Henry Enfield Roscoe

Spectrum analysis

six lectures, delivered in 1868, before the Society of Apothecaries of London : with appendices ...

Henry Enfield Roscoe

Spectrum analysis
six lectures, delivered in 1868, before the Society of Apothecaries of London : with appendices ...

ISBN/EAN: 9783741173622

Manufactured in Europe, USA, Canada, Australia, Japa

Cover: Foto ©berggeist007 / pixelio.de

Manufactured and distributed by brebook publishing software (www.brebook.com)

Henry Enfield Roscoe

Spectrum analysis

SPECTRA OF THE METALS OF THE ALKALIES & ALKALINE EARTHS.

From the Drawings of BUNSEN & KIRCHHOFF.

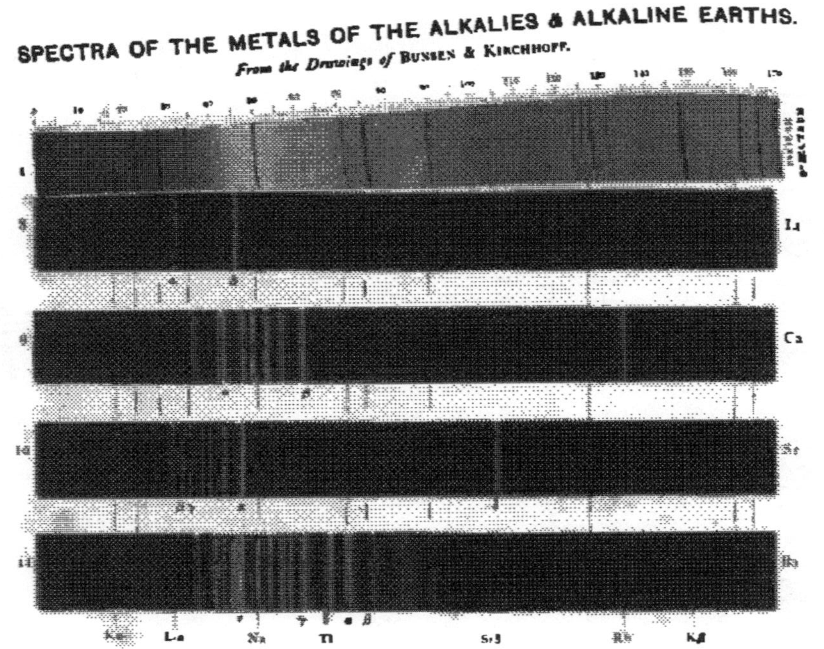

SPECTRUM ANALYSIS.

SIX LECTURES,

DELIVERED IN 1868, BEFORE THE SOCIETY OF APOTHECARIES OF LONDON.

BY

HENRY E. ROSCOE, B.A. PH.D. F.R.S.

PROFESSOR OF CHEMISTRY IN OWENS COLLEGE, MANCHESTER.

WITH APPENDICES, COLOURED PLATES, AND ILLUSTRATIONS.

SECOND EDITION.

London:

MACMILLAN AND CO.

1870.

PREFACE TO THE SECOND EDITION.

In the year which has elapsed since the first publication of these Lectures on Spectrum Analysis, much has been done by the active workers in this branch of science to extend our knowledge, especially concerning the physics of the sun. I have, therefore, found it necessary to rewrite almost the whole of the portion of the book relating to celestial chemistry, introducing into the text the latest discoveries of Huggins, Lockyer, Janssen, and Zöllner, which in the former edition were but imperfectly given. It has also been found expedient to re-arrange the subject-matter of Lectures III. and IV., and to add thereto a short description of investigations in several collateral branches of science. The Appendices have been enriched by abstracts of the original communications made to learned Societies by the investigators themselves. Many new illustrations have been introduced into the text, and no pains have been spared to render the work as complete a record as possible of the present state of the subject.

H. E. Roscoe.

April 1870.

PREFACE TO THE FIRST EDITION.

In publishing the following Lectures I have endeavoured to preserve the elementary character which they naturally assumed in delivery, thinking it best to give further detail in a series of Appendices. If the book thus assumes less of the character of a complete treatise than might be desirable, it gains in value for the general reader, inasmuch as the science of Spectrum Analysis is at present in such a rapid state of growth that much of the subject is incomplete, and, therefore, necessarily unsuited to the public at large. I hope, however, that the addition of many extracts from the most important Memoirs on the subject may prove interesting to all, as it will certainly be useful to those specially engaged in scientific inquiry, as indicating the habits of exact research and accurate observation by which alone such striking results have been attained. For the permission to reproduce exact copies of Kirchhoff's, Angström's, and Huggins' maps, together with the Tables of the positions of the dark solar and bright metallic lines, I have to thank the above-named gentlemen. These maps will render the work valuable to the student for a reference, whilst the chromolithographic plates of the spectra of the metals of the alkalies and alkaline earths, and of the spectra of the stars, nebulæ, and non-metallic elements, serve to give some idea of the peculiar beauty of the real phenomena thus represented.

Since last summer, when these Lectures were delivered, our knowledge of the constitution of the sun especially has made giant strides; and although I have been unable to introduce these newest facts into the text of the Lectures, I have still brought forward the most important of these discoveries in the Appendices to Lecture V.

As the latest news on this subject, I may mention the arrangement contrived by Mr. Huggins, by which the wonderful changes of the red solar prominences can all be viewed at once; changes so enormously rapid that Mr. Lockyer has observed one of these red solar flames, 27,000 miles in length, disappear altogether in less than ten minutes. Mr. Lockyer has also succeeded in seeing in the flames the red (C) line of hydrogen, as well as the line in the violet, which he finds corresponds to the line marked 2796 on Kirchhoff's map, and not, as was supposed, identical with Fraunhofer's line G.

At the end of the volume will be found a tolerably complete List of Memoirs forming the literature on the subject.

My thanks are due to Mr. J. D. Cooper and Mr. Collings for the very great care which they have bestowed upon the illustrations, and especially upon the difficult task of reproducing Kirchhoff's maps in four tints.

H. E. R.

MANCHESTER, *April* 1869.

CONTENTS.

LECTURE I.

LECTURE II.

LECTURE III.

LECTURE IV.

LECTURE V.

LECTURE VI.

LIST OF THE PRINCIPAL MEMOIRS, ETC. ON SPECTRUM ANALYSIS.

LIST OF ILLUSTRATIONS.

PLATES.

ON

SPECTRUM ANALYSIS.

LECTURE I.

Introduction.—Newton's Discovery of the Composition of White Light, 1675.—Properties of Sun Light.—Heating Rays.—Luminous Rays.—Chemically Active Rays.—The Solar Spectrum.—Position of Maxima.—Illustrations of these Radiations.—Means of obtaining a Pure Spectrum.—Fraunhofer's Lines.—Planet and Moon Light.—Star Light.

APPENDIX A.—Extracts from "Newton's Opticks."

APPENDIX B.—Burning Magnesium Wire a Source of Light for photographic purposes.

APPENDIX C.—On the Chemical Action of the constituent parts of Solar Light.

AMONGST all the discoveries of modern science none has deservedly attracted more attention, or called forth more general admiration, than the results of the application of Spectrum Analysis to chemistry. Nor is this to be wondered at when we remember that a new power has thus been placed in the hands of the chemist, enabling him to detect the presence of chemical substances with a degree of delicacy and accuracy hitherto unheard of, and thus to obtain a far more intimate knowledge of the composition of terrestrial matter than he formerly enjoyed. So valuable a means of research has this new process of analysis proved itself to be that since its first establishment, some seven short years ago, no less than four new chemical elements have by its help been discovered.

Not only, however, have we to consider the importance and interest which attaches to the subject as evidenced by the discovery of these new elementary bodies, but we are forced to admit that by the application of the simple principles of spectrum analysis the chemist is able to overstep the narrow bounds of our planet, and, extending his intellectual powers into almost unlimited space, to determine, with as great a degree of certainty as appertains to any conclusion in physical science, the chemical composition of the atmosphere of the sun and far distant fixed stars. Nay, he has even succeeded in penetrating into the nature of those mysteries of astronomy, the nebulæ; and has ascertained not only the chemical composition, but likewise the physical condition, of these most distant bodies.

It does, indeed, appear marvellous that we are now able to state with certainty, as the logical sequence of exact observations, that bodies common enough on this earth are present in the atmosphere of the sun, at a distance of ninety-one millions of miles, and still more extraordinary that in the stars the existence of such metals as iron and sodium should be ascertained beyond a shadow of doubt. We thus see that the range of inquiry which the subject opens out is indeed vast, and it is well to bear in mind that as the discoveries in this branch of science are so recent, they are necessarily incomplete, so that we must expect to meet with many facts and observations which still stand alone, and require further investigation to bring them into harmony with the rest. The advance in these new fields of research is, however, so rapid that, as time rolls on, and our range of knowledge widens, new facts quickly come to support the hitherto unexplained phenomena, and thus our theory becomes more and more complete. It will be my duty in the Lectures which I have the honour of delivering in this Hall to endeavour to explain to you that these results, apparently as marvellous as the discovery of the elixir vitæ or the philosopher's stone, are the plain and necessary deductions from exact and laborious experiment, and to show how two German philosophers, quietly working in their laboratories in Heidelberg, obtained this startling insight into the processes of creation.

The only means of communication which we possess with the sun, planets, or far distant stars, or by which we can ascertain anything respecting their chemical constitution, is by means of the life-supporting radiation which they pour down upon the earth, producing the effects which we call light and heat. It will, therefore, be our business, in the first place, to investigate the composition of the radiations which these bodies give off, and next gradually to notice, as our field of observation enlarges, the applications to which the properties of the light thus emitted lead us. One cannot help regretting that when, as at present, the sunlight is shining so brightly, we are unable to utilize it, and illustrate by experiments made with the sunlight itself the points which we wish to explain. In this climate, however, even if we could conveniently do it, the sun shines so intermittently, and it is so doubtful if we can have it just when it is required, that we have to make use of other means, especially of this bright light of the electric arc, which, if less perfect than sunlight, is more under our control.

In the year 1675, Sir Isaac Newton presented to the Royal Society his memorable treatise on Optics.[1] In this treatise, which contains a large number of experimental and theoretical investigations, one point especially attracts our attention: it is the discovery of the decomposition of white light. We have here (Fig. 1) a facsimile of the drawing illustrating Newton's experiment on this subject. He heads the first paragraph in his memoir,

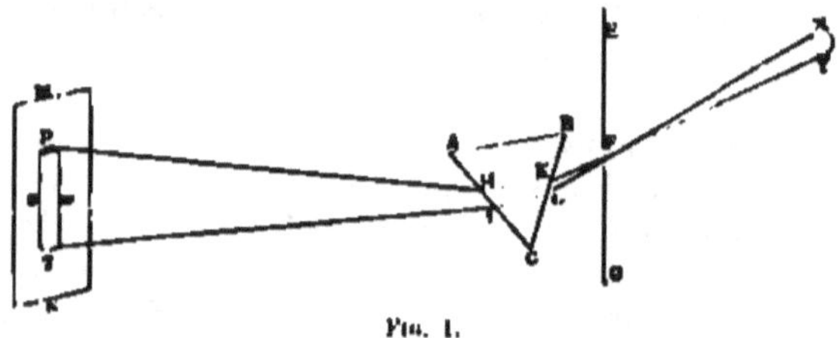

FIG. 1.

written in the year 1675, with the words, "lights which differ in colour differ also in refrangibility." Newton allowed the sun to shine through a round hole (Fig. 1, F) in a shutter, and he then examined the character of this light by means of a triangular piece of glass (A B C) called a prism. He found that the white light, after passing through the prism, was bent or refracted out of its course, and split up into a coloured band (P T) which, when received on the white screen (M N), exhibited all the colours of the rainbow in regular succession, passing from red through all the shades of orange, yellow,

[1] For extracts from "Newton's Opticks," see Appendix A.

green, and blue, to violet. Newton termed this coloured band the Solar Spectrum, and came to the conclusion that the light of the sun consists of rays of different degrees of refrangibility. He also showed that all the various portions of this coloured band, when again brought together, produce upon the eye the effect of white light. This experiment (represented in Fig. 2), Newton performed by simply allowing the light passing through the round hole (F) in the shutter to fall on a prism (A B C), producing the solar spectrum, and then,

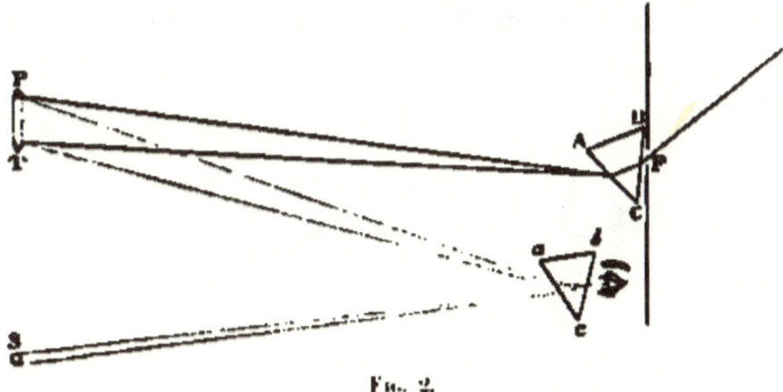

Fig. 2.

on looking at this coloured band through another prism ($a\,b\,c$) placed in the same direction, instead of seeing a coloured band he observed a spot of white light, thus showing that the whole of these differently coloured rays, when brought together by means of the second prism, produce on the eye the effect of white light. Here we have the intensely white electric light, and by means of these two prisms you observe that I can split the light up into its various constituent parts, and we obtain this splendid coloured band, the spectrum of the electric arc. Now I shall endeavour to show you the second effect which Newton observed. For this purpose I have only to reverse the position of one of the prisms; for if I allow this band of coloured light emitted from the first prism to pass through the second prism, placed in the opposite direction to the first, I shall again bring these coloured rays together, the second prism neutralizing altogether the effects of the first, and we obtain the bright white image of the slit.

I should like next to show you a third part of the experiment made by Newton. If I cut off, by means of this screen, a portion of the spectrum which has passed through the first prism, say the yellow or the green or

FIG. 3.

the red, you will see that another refraction through the second prism does not alter this coloured light. Here, for instance, I take the green, and bring my prism into a proper position: we shall find that I only get the same green light on the screen behind, proving that the green ray cannot again be split up by further refraction. The mode in which Newton performed this experiment is seen in Fig. 3. The green rays (g) in the spectrum (de), when refracted through a second prism (abc), appear as a green band (N M) on the second screen.

Now the reason why all these different coloured lights, when they reach the eye, produce the effect of white light, is a physiological question which we cannot explain. We find that not only do all the colours of the spectrum, when they are brought upon the eye, produce the effect of white light, but that several mixtures of only a few out of all these different colours have the same power of producing upon the eye the effect of white light.

Thus Helmholtz and Maxwell have shown that the

FIG. 4.

following mixtures of two complementary colours when brought together into the eye produce the effect of whiteness:—violet and greenish yellow; indigo and yellow; blue and orange; greenish blue and red. This I may readily illustrate to you by taking a revolving disc (Fig. 4), upon which segments of various colours have been painted. When I turn this disc—which you observe contains, to begin with, all the colours of the spectrum in due proportion—and then, when it is revolving, exhibit it to you by the light of burning magnesium, you will notice that the effect produced upon the eye is that of a uniformly white disc. Let me now substitute for this painted disc another one, which only contains the three colours red, yellow, and blue. You will still perceive that when quickly revolved the disc appears white; and so I might substitute any of the above mixtures of colours, all of which produce on the eye the sensation of white light by the rapid succession of the images of the different colours, the effect of one of which not having time to disappear from the retina before that of the other comes into play. If I illuminate the rotating disc by means of an instantaneous electric spark, you will see that all the colours of the disc become noticeable at once. This is because the period of illumination is so exceedingly short that the disc appears as if stationary. It was indeed at one time supposed that the various shades of colour in the solar spectrum were produced by an overlapping, as it were, of three distinct coloured spectra, one red, the second yellow, and the third a blue spectrum, the maxima of which are situated at different points, that of the red and blue at the extremes, and that of the yellow in the middle of the visible spectrum. This theory of Brewster's has, however, been proved to be fallacious, for Helmholtz has shown that the green ray, for example, is not made up of blue and yellow light superposed, and we cannot separate anything else but green out of it. Hence we conclude that each particular ray has its own peculiar colour, and that light of each degree of refrangibility is monochromatic. But, on the other hand, although physically, and in the actual spectrum, there is no such thing as a superposition, or overlapping of different spectra, yet it is very likely, nay, more than likely, that the retina is mainly sensitive to three impressions, viz. red, yellow, blue; in fact it appears that there are some nerves especially sensitive to red, others to yellow, and others again to blue light, whilst the impressions of the other tints are obtained by the joint impressions produced on these three classes of nerves. This theory, indeed, was proposed so long ago as the beginning of the century by our celebrated countryman, Thomas Young, and quite recently it has been proved by Max Schulze, that in the eyes, not indeed as yet of man, but of certain animals, there exist differences which are observable in the nerve ends situated at the back of the retina: some of these end in little red drops, some of them in yellow drops, and some of them in colourless ones. The nerves whose ends contain the little red drops are more sensitive to red colour than the others; and so those containing yellow drops are more sensitive to the yellow colour; and in this way we believe that the peculiar effects which we observe in the mixtures of colour may be explained.

In noticing the physical properties of the variously coloured light obtained when the sunlight or white light is decomposed into its constituent parts, we must remember, in the first place, that light is due to the undulations of the elastic medium pervading all space, to which physicists have given the name of luminiferous ether. As the undulations of the particles of water, causing the waves of the sea, differ in length and intensity of vibration, sometimes being minute and shallow like the ripples on the surface of a pond, sometimes rising and falling into the gigantic crests and valleys of the storm-ridden ocean, so also the undulations of the ether producing light differ in amplitude and intensity, giving rise to the different effects of colour to which we have referred.

Let us now compare the power of the ear and the eye, the one to receive the vibrations of the air called sound, and the other the vibrations of the ether termed light.

The human ear has the power of distinguishing variations in sounds differing widely in wave-length and in rapidity of vibration. Thus, for instance, the musical note, the deepest in the bass of all those that can be heard by the human ear, is produced by the regular succession of impulses occurring about 16 times in each second, whereas the highest note which is perceptible to the ear is caused by about 4,000 vibrations per second. Hence the range of audible notes extends over about 11 octaves. Let us next examine what is the range to which the eye is sensitive. It is found that we can observe, ordinarily, in the solar spectrum from the position indicated by a fixed dark line in the red portion termed A (see Frontispiece of the Solar Spectrum) to a line H in the violet or most refrangible portion of the visible spectrum. Now the difference in the number of the vibrations from A to H is but slight; it is indeed not one octave. The length of the undulation of each of these waves is excessively small, whilst the number of the vibrations of the ether which take place every second is enormously large. By certain methods of exact measurement, physicists have been enabled to determine with accuracy the length and the duration of these various waves of light; and they have come to the conclusion that the wave causing this red ray which is only just visible to the eye has a length of the $\frac{291}{10000000}$th part of an inch, and that in one second of time no less than

458 millions of millions of these vibrations occur—whereas at the line H in the violet the length of the wave is $\frac{1458}{100000000}$th part of an inch, and the number of vibrations is 727 millions of millions per second. Hence the difference is only from 458 to 727, and you see that the rapidity of the vibration at the one point is not twice as great as it is at the other, and we are correct in stating that we can hear about 11 octaves, but that we cannot see a single octave.

Nevertheless, although they do not produce on the retina the impression which we call light, there are rays extending both beyond the visible red and beyond the visible blue. By certain devices we can make ourselves aware of the existence of these invisible rays, which play a most important part in the nature of the solar light.

If we observe the effects which are produced by the different rays constituting the visible portion of the solar spectrum, we find, to begin with, that those rays which mainly produce heating effects are situated at the red end of the solar spectrum. We learn that the maximum heating effect is produced at a point beyond that at which we can see any red light. The maximum of the luminous rays as affecting the eye exists in the yellow. Passing on from the red towards the violet portion, we find, by means of a thermo-pile and a delicate galvanometer, that the quantity of heating effect produced in the yellow and green portions of the spectrum gradually diminishes, and sinks down to a very insignificant amount in the violet part of the band. In the blue and violet portion of the spectrum, so slightly endowed with heating power, we have, however, to notice the existence of a new and striking peculiarity, that of producing chemical action; that is, of causing the combination and decomposition of certain chemical substances, as for instance the decomposition and blackening of silver salts, upon which the art of photography is based.

It is to Sir W. Herschel that we are indebted for the first notice, in the year 1800, of the fact of the heating rays existing especially in the red portion of the spectrum, whilst Ritter and Scheele at the early part of this century observed the peculiar power possessed by the blue, violet, and ultra-violet rays to blacken silver salts.

In Fig. 5 we have a graphical representation of the varying intensity of the heating, luminous, and chemically active rays in the various parts of the solar spectrum.

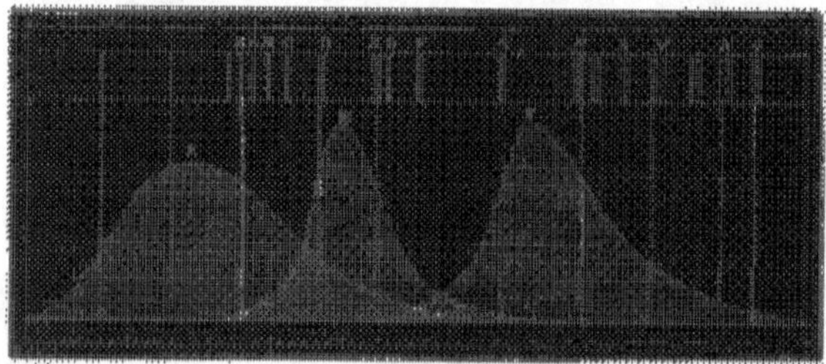

FIG. 5.

The figure exhibits three curves, A, B, and C, showing the distribution of these three actions produced by the rays of the solar spectrum, whose position is given in the upper part of the diagram, the differently coloured portions being indicated by certain fixed lines to which letters are attached, beginning with A in the red, and ending with O in the ultra-violet portion of the spectrum.

This curve A (Fig. 5) represents the intensity of the heating power of the spectrum. You observe how far it extends, a long way beyond the visible red, which ends a little to the left of the line A. In fact it has been noticed, by those who have made careful measurements, that half the rays of heat reaching the earth from the sun are invisible, and I shall hope to show you directly the effect of these invisible heating rays.

In the part shaded with vertical lines only, there is no perceptible luminous or chemical activity; in that shaded only with horizontal lines there is nothing but chemical action. From the beginning to the end of the luminous spectrum shaded with oblique lines two, at least, of these three forms of action exist simultaneously. The intensity of each at any point of the spectrum is measured by the vertical line drawn from the point on the base line to meet its proper curve.

Whilst noticing these peculiar properties of the different rays we must carefully remember that there really is no difference in kind between those rays which are called heating rays, those which are called light rays, and those which are called chemically active rays. These differ one from the other in exactly the same way that the visible yellow rays differ from the green rays, or as the green rays differ from the blue; only in wave-length and intensity of vibration. In any particular portion of the spectrum we cannot separate the rays of light and leave the rays of heat behind; we cannot, for instance, separate out the yellow rays of light from the yellow portion of the spectrum, and leave behind any rays of heat of the same degree of refrangibility. But, as I shall show you, we can separate from the whole radiation the luminous rays, and with them the heating effect of those luminous rays, and still leave the dark or invisible rays of heat of lower refrangibility.

I will endeavour to prove to you, in the first place, the existence of these dark heating rays of really invisible light. We see that the maximum of these rays is placed beyond the visible red. This may be clearly ex-

FIG. 6.

hibited with the electric light in the beautiful experiment by which Dr. Tyndall first accomplished the separation of which I have just spoken. I have for this purpose placed in the dark box (B, Fig. 6) an electric lamp (L), which gives us a very bright light, and by means of this mirror (M) I can bring the rays to a focus at any desired point. Here is a cell (C) which Dr. Tyndall very expressively calls a ray-filter, by which I can filter out the whole of the luminous rays, by passing them through this opaque solution of iodine in disulphide of carbon, whilst the invisible heating rays are transmitted, and will soon render themselves evident to you. A current of cold water circulates through a double jacket on the outside of the cell, to keep the volatile disulphide cool. Now I think you will observe that no rays of light come through, but if I take a piece of black paper, and place it in the focus of our mirror, you see the paper is ignited, owing to the presence of these dark heating rays. I may now do the same with a piece of blackened platinum (P, Fig. 6); you see that this also is heated to redness. I can show you this again in a variety of forms. Here is some gunpowder strewed on this paper; you observe that it at once explodes when brought into the focus of the dark rays. Here I have some blackened gun-cotton, which instantly catches fire. I may vary the experiment by lighting a cigar; and here you see the brilliant scintillations of charcoal burning in oxygen, having been heated up to the temperature of ignition in the focus of the dark rays. Dr. Tyndall has measured the proportional amount of the entire heating rays which, pouring forth from this incandescent carbon, has passed through this dark filter, and he has found that this consists of $\frac{7}{8}$ of the whole amount; so that only $\frac{1}{8}$ of the radiation is really visible.

Understanding then the existence in this ultra-red of a large amount of heating rays, let us pass now to consider the properties of the light which is given off at the opposite or blue end of the spectrum, which I have called the chemically active rays. Allow me to show you an experiment to prove that it is in the blue portion of the spectrum that we have essentially these chemically active rays. In order to render the illustration more perfect, I will first make an experiment with reference simply to white light, to show you that the brilliant light which is emitted by this burning magnesium, and is almost too dazzling for the eye to

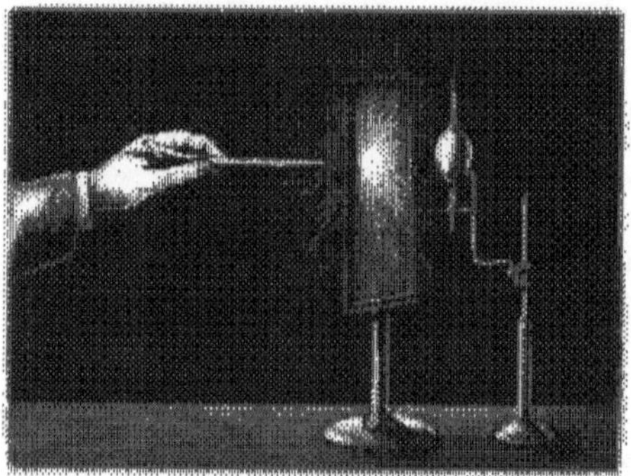

Fig. 7.

bear, contains a very large proportion of the rays which we are about to investigate.[1]

I have here a thin glass bulb containing a mixture of equal volumes of two gases, chlorine and hydrogen. These gases when exposed to a bright light combine together, and form hydrochloric acid gas. If I were to throw this bulb out into the sunlight, so rapid would be

[1] See Appendix D for the measurement of the chemical intensity of magnesium light.

the combination, and so great the consequent evolution of heat and sudden expansion, that this little bulb would instantly be shattered into a thousand fragments. Almost as sudden an effect will be produced if I simply burn a bit of magnesium wire in the neighbourhood of the bulb (Fig. 7); it explodes with a pretty loud report, the bulb is shattered, and the gases have been combined by virtue of the blue rays contained in this kind of

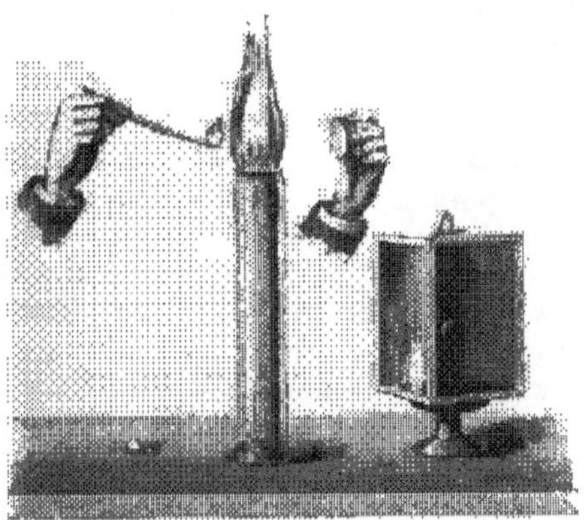

Fig. 8.

light. I will next show you that it is really the blue rays which thus act chemically. This lantern (Fig. 8) contains panes of different coloured glass,—here a white one, there a yellow one; here a red one, there a blue one. I am going to put another of these little bulbs filled with chlorine and hydrogen in the inside of this lantern, and then I will produce, not by magnesium wire, but by another means, a very bright blue light, a light which contains these chemically active rays in great quantity. I will first allow this blue light to shine upon the bulb through the red pane of glass. Here I produce a very bright flame, by throwing some carbon disulphide into a tall cylinder full of nitric oxide gas, and igniting the mixture. There you have the bright flash, but you have noticed no explosion of the bulb, for all the chemical rays have been held back: filtered off by this red glass, they cannot pass through; and the consequence is, there has been no action on my bulb. I will now allow another of these flashes of light to pass through the blue glass, which being of course transparent to the blue rays my little bulb will be shattered into a fine powder, as you observe. Here then we have ascertained by experiment that the blue rays act chemically, whilst the red rays produce heating effects.

This sensitive mixture of chlorine and hydrogen, which, as you have seen, explodes when the chemical activity of the light is great, may be used as a most delicate means of measuring the amount of light of a less intensity. The combination of the gases then occurs slowly, and may be rendered evident by allowing the hydrochloric acid thus formed to be absorbed by water, when the consequent diminution in bulk of the gas accurately represents the chemical action effected.

The varying intensity of the chemically active rays in different parts of the solar spectrum has been carefully measured by means of this sensitive mixture of chlorine and hydrogen gases.[1] The accompanying figure (Fig. 9) exhibits the chemical action effected by the various portions of the spectrum on the sensitive mixture for one particular zenith distance of the sun. The lines marked

[1] See Appendix C for description of method.

with the letters of the alphabet from A to W, at the bottom of the figure, represent the fixed dark lines which exist in the solar spectrum, of which I shall have much to say in the subsequent lectures. They serve as a sort of landmarks by which to ascertain the position of any given point in the spectrum. The greatest amount of chemical action is noticed between the line in the indigo marked G, and that in the violet marked H. In the direction of the red end of the spectrum, the action becomes imperceptible

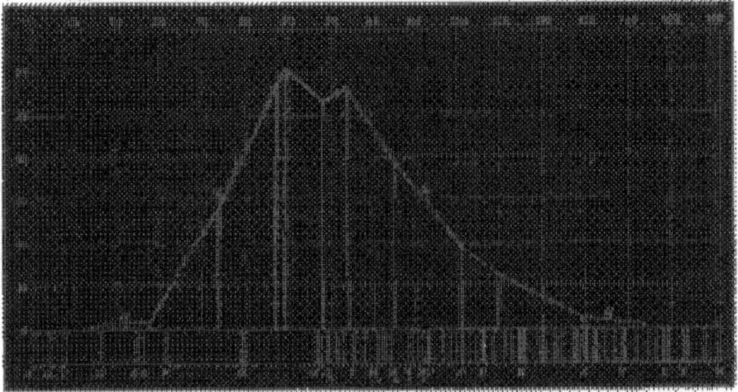

FIG. 6.

about D, in the orange (the maximum of visible illumination); whilst towards the other end of the spectrum the action was found to extend as far as the line marked U, or to a greater distance beyond the line H in the violet than the total length of the ordinary visible spectrum.

This same fact may again be illustrated by showing that I can photograph with these blue rays, whereas I fail to produce the same effect with the red rays. I will coat a plate with collodion, and then darken the room, with the exception of this yellow mono-chromatic flame, produced by the volatilization of soda salts, which is incapable of acting chemically, and with which we may work without at all affecting our photographic plate. Now I have coated a plate with collodion, and sensitised it in the silver bath. I shall next expose this to the action of the light of the spectrum of the electric lamp. Let me first show you that I have here (Fig. 10) a negative

FIG. 10.

photograph, of which I am about to take a print by means of the blue rays of the electric lamp. You will observe that there are two figures upon the negative, one marked V and the other marked R: these letters being intended to signify Violet and Red. The one figure marked V, I propose to place in front of my sensitised plate in the blue or violet ray, and the one marked R I shall open in the red ray, and I hope to be able to produce a chemical effect on that portion of my sensitised plate which has been exposed to the blue, whilst we shall get no corresponding effect on the portion exposed to the red ray. I next place my plate with its face downwards on the negative; we now start our electric lamp, using a small spectrum in order to have the action rather more distinct. I then expose half my plate in the red rays for about twenty seconds, and afterwards expose the other half, with the V upon it, for

FIG. 11.

about the same length of time, to the violet light. I will now develop and wash the photograph, and throw the image produced on the screen, when you will observe (Fig. 11) a very marked difference between the two halves, the one showing that no action has been produced on the sensitive plate exposed to the red rays, and the other giving us the perfect picture with the V upon it.

These chemically active rays which the sun constantly

pours down upon our earth, doubtless exert a most important influence on the fauna and flora of a country, and it becomes a matter of some importance to measure accurately their varying intensity with the changing seasons, and at different parts of the globe. It may not be uninteresting if I point out some of the results which have already been obtained by measurements of this kind, made by a method proposed by Professor Bunsen and myself, and depending on the darkening of a sensitive paper prepared with chloride of silver. The principles upon which the method is based are: (1) that a photographic paper can be prepared which always shall possess a constant degree of sensitiveness; and (2) that the darkening effect produced by the light on this paper is constant when the product of the intensity of the light into the time of exposure is also constant. Measurements according to this method have been carried out at Kew Observatory, and an interesting series of curves obtained, showing the variation of the chemical action of the daylight with the seasons. The curves (Fig. 12) show the rise and fall of monthly chemical intensity with the hour of the day, from 6 A.M. to 6 P.M., for the year 1866. We see from these curves that the maximum amount of chemical action occurs at 12 o'clock, and that this action is equal at hours equidistant from noon; we learn also that if the chemical intensity in July 1866 be represented by the number 100, that in January is represented by 14; or in mid-summer the chemical intensity of the sun is more than seven times as great as in mid-winter.[1]

Still more interesting is it to ascertain the distribution of the chemically active rays on the earth's surface, about

[1] (Roscoe) Bakerian Lecture, Phil. Trans. part ii. 1865, p. 605; and also (Roscoe) Phil. Trans. 1867, p. 555.

which we as yet know but little. According to the vague observations of photographers, it would appear that in advancing from England towards the equator the difficulty of obtaining good pictures is increased, and more time is said to be needed to produce the same effect on the photographic plate under the full blaze of a tropical sun than in the gloomier atmosphere of London. It has likewise been stated that in Mexico, where the light is very intense, from twenty minutes to half an hour was required to produce photographic effects which in England occupy only a minute; and it is said that

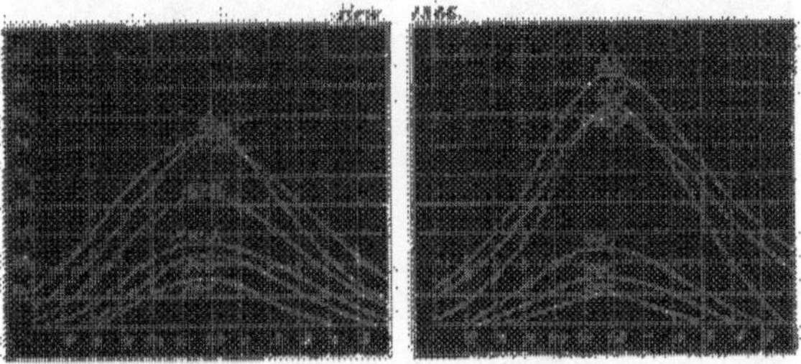

FIG. 12.

travellers engaged in copying the antiquities of Yucatan have on several occasions been obliged to abandon the use of the photographic camera, and have had to take to their sketch-books! In order to test the validity of these statements, it becomes a matter of great importance to determine directly the intensity of the chemically active rays in the tropics; and, thanks to the zeal and ability of my friend Dr. T. E. Thorpe, I am able to show you the results of such measurements, made at Pará, situated nearly under the equator, in the northern province of the Brazils, and lying on a branch of the Amazons. Owing to the rainy season having commenced when the experiments were made, the changes in the chemical intensity as observed from hour to hour, and even from minute to minute, are very sudden and remarkable, and are well seen on the curves (Figs. 13 and 14); and these, compared with the dotted lines below indicating the corresponding action on the same day at Kew, show the enormous variation in chemical intensity

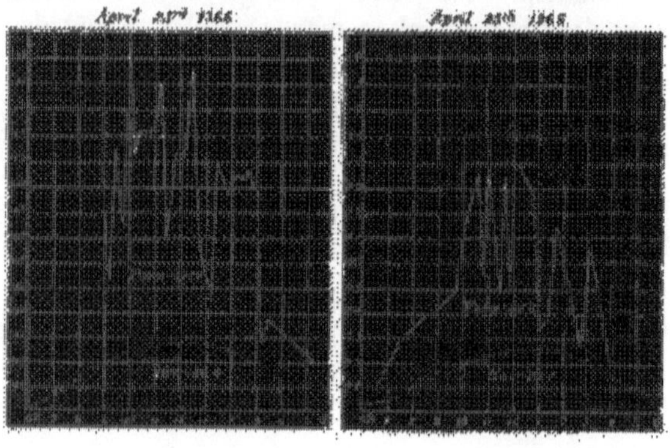

Fig. 13. Fig. 14.

which occurs under a tropical sun in the rainy season. Regularly every afternoon, and frequently at other hours of the day, enormous thunder-clouds obscure the sky, and, discharging their contents in the form of deluging rain, reduce the chemical action nearly to zero. The storm quickly passes over, and the chemical intensity rapidly rises to its normal value. By comparing the curves for Pará and Kew on the same days, we obtain some idea of the energy of chemical action at the tropics, and it is at once evident that the alleged failure of the photographer cannot at any rate be ascribed to a diminution in the sun's chemical intensity, which in the month of April 1866 was nearly seven times as great at Pará as at Kew.

I have in conclusion to point out to you that the solar spectrum differs in certain respects from that beautiful spectrum of the electric arc with which we have been working; and it differs in this way, that the solar spectrum consists, not of a continuous band, passing without break or interruption from red to violet, through all the shades of colour which we know as the rainbow tints, but that in the solar spectrum we find, interspersed between these certain dark lines which we may regard as shadows in the sunlight, spaces where certain rays are absent. The first person who observed these dark lines was Dr. Wollaston.[1] Newton did not observe them, and for the good reason that he allowed the light to fall on the prism from a round hole in the shutter. In this way he did not obtain what is termed a pure spectrum, but a series of spectra, one overlapping the other, owing to the light coming through different parts of the round hole. If he had allowed the light to pass through a fine vertical slit, and if this slit of light, if we may use such a term, had then fallen upon the prisms, placed so that the edge of the refracting angle is parallel to the slit, he would have observed that the solar spectrum was not continuous, but broken up by permanent dark lines. Dr. Wollaston, making use of a fine slit of light, discovered these fixed dark lines in the solar spectrum.

[1] Phil. Trans. 1802, p. 378.

I invite your attention to the drawings of these lines in this very imperfect sketch of the solar spectrum (see coloured diagram on Frontispiece). These dark lines are always found in the same position in the sunlight, whether you take direct, diffused, or reflected sunlight. The exact mapping and observation of these lines in the solar spectrum is a matter of as great importance to astronomy and to physical science generally as the mapping of the stars in the heavens, because by knowing exactly the position of these dark lines in the solar spectrum we can ascertain that iron, sodium, and other well-known substances exist in the solar atmosphere.

I will now show you a diagram illustrative of this fact, and remind you that we are indebted for the first careful examination of these lines to a German optician, Fraunhofer, whose name has been attached to these lines. Fraunhofer mapped no less than 576 of these lines in the year 1814.[1] This is an exact copy of his map (Fig. 15). On the left is the red and on the right is the violet portion of the visible spectrum. You observe how this spectrum is shaded. Notice, if you please, the immense number of lines which intersect and almost appear to make the solar spectrum dark. Fraunhofer employed the letters of the alphabet to designate some of the principal lines, beginning with A in the red and passing over to H in the violet. Many of these lines are as fine as the finest spider's web, so that they occupy but a small portion of the whole area of the spectrum—that is, the portion which is filled with light is far greater than the portion filled with these shadows, although the number of these shadows is so exceedingly

[1] Denkschriften der Münchener Akademie, 1814.

large. The curve in the upper part of the figure gives

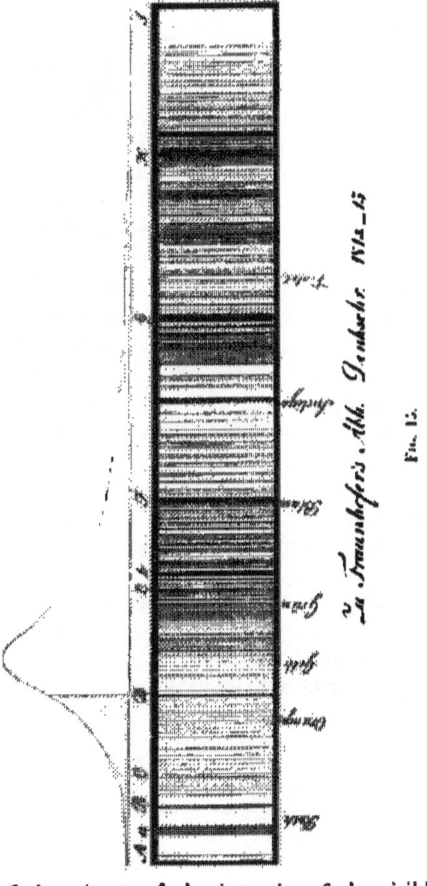

Fig. 13.

Fraunhofer's estimate of the intensity of the visible rays

at different parts of the spectrum, and this corresponds closely to the curve B in Fig. 5.

Fraunhofer first ascertained that these lines are present in every kind of sunlight, and that moonlight, as well as the light of the planet Venus, exhibits the same dark lines. He likewise measured the refractive indices of these lines,—that is, determined their relative positions in the solar light; and he found that the relative distances between any given lines remained constant, whether he took direct sunlight, or sunlight reflected from the moon or planets.

Another important observation was made by Fraunhofer,—namely, that the light from the fixed stars, which, as you know, are self-luminous, also contains dark lines, but different lines from those which characterise the sunlight, the light of the planets, and that of the moon; and hence in 1814 Fraunhofer came to this remarkable conclusion: that whatever produced these dark lines—and he had no idea of the cause—was something which was acting beyond and outside our atmosphere, and not anything produced by the sunlight passing through the air.

This conclusion of Fraunhofer has been borne out by subsequent investigation, and the observations upon which it was based may truly be said to have laid the foundation-stone of solar and stellar chemistry.

LECTURE I.—APPENDIX A.

EXTRACTS FROM "NEWTON'S OPTICKS," 1675.

BOOK I. PART I.

PROP. I. THEOR. 1.—*Lights which differ in colour, differ also in degrees of refrangibility.*

THE PROOF BY EXPERIMENTS.

Exper. 1.—I took a black oblong stiff paper terminated by parallel sides, and, with a perpendicular right line drawn across from one side to the other, distinguished it into two equal parts. One of these parts I painted with a red colour and the other with a blue. The paper was very black, and the colours intense and thickly laid on, that the phenomenon might be more conspicuous. This paper I viewed through a prism of solid glass, whose two sides through which the light passed to the eye were plane and well polished, and contained an angle of about 60°: which angle I call the refracting angle of the prism. And whilst I viewed it, I held it and the prism before a window in such manner that the sides of the paper were parallel to the prism, and both those sides and the prism were parallel to the horizon, and the cross line was also parallel to it; and that the light which fell from the window upon the paper made an angle with the paper, equal to that angle which was made with the same paper by the light reflected from it to the eye. Beyond the prism was the wall of the chamber under the window covered over with black cloth, and the cloth was involved in darkness that no light might be reflected from thence, which in passing by the edges of the paper to the eye might mingle itself with the light of the paper, and obscure the phenomenon thereof. These things being thus ordered, I found that if the refracting angle of the prism be turned upwards, so that the paper may seem to be lifted upwards by the refraction, its blue half will be lifted higher by the refraction than its red half. But if the refracting angle of the prism be turned downward, so that the paper may seem to be carried lower by the refraction, its blue half will be carried something lower thereby than its red half. Wherefore in both cases the light which comes from the blue half of the paper through the prism to the eye, does in like circumstances suffer a greater refraction than the light which comes from the red half, and by consequence is more refrangible.

Exper. 2.—About the aforesaid paper, whose two halves were painted over with red and blue, and which was stiff like thin pasteboard, I lapped several times a slender thread of very black silk, in such manner that the several parts of the thread might appear upon the colours like so many black lines drawn over them, or like long and slender dark shadows cast upon them. I might have drawn black lines with a pen, but the threads were smaller and better defined. This paper thus coloured and lined I set against a wall perpendicularly to the horizon, so that one of the colours might stand to the right hand, and the other to the left. Close before the paper at the confine of the colours below, I placed a candle to illuminate the paper strongly: for the experiment was tried in the night. The flame of the candle reached up to the lower edge of the paper, or a very little higher. Then at the distance of six feet and one or two inches from the paper upon the floor I erected a glass lens four inches and a quarter broad, which might collect the rays coming from the several points of the paper, and make them converge towards so many other points at the same distance of six feet and one or two inches on the other side of the lens, and so form the image of the coloured paper upon a white paper placed there, after the same manner that a lens at a hole in a window casts the images of objects abroad upon a sheet of white paper in a dark room.

The aforesaid white paper, erected perpendicular to the horizon and to the rays which fell upon it from the lens, I moved sometimes towards the lens, sometimes from it, to find the places where the images of the blue and red parts of the coloured paper appeared most distinct. Those places I easily knew by the images of the black lines which I had made by winding the silk about the paper. For the images of those fine and slender lines (which by reason of their blackness were like shadows on the colours) were confused and scarce visible, unless when the colours on either side of each line were terminated most distinctly. Noting therefore, as diligently as I could, the places where the images of the red and blue halves of the coloured paper appeared most distinct, I found that where the red half of the paper appeared distinct, the blue half appeared confused, so that the black lines drawn upon it could scarce be seen; and on the contrary, where the blue half appeared most distinct, the red half appeared confused, so that the black lines upon it were scarce visible. And between the two places where these images appeared distinct there was the distance of an inch and a half: the distance of the white paper from the lens, when the image of the red half of the coloured paper appeared most distinct, being greater by an inch and a half than the distance of the same white paper from the lens when the image of the blue half appeared most distinct. In like incidences therefore of the blue and red upon the lens, the blue was refracted more by the lens than the red, so as to converge sooner by an inch and a half, and therefore is more refrangible. . . . .

Scholium.—The same things succeed notwithstanding that some of the circumstances be varied; as in the first experiment when the prism and paper are any ways inclined to the horizon, and in both when coloured lines are drawn upon very black paper. But in the description of these experiments, I have set down such circumstances by which either the phenomenon might be rendered more conspicuous, or a novice might more easily try them, or by which I did try them only. The same thing I have often done in the following experiments; concerning all which this one admonition may suffice. Now from these experiments it follows not that all the light of the blue is more refrangible than all the light of the red; for both lights are mixed of rays differently refrangible, so that in the red there are some rays not less refrangible than those of the blue, and in the blue there are some rays not more refrangible than those of the red; but these rays in proportion to the whole light are but few, and serve to diminish the event of the experiment, but are not able to destroy it. For if the red and blue colours were more dilute and weak, the distance of the images would be less than an inch and a half; and if they were more intense and full, that distance would be greater, as will appear hereafter. These experiments may suffice for the colours of natural bodies. For in the colours made by the refraction of prisms this proposition will appear by the experiments which are now to follow in the next proposition.

PROP. II. THEOR. 2.—*The light of the sun consists of rays differently refrangible.*

THE PROOF BY EXPERIMENTS.

Exper. 3.—In a very dark chamber at a round hole about one third part of an inch broad made in the shut of a window I placed a glass prism, whereby the beam of the sun's light which came in at that hole might be refracted upwards towards the opposite wall of the chamber, and there form a coloured image of the sun. The axis of the prism (that is, the line passing through the middle of the prism from one end of it to the other end parallel to the edge of the refracting angle) was in this and the following experiments perpendicular to the incident rays. About this axis I turned the prism slowly, and saw the refracted light on the wall or coloured image of the sun first to descend, and then to ascend. Between the descent and ascent when the image seemed stationary, I stopped the prism, and fixed it in that posture, that it should be moved no more. For in that posture the refractions of the light at the two sides of the refracting angle, that is at the entrance of the rays into the prism, and at their going out of it, were equal to one another. So also in other experiments, as often as I would have the refractions on both sides the prism to be equal to one another, I noted the place where the image of the sun formed by the refracted light stood still between its two contrary motions, in the common period of its progress and regress; and when the image fell upon that place, I made fast the prism. And in this posture, as the most convenient, it is to be understood that all the prisms are placed in the following experiments, unless where some other posture is described. The prism therefore being placed in this posture, I let the refracted light fall perpendicularly upon a sheet of white paper at the opposite wall of the chamber, and observed the figure and dimensions of the solar image formed on the paper by that light. This image was oblong and not oval, but terminated with two rectilinear and parallel sides, and two semicircular ends. On its sides it was bounded pretty distinctly, but on its ends very confusedly and indistinctly, the light there decaying and vanishing by degrees. The breadth of this image answered to the sun's diameter, and was about two inches and the eighth part of an inch, including the penumbra. For the image was eighteen feet and a half distant from the prism; and at this distance that breadth, if diminished by the diameter of the hole in the window-shut, that is by a quarter of an inch, subtended an angle at the prism of about half a degree, which is the sun's apparent diameter. But the length of the image was about ten inches and a quarter, and the length of the rectilinear sides about eight inches, and the refracting angle of the prism whereby so great a length was made was 64°. With a less angle the length of the image was less, the breadth remaining the same. If the prism was turned about its axis that way which made the rays emerge more obliquely out of the second refracting surface of the prism, the image soon became an inch or two longer, or more; and if the prism was turned about the contrary way, so as to make the rays fall more obliquely on the first refracting surface, the image soon became an inch or two shorter. And therefore, in trying this experiment, I was as curious as I could be, in placing the prism by the above-mentioned rule exactly in such a posture that the refractions of the rays at their emergence out of the prism might be equal to that at their incidence on it. This prism had some veins running along within the glass from one end to the other, which scattered some of the sun's light irregularly, but had no sensible effect in increasing the length of the coloured spectrum. For I tried the same experiment with other prisms with the same success; and particularly with a prism which seemed free from such veins, and whose refracting angle was 62½°. I found the length of the image 9¾ or 10 inches at the distance of 18½ feet from the prism, the breadth of the hole in the window-shut being ¼ of an inch, as before. And because it is easy to commit a mistake in placing the prism in its due posture, I repeated the experiment four or five times, and always found the length of the image that which is set down

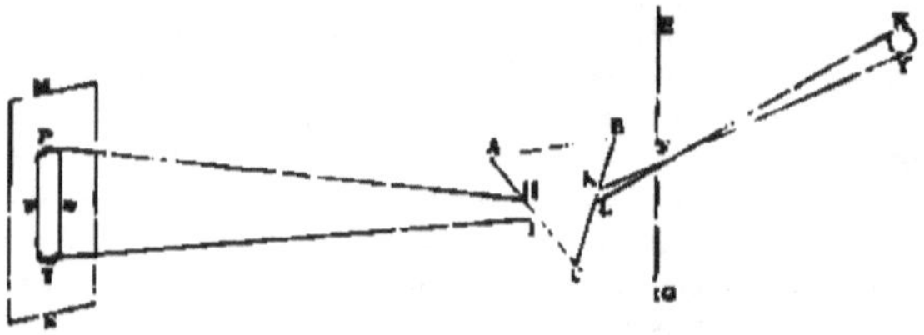

FIG. 16.

above. With another prism of clearer glass and better polish, which seemed free from veins, and whose refracting angle was 63½°, the length of this image at the same distance of 18½ feet was also about 10 inches, or 10⅛. Beyond these measures for about ¼ or ⅓ of an inch at either end of the spectrum the light of the clouds seemed to be a little tinged with red and violet, but so very faintly, that I suspected that tincture might either wholly or in great measure arise from some rays of the spectrum scattered irregularly by some inequalities in the substance and polish of the glass, and therefore I did not include it in these measures. Now the different magnitude of the hole in the window-shut, and different thickness of the prism where the rays passed through it, and different inclinations of the prism, to the horizon, made no sensible changes in the length of the image. Neither did the different matter of the prisms make any: for in a vessel made of polished plates of glass cemented together in the shape of a prism and filled with water, there is the like success of the experiment according to the quantity of the refraction. [After giving a rigorous proof that the rays in different parts of the spectrum are differently refracted, Newton proceeds.]

This image or spectrum P T was coloured, being red at its least refracted end T, and violet at its most refracted end P, and yellow, green, and blue in the intermediate spaces, which agrees with the first proposition, that lights which differ in colour do also differ in refrangibility. The length of the image in the foregoing experiments I measured from the faintest and outmost red at one end, to the faintest and outmost blue at the other end, excepting only a little penumbra, whose breadth scarce exceeded a quarter of an inch, as was said above.

PROP. III. THEOR. 3.—*The sun's light consists of rays differing in reflexibility, and those rays are more reflexible than others which are more refrangible.*

This is manifest by the ninth and tenth experiments, for in the ninth experiment, by turning the prism about its axis until the rays within it, which in going out into the air were refracted by its base, became so oblique to that base as to begin to be totally reflected thereby; those rays became first of all totally reflected which before at equal incidences with the rest had suffered the greatest refraction. And the same thing happens in the reflexion made by the common base of the two prisms in the tenth experiment.

BOOK I. PART 2.

PROP. V. THEOR. 4.—*Whiteness and all grey colours between white and black may be compounded of colours, and the whiteness of the sun's light is compounded of all the primary colours mixed in a due proportion.*

Exper. 11.—Let the sun's coloured image P T (Fig. 17) fall upon the wall of a dark chamber, as in the third experiment of the first book, and let the same be viewed through a prism *a b c*, held parallel to the prism A B C, by whose refraction that image was made, and let it now appear lower than before, suppose in the place S over against the red colour T. And if you go near to the image P T, the spectrum S will appear oblong and coloured

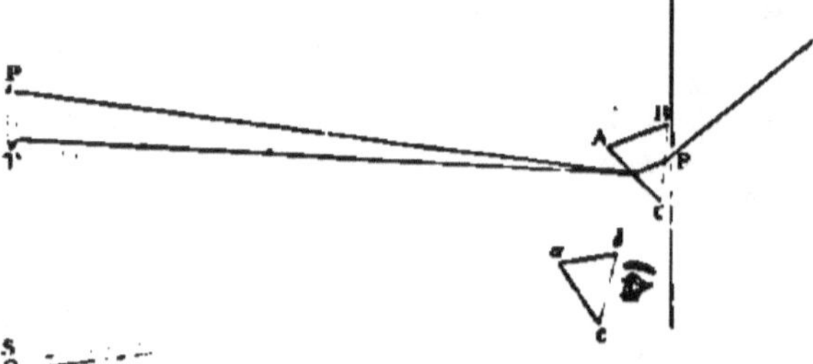

FIG. 17.

like the image P T; but if you recede from it, the colours of the spectrum S will be contracted more and more, and at length vanish, that spectrum S becoming perfectly round and white: and if you recede yet farther, the colours will emerge again, but in a contrary order. Now that spectrum S appears white in that case when the rays of several sorts which converge from the several parts of the image P T, to the prism *a b c*, are so refracted unequally by it, that in their passage from the prism to the eye they may diverge from one and the same point of the spectrum S, and so fall afterwards upon one and the same point in the bottom of the eye, and there be mingled.

And farther, if the comb be here made use of, by whose teeth the colours at the image PT may be successively intercepted, the spectrum s when the comb is moved slowly will be perpetually tinged with successive colours; but when, by accelerating the motion of the comb, the succession of the colours is so quick that they cannot be severally seen, that spectrum s, by a confused and mixed sensation of them all, will appear white.

PROP. II. THEOR. 2.—*All homogeneal light has its proper colour, answering to its degree of refrangibility, and that colour cannot be changed by reflexions or refractions.*

In the experiments of the fourth proposition of the first book, when I had separated the heterogeneous rays from one another, the spectrum *p t* formed by the separated rays did, in the progress from its end *p*, on which the most refrangible rays fell, unto its other end *t*, on which the least refrangible rays fell, appear tinged with this series of colours, violet, indigo, blue, green, yellow, orange, red, together with all their intermediate degrees in a continual succession, perpetually varying. So that there appeared as many degrees of colours as there were sorts of rays differing in refrangibility.

Exper. 5.—Now, that these colours could not be changed by refraction, I knew by refracting with a prism sometimes one very little part of this light, sometimes another very little part, as is described in the twelfth experiment of the first book (see Fig. 18). For by this refraction the colour of the light was never changed in the least. If any part of the red light was refracted, it remained totally of the same red colour as before. No orange, no yellow, no green or blue, no other new colour was produced by that refraction. Neither did the colour any way change by repeated refractions, but continued always the same red entirely as at first. The like constancy and immutability I found also in the blue, green, and other colours. So also if I looked through a prism upon any body illuminated with any part of this homogeneal light, as in the fourteenth experiment of the first book is described, I could not perceive any new colour generated this way. All bodies illuminated with compound light appear through prisms confused (as was said above), and tinged with various new colours, but those illuminated with homogeneal light appeared through prisms neither less distinct, nor otherwise coloured, than when viewed with the naked eyes. Their colours were not in the least changed by the refraction of the interposed prism. I speak here of a sensible change of colour: for the light which I here call homogeneal, being not absolutely homogeneal, there ought to arise some little change of colour from its heterogeneity. But if that heterogeneity was so little as it might be made by the said experiments of the fourth proposition, that change was not sensible, and therefore in experiments, where sense is judge, ought to be accounted none at all.

Exper. 6.—And as these colours were not changeable by refractions, so neither were they by reflexions. For all white,

FIG. 18.

grey, red, yellow, green, blue, violet bodies, as paper, ashes, red-lead, orpiment, indigo, bise, gold, silver, copper, grass, blue flowers, violets, bubbles of water tinged with various colours, peacocks' feathers, the tincture of *Lignum nephriticum*, and such like, in red homogeneal light appeared totally red, in blue light totally blue, in green light totally green, and so of other colours. In the homogeneal light of any colour they all appeared totally of that same colour, with this only difference, that some of them reflected that light more strongly, others more faintly. I never yet found any body which by reflecting homogeneal light could sensibly change its colour.

From all which it is manifest, that if the sun's light consisted of but one sort of rays, there would be but one colour in the whole world, nor would it be possible to produce any new colour by reflexions and refractions, and by consequence that the variety of colours depends upon the composition of light.

DEFINITION.

The homogeneal light and rays which appear red, or rather make objects appear so, I call rubrific or red-making; those which make objects appear yellow, green, blue, and violet, I call yellow-making, green-making, blue-making, violet-making, and so of the rest. And if at any time I speak of light and rays as coloured or endued with colours, I would be understood to speak not philosophically and properly, but grossly, and accordingly to such conceptions as vulgar people in seeing all these experiments would be apt to frame. For the rays, to speak properly, are not coloured. In them there is nothing else than a certain power and disposition to stir up a sensation of this or that colour. For as sound in a bell or musical string, or other sounding body, is nothing but a trembling motion, and in the air nothing but that motion propagated from the object, and in the sensorium 'tis a sense of that motion under the form of a sound; so colours in the object are nothing but a disposition to reflect this or that sort of rays more copiously than the rest, in the rays they are nothing but their dispositions to propagate this or that motion into the sensorium, and in the sensorium they are sensations of those motions under the forms of colours.

APPENDIX B.

BURNING MAGNESIUM WIRE, A SOURCE OF LIGHT FOR PHOTOGRAPHIC PURPOSES.[1]

Another interesting practical application of our knowledge concerning the properties of the kind of light which certain bodies emit when heated, is the employment of the light evolved by burning magnesium wire for photographic purposes. The

[1] Professor Roscoe on Spectrum Analysis, Royal Institution of Great Britain Proceedings, May 6, 1864.

spectrum of this light is exceedingly rich in violet and ultra-violet rays, due partly to the incandescent vapour of magnesium, and partly to the intensely-heated magnesia formed by the combustion. Professor Bunsen and the speaker in 1859 determined the chemically active power possessed by this light, and compared it with that of the sun; and they suggested the application of this light for the purpose of photography. They showed[1] that a burning surface of magnesium wire, which, seen from a point at the sea's level, has an apparent magnitude equal to that of the sun, effects on that point the same chemical action as the sun would do if shining from a cloudless sky at a height of 9° 53′ above the horizon. On comparing the visible brightness of these two sources of light it was found that the brightness of the sun's disc, as measured by the eye, is 524·7 times as great as that of burning magnesium wire, when the sun's zenith distance is 67° 22′; whilst at the same zenith distance the sun's *chemical* brightness is only 36·6 times as great. Hence the value of this light as a source of the chemically active rays for photographic purposes becomes at once apparent.

Professor Bunsen and the speaker state, in the memoir above referred to, that "the steady and equable light evolved by magnesium wire burning in the air, and the immense chemical action thus produced, render this source of light valuable as a simple means of obtaining a given amount of chemical illumination; and that the combustion of this metal constitutes so definite and simple a source of light for the purpose of photochemical measurement, that the wide distribution of magnesium becomes desirable. The application of this metal as a source of light may even become of technical importance. A burning magnesium wire of the thickness of 0·297 millimetre evolves, according to the measurement we have made, as much light as 74 stearine candles of which five go to the pound. If this light lasted one minute, 0·987 metre of wire, weighing 0·120 gramme, would be burnt. In order to produce a light equal to 74 candles burning for ten hours, whereby about 20lbs. of stearine are consumed, 72·2 grammes (2½ ounces) of magnesium would be required. The

[1] Phil. Trans. 1859, p. 920.

magnesium wire can be easily prepared by forcing out the metal from a heated steel press having a fine opening at bottom: this wire might be rolled up in coils on a spindle, which could be made to revolve by clockwork, and thus the end of the wire, guided by passing through a groove or between rollers, could be continually pushed forward into a gas or spirit-lamp flame in which it would burn."

It afforded the speaker great pleasure to state that the foregoing suggestion had now been actually carried out. Mr. Edward Sonstadt has succeeded in preparing magnesium on the large scale, and great credit is due to this gentleman for the able manner in which he has brought the difficult subject of the metallurgy of magnesium to its present very satisfactory position.

Some fine specimens of crude and distilled magnesium weighing 3 lb. were exhibited as manufactured by Mr. Sonstadt's process, by Messrs. Mellor and Co. of Manchester.

The wire is now to be had at the comparatively low rate of 3*d.* per foot; and half an inch of the wire evolves on burning light enough to transfer a positive image to a dry collodion plate; whilst by the combustion of 10 grains a perfect photographic portrait may be taken; so that the speaker believed that for photographic purposes alone the magnesium light will prove most important. The photo-chemical power of the light was illustrated by taking a portrait[1] during the discourse. In doing this the speaker was aided by Mr. Brothers, photographer, of Manchester, who was the first to use the light for portraiture.

APPENDIX C.

ON THE CHEMICAL ACTION OF THE CONSTITUENT PARTS OF SOLAR LIGHT.[2]

The chemical action effected by the several portions of the solar spectrum depends not only upon the nature of the refracting body, but also upon the thickness of the column of air through which the light has to pass before decomposition. In

[1] Of Professor Faraday. [2] Bunsen and Roscoe, Phil. Trans. 1859.

the following experiments we have employed prisms and lenses of quartz, cut by Mr. Darker of Lambeth, instead of glass prisms, which, as is well known, absorb a large portion of the chemically active rays. In order to render our experiments as free as possible from the irregularities arising from variation in the atmospheric absorption, the observations were made quickly one after the other, so that the zenith distance of the sun altered but very slightly.

A perfectly cloudless day was chosen for these observations, and the direct sunlight reflected from the speculum mirror of a Silbermann's heliostat through a narrow slit into our dark room. The spectrum produced by the rays passing through two quartz prisms and a quartz lens fell upon a white screen, which was covered with a solution of sulphate of quinine to render the ultra-violet rays and the accompanying dark lines visible. In this screen a narrow slit was made, through which the rays from any wished-for portion of the spectrum could be allowed to pass, so as to fall directly upon the insolation vessel,[1] situated at the distance of from four to five feet. A finely-divided millimetre scale was also placed on the screen, by means of which the distance between the Fraunhofer lines could be accurately measured, and the portion of light employed thus exactly determined.

In order to recognise with accuracy the various portions of the spectrum, we employed a map of the dark lines prepared by Mr. Stokes, which he most kindly placed at our disposal. The figure (Fig. 9) contains a copy of Mr. Stokes's map, with the distance measured by him, and letters given according to his notation. We have divided the space between the letter A in the red to the last ray Stokes observed, W in the lavender rays, into 160 equal parts, and we represent the position and breadth of the bundle of rays which effected a given action upon the insolation vessel as follows:—If a bundle of rays lying between

[1] This vessel was filled with the sensitive mixture of chlorine and hydrogen gases together with water. The chemically active rays effected a union of the gases, and the resulting hydrochloric acid gas being absorbed by the water, gave a diminution of volume, directly proportional to the intensity of the acting chemical rays.

the abscissæ 20·5 and 34 in Fig. 9, page 18, had to be represented, we should call the edge of the bundle towards A, ½ DE, and that towards W, ½ FG, whilst the middle of the portion of the spectrum, which produces the action, we call "½ DE to ½ FG." The breadth of this bundle of rays in which the insolation vessel was completely bathed was $\frac{5}{108}$ of the total length of the spectrum.

The following table gives the direct results of a series of observations made by perfectly cloudless sky at Heidelberg, on the 14th of August, 1857, under a barometric pressure of 0·7494 m. The first column gives the numbers of the observations in the order in which they were made; Column II. the times of observation in true solar time; Column III. the portion of spectrum under examination; and Column IV. the action corresponding to this portion.

I.	II.	III.	IV.
	H. M.		
1	10 54 A.M.	From ½ GH to I . .	48·80
2	10 58 A.M.	From ½ DE to E . .	1·27
3	11 4 A.M.	From C to ½ DE . .	0·47
4	11 8 A.M.	From N_1 to ½ QR . .	18·28
5	11 13 A.M.	From ½ RS to ½ ST .	2·03
6	11 41 A.M.	From ½ ST to ½ UV .	1·27
7	11 47 A.M.	From ½ N_4Q to ½ RS .	11·73
8	11 50 A.M.	From ½ ST to ½ UV .	1·02
9	11 54 A.M.	From ½ IM_1 to N_4 . .	37·87
10	11 57 A.M.	From I_1 to ½ IM_1 . .	57·42
11	0 1 P.M.	From I_1 to ½ IM_1 . .	52·30
12	0 4 P.M.	From ½ GH to H . .	61·38
13	0 7 P.M.	From ½ FG to G . .	27·61
14	0 16 P.M.	From ½ FG to G . .	28·74
15	0 20 P.M.	From ½ DE to F . .	1·39
16	0 25 P.M.	From ½ N_4Q to ½ RS .	13·19
17	0 32 P.M.	From ½ N_4Q to ½ RS .	12·11
18	0 40 P.M.	From G to ½ GH . .	53·76
19	0 42 P.M.	From ½ GH to H . .	58·74
20	0 45 P.M.	From ½ GH to I . .	53·9

If the refraction of the unit amount of incident light which is reflected from the mirror of the heliostat at the commencement and at the end of the series of the experiments be calculated, we get the numbers 0·644 and 0·642, which differ so slightly that the variations brought about by the reflection may be neglected without overstepping the observational errors. At the times of observation on the 14th of August, 1857, the sun's zenith distance was as follows:—

At 10h. 54m. A.M. . . . 37° 35′
At 0 0 A.M. . . . 35 13
At 0 45 P.M. . . . 36 16

The chemical intensity of the sun's rays at these various periods may be calculated by formula (14). They are in the proportion of the numbers 1·002, 1·000, and 1·016. Although the differences between these numbers are but small, we have reduced all the observations to that chemical action which would have been observed if they had all been made at 12h. 0m. A.M. upon the day in question. The following table contains the numbers thus reduced, the mean value having been taken of those observations which occur more than once:—

No.	True solar time	Position in the spectrum.	Relative chemical action.
	H. M.		
1	10 54 A.M.	From ½ GH to I . .	52·7
2	10 58 A.M.	From ½ DE to E . .	1·3
3	11 4 A.M.	From C to ½ DE . .	0·5
4	11 8 A.M.	From N_1 to ½ QR . .	18·9
5	11 13 A.M.	From ½ RS to ½ ST .	2·1
6	11 11 A.M.	From ½ ST to ½ UV .	1·2
7	11 47 A.M.	From ½ N_1Q to ½ RS .	12·5
8	11 54 A.M.	From ½ IM, to N_1 . .	38·6
9	0 1 P.M.	From H_1 to ½ IM, .	55·1
10	0 4 P.M.	From ½ GH to H . .	60·5
11	0 16 P.M.	From ½ FG to G . .	28·4
12	0 20 P.M.	From ½ DE to F . .	1·4
13	0 40 P.M.	From G to ½ GH . .	54·5

The lines *a a a a* (Fig. 9, page 18) give a representation of the relative chemical action which the various parts of the spectrum, the rays of which have only passed through air and quartz, effect on the sensitive mixture of chlorine and hydrogen. It is seen that this action attains many maxima, of which the largest lies by ½ GH to H, and the next at I, and also that the action diminishes much more regularly and rapidly towards the red than towards the violet end of the spectrum.

The sun, when it was employed for these experiments, was 35° 13' removed from the zenith. If the atmosphere were throughout of the density corresponding to 0·76 m. and 0° C., the perpendicular height which, during our experiment, it would have possessed, is

$$\frac{0{\cdot}7494}{0{\cdot}000095081} = 7{,}881 \text{ metres.}$$

The depth of atmosphere through which the rays had to pass in this experiment was, however,

$$\frac{7881}{\cos 35^\circ\ 13'} = 9{,}647 \text{ metres.}$$

We have stated in one of our previous communications,[1] that the solar rays which at different hours of the day pass through the same column of chlorine are altered in a very different manner. This shows that rays of different chemical activity are absorbed in very different ways by the air. The above results are therefore only applicable for sunlight which has passed through a column of air, measured at 0·76 m. and 0° C. of 9,647 metres in thickness. For rays which have to pass through a column of air of a different length from this, the chemical action of the various constituents of the spectrum must be different. The order and degree in which the chemical rays are absorbed, may be obtained by repeating the observations according to the above method from hour to hour during a whole day. Such a series of experiments we have unfortunately as yet been unable to execute, owing to the variability of the weather in our latitudes. One very imperfect series of observations we can, however, quote, and they suffice to show that the relation between the chemical action of the spectral colours is perceptibly altered when the thickness of air through which the rays pass changes from 9,647 to 10,735 metres.

These experiments were likewise made on August 14th, 1857, in the short space of time from 9h. 44m. to 10h. 10m. A.M., and gave the following numbers reduced to the zenith distance (42° 46'), corresponding to 10h. 0m. A.M. They were, however, made with a bundle of rays of a different thickness from the

[1] Phil. Trans. 1857, p. 617, &c.

former experiments, and therefore cannot be compared with those.

No.	Time.	Portion of Spectrum	Relative chemical action.
	H. M.		
1	9 44 A.M.	From ½ GH to I . .	14·6
2	9 48 A.M.	From N_2 to H_γ . . .	10·1
3	9 54 A.M.	From $\frac{1}{10}$ H_1S to ½ ST .	2·4
4	9 59 A.M.	From ½ ST to U . .	0·0
5	10 4 A.M.	From G to ½ GH . .	13·0
6	10 8 A.M.	From F to ½ FG . .	7·1
7	10 11 A.M.	From *b* to ½ FG . .	3·2
8	10 15 A.M.	From ½ DE to ½ EF .	0·4

From this it is seen that the relation of the chemical action of the spectrum from the line F to the line H undergoes a considerable alteration when the rays have to pass through a column of air 10,735 metres in height instead of 9,647 metres.

An extended series of measurements of the chemical action of the several portions of the solar spectrum under various conditions of atmospheric extinction may prove of great interest, if, as we can now scarcely doubt, the solar spots appear at regular intervals, and our sun belongs to the class of fixed stars of variable illuminating power. It is possible that such observations, made during the presence and during the absence of the solar spots, may give rise to some unlooked-for relations concerning the singular phenomena occurring on the sun's surface. Whether, however, the atmospheric extinction can ever be determined with sufficient accuracy to render visible the alteration in the light which probably occurs with the spots, is a question which can only be decided by a series of experimental investigations which must extend far beyond the scope of any single observer.

LECTURE II.

Continuous Spectrum of Incandescent Solids.—Effect of Increase of Heat.—Broken Spectrum of glowing Gases.—Application to Chemical Analysis.—Spectra of the Elementary Bodies.—Construction of Spectroscopes.—Means of obtaining Substances in the State of glowing Gas.—Examination of the Spectra of Coloured Flames.—Spectra of the Metals of the Alkalies and Alkaline Earths.—Delicacy of the Spectrum Analytical Method.

APPENDIX A.—Description of the Spectrum Reactions of the Salts of the Alkalies and Alkaline Earths.

APPENDIX B.—Bunsen and Kirchhoff on the Mode of using a Spectroscope.

APPENDIX C.—Bunsen on a Method of mapping Spectra.

IN the last lecture I pointed out to you some of the chief properties of the light with which we are now, I am glad to say, illumined—the light of the sun. I explained that the white sunlight can be divided up into a large number of different coloured rays by means of the prism; that these differently refrangible rays possess different properties, that we find the heating rays chiefly situated at the red end, or in the least refrangible part. I showed that we could separate out by certain means the light rays from the less refrangible ultra-red rays, and obtain at the dark focus of these rays the phenomena of incandescence and of combustion, showing that these rays, which do not affect the eye, are capable when brought together of producing ignition. We also saw that at and beyond the other end, the blue end, of the spectrum we have the rays termed the chemically active rays, and that these rays are capable of effecting chemical change.

We proceed to-day in the examination of the action of heat upon terrestrial matter in so far as it evolves light. The question may very properly be asked, "What has all this to do with chemical analysis?" It might be said, "It is true you have pointed out the difference between the various parts of the solar spectrum; but how is this connected with the analysis which we expect to be told about—with the method by means of which chemical substances may be detected or examined with a degree of accuracy beyond anything that has hitherto been attained?" In order to enable you to answer this question, let us begin by examining the action of heat upon terrestrial matter, and, in the first place, upon solid bodies. I have here the means of heating a long piece of platinum wire, first of all to redness, and by diminishing its length I shall be able to increase the temperature of the wire gradually until I raise it to the melting point of platinum. The first thing we observe when a solid body, such as this wire, is heated, is that it becomes red hot; and that as we increase the temperature, the light which it gives off increases in refrangibility, so that it ends by emitting light of every degree of refrangibility. I cannot show you on the screen the spectrum which this heated wire yields, simply because the intensity of light which it emits is insufficient for the purpose; but if I were to allow the light to fall into my eye through a prism, I should see that the red rays become first visible, and that then a gradual increase in the refrangibility of the light occurs, and that successively yellow, green, blue, and violet rays will be emitted as the temperature is increased up to a white heat, when all the rays of light are given off.

I will endeavour to render this fact visible to you in a rougher way by heating the wire gradually up to whiteness, and allowing the light to pass through these coloured glasses placed between you and the wire. At first, when it is red-hot, the glowing wire will be visible only through the red glass, none of the rays are able to pass through the blue glass; or in other words, there is no blue light given off: when the temperature is increased, blue rays begin to be given off, and these can pass through the blue glass, as you now plainly see when I raise the temperature of the wire. Here I can increase the temperature of the wire until we get at a point at which I have no doubt you will be able to see that the blue rays are emitted, and if I continue this and go on until the wire becomes intensely white-hot, you will see it through this blue glass perfectly well.

Such then is the action of increased temperature upon *solid* bodies. If I had taken any other substance which I could have heated in the same way, I should have produced the same effect: for it has been found that all solid and liquid substances act in this same way with regard to increase of heat; they all begin to be visibly hot at the same temperature, and the spectrum thus produced is in every case a continuous one.[1] I may remind you that this is the case by again throwing on the screen the spectrum of the white-hot carbon points

[1] This law was discovered by Draper (Phil. Mag. 1847). The only known exception to this law is glowing solid Erbia, whose spectrum exhibits bright lines; see Appendix E. to Lecture IV.

heated in the electric arc. Here we have this grand continuous band of light. The arrangements for producing this are simple enough. We require to connect the terminal wires from about sixty pairs of Grove's cells with the carbon electrodes of a Duboscq's lamp (E) contained inside this lantern. The light passes through a narrow vertical slit (s, Fig. 19), and by means of the moveable lens (L) a distinct image of the slit is thrown upon the screen (W W). A hollow prism filled with bisulphide of carbon (p_1) is now introduced at the distance of about two

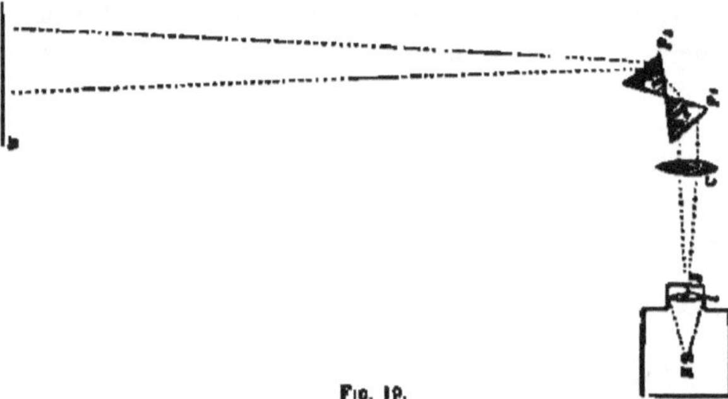

FIG. 19.

feet from the lens; next the lamp, with the arm carrying lens and prism, is turned round until the coloured band falls upon the screen, and the prism then adjusted to the angle of minimum deviation for the yellow rays. A second prism (p_2) is then interposed, and the lamp and arm again turned so as to allow the lengthened spectrum to fall on the screen. A drawing of lamp, lens, and prisms, thus placed, is shown in Fig. 20.

How does the case stand with respect to that important form of matter termed the gaseous? Do gases when they become incandescent all emit the same kind of light, like solids, or does each chemically different gas emit a characteristic and peculiar kind of light? I purpose now to show you that every different chemical element in the

FIG. 20.

state of gas, when heated until it becomes luminous, gives off a peculiar light, so that the spectrum of every element in the state of glowing gas is totally different from that of any solid body, inasmuch as, instead of giving a continuous spectrum, it presents a broken or discontinuous one containing bright bands or lines, indicative of the presence of the particular elementary gas in question.[1] I will illustrate this fundamental difference to you by means of the following experiment. It has long been known to chemists that certain substances have the power when brought into a colourless flame of producing peculiar tints. Thus, for instance, if we bring various bodies into the flame, such as the alkalies soda and potash, we observe that the flame becomes coloured in the first case of a bright yellow, and in the second of a pale violet tint; whilst the salts of strontium colour the flame crimson, and those of barium produce a green tint, and calcium compounds impart a red colour to flame. Here we have the beautiful non-luminous gas flame produced by the combustion of coal gas mixed with air, in what we know as the Bunsen burner. The air and gas mix in the chimney, the gas issuing from a jet at the centre of the foot, and the air entering by the holes at the side; the mixture burns with a light blue flame, which we can tinge with the peculiar colours of the alkalies by bringing a small fused bead of salt into the outer mantle of the flame on the loop of thin platinum wire (Fig. 21). Here is another substance called lithium; if we bring the slightest trace of this lithium salt into the flame, you perceive the magnificent crimson tint which it at once imparts to the flame: whilst in these other burners we see the colours due to the salts of potassium, calcium, strontium, and barium.

A most important observation has now to be made, namely, that all the salts of sodium give off this yellow light when brought into the flame; so, too, all the lithium

[1] Under peculiar circumstances certain incandescent gases give continuous spectra; see Appendix C. to Lecture IV.

compounds tint the flame crimson; and this property of emitting a peculiar kind of light is one of the means by which the presence of these various chemical substances can be detected. Here I will produce a peculiar blue flame by a substance which differs entirely from the foregoing in properties, viz. the non-metallic element selenium: it is a very volatile substance, and the blue flame lasts only for a short time. Further on we have

FIG. 21.

the characteristic green colours communicated to the flame by salts of copper and boracic acid.

I will next show you the same thing in other ways; for instance, I can here produce a much larger flame, and show you the colour of the same salts. I have a large gas burner which, when urged by this blowpipe, gives us a colourless flame three feet long. If I hold in this flame pieces of pumice-stone moistened with solutions of the chlorides of sodium, potassium, lithium, barium, strontium, and calcium, the colours imparted by these substances will be rendered evident. Again, I have another illustration in these gun-papers, which have been soaked in solutions of the chlorates of these metals and then dried. The combustion is rather quick, but by reflection on the white screen their peculiar colours come out well. Here you have the violet potash tint; here the bright green colour characteristic of the barium compounds. The common fireworks of the stage are further illustrations of the peculiar colours produced by certain chemical substances. I may imitate the red fire by igniting some chlorate of strontium in coal gas; we must melt the salt and then plunge it into the jar of burning coal gas, when we get this splendid combustion of oxygen in coal gas, coloured crimson by the ignited vapour of strontium salt.

We have already seen that the quality of the light emitted by solid bodies varies with difference of temperature. The quality of the light emitted by gaseous bodies, however, with certain exceptions—about which I shall have to speak subsequently—does not vary under change of temperature. Here I have the means of igniting some sodium salt at various temperatures. There, in the first place, is the bluish flame of burning sulphur, one of the coldest flames we can obtain, the temperature being about 1,820° Centigrade; then I next ignite the flame of burning carbon disulphide, having a temperature of 1,295° C. Here we see the flame of coal gas burning mixed with air: if I cut off the air, we get the common luminous flame of coal gas; but if I allow the air to mix with it before it burns, then we have this beautiful non-luminous flame. The temperature of this flame has been calculated to be 2,350° C. Here I have another jet, from which burns the blue flame of carbonic oxide gas, the body which produces the blue lambent flame frequently seen in coal fires: the temperature of this flame is somewhat higher, and has been calculated at 3,000°. If I bring a little common salt (sodium chloride) into these flames, you observe that in all cases we get them coloured yellow. I have here a hydrogen flame, of which the temperature is 3,259° C., and you see, when we bring the sodium salt into it, we have the same yellow colour produced; in other words, we cannot get sodium vapour either red-hot or blue-hot, it always remains yellow-hot; that is to say, the first moment that the sodium vapour becomes luminous, it gives off this particular and peculiar yellow light, and if we heat it more, the effect is not to alter the refrangibility of the rays, but merely to increase their intensity.

As a further illustration, we have this oxyhydrogen flame, of which the temperature is said to be about 6,000° C. If I bring a piece of soda into it, the effect is intense ignition; but still there is only the yellow light, no blue light. This indicates to us that when a body becomes gaseous, the light which it gives off is of a particular kind, and does not alter when we increase the temperature. One other experiment will indicate this to you still more fully, and this I can make by means of the electric spark, which I have here the means of producing. The temperature of this electric spark is so high that it has never been measured, but it is certainly infinitely higher even than the temperature of the oxyhydrogen flame. Still, if I bring this piece of sodium salt into the electric spark, I find that the same thing occurs—I get the same yellow-coloured light; and if I take some other substance, such as lithium, the permanent red colour which lithium vapour gives off will be clearly seen.

Now the methods by means of which we can obtain bodies in the state of luminous gas vary with the nature of the substance, but I would beg you to understand that the property which we have noticed with regard to sodium and the other alkalies is not confined to those bodies which have the power of being volatilized in such a flame as I have burning before me. This property belongs to matter in general; it belongs to every chemical element; and if we can by any method get the vapour of a chemical element so hot as to become luminous, we find that the light emitted by it is peculiar to itself, and is distinctive of that special body, whether under the ordinary circumstances the element be gaseous, solid, or liquid. Hence you see that we have at last reached the principles upon which the science of spectrum analysis is based, by means of which we can detect the presence of any of the elementary bodies when they can be obtained in this state of glowing gas.

We must now pass on to the consideration of the various methods by which the elements can be obtained as luminous gases.

I purpose to confine our attention in this lecture to the method by which we can detect the presence of the metals of the alkalies and alkaline earths. Let me, however, first point out to you the kind of spectrum which we obtain when we look at any one of these variously-coloured flames through a prism or spectroscope, the construction of which we will now briefly consider.

The simplest form of spectroscope which Bunsen first adopted is represented in Fig. 22. It consists of a common hollow prism (F) placed in a box; a telescope (C) is fixed at one side of the box, and a slit is placed at one end of a tube having a lens at the other end, in order to obtain a pure spectrum, and to render the rays parallel; and this

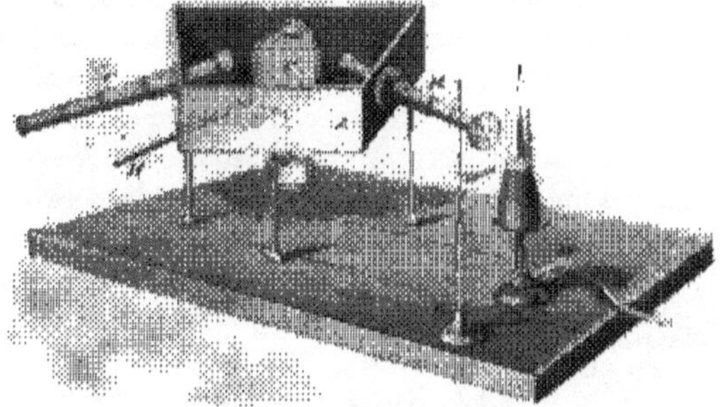

FIG. 22.

collimator (B) is fixed at the other side of the box. The substance to be examined is placed in the non-luminous Bunsen's flame, and the light passing through the slit falls upon the prism, and having been split up into its

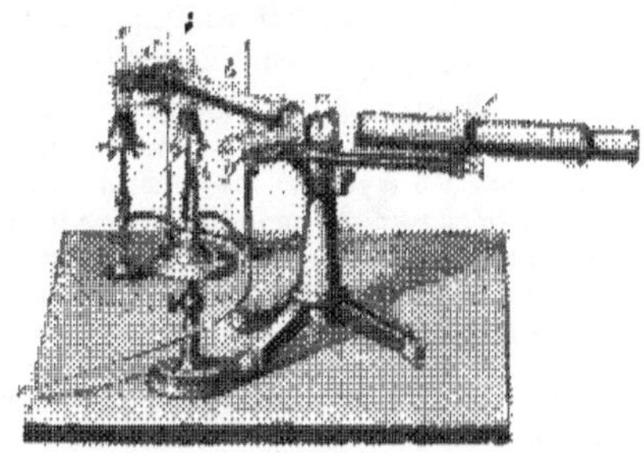

FIG. 23.

constituent parts, the differently-coloured rays pass through this telescope, are magnified, and then fall upon the retina. In Fig. 23 we have the more perfect form of the instrument represented, as made by Steinheil of Munich.[1] With this we are enabled to use two flames, and the apparatus is so arranged that we can see the two spectra placed one above the other. The object of this superposition of the spectra is evident: it is to enable us to see whether the substance under examination really is the body which it is supposed to be. For instance, putting a small quantity of the substance we know to contain sodium in this flame, we place a substance supposed to contain sodium in the other flame, and then by means of a small reflecting prism placed on the end of the slit we have the spectra of these two flames sent into the telescope one above the other, so that we see at the same time the spectrum of the pure sodium and the spectrum *supposed* to be that of sodium; and we can readily observe whether the lines coincide. If they coincide, and the two spectra have these lines exactly continuous one below the other, then we are quite certain that sodium, or any other substance which we may have been investigating, is

FIG. 24.

[1] This instrument consists of a prism (*a*) fixed upon a firm iron stand, and a tube (*b*) carrying the slit (*d*), seen on an enlarged scale in Fig. 24, through which the rays from the coloured flames (*e* and *e*) fall upon the prism, being rendered parallel by passing through a lens. The light having been refracted, is received by the telescope (*f*), and the image magnified before reaching the eye. The rays from each flame are made to pass into the telescope (*f*); one set through the upper uncovered half of the slit, the other by reflection from the sides of the small prism (*c*), Fig. 24, through the lower half; thus bringing the two spectra into the field of view at once, so as to be able to make any wished-for comparison of the lines. The small luminous gas flame (*h*), Fig. 23, is placed so as to illuminate a fixed scale contained inside the tube (*g*): this is reflected from the surface of the prism (*a*) into the telescope, and serves as a means of measuring the position of the lines.

present. Another arrangement for facilitating the comparison of spectra consists in the illuminated millimetre scale contained in the tube g (Fig. 23), a magnified reflection of which is thrown into the telescope from the surface of the prism. The illuminated scale is thus seen between the two superimposed spectra, and the position of any line or lines can be accurately determined. The further arrangements—mechanical and optical—of these inrtruments I need hardly trouble you with in detail. I have here a variety of spectroscopes kindly lent to me by the maker, Mr. Browning; one with one, one with two, one with three, and one with four prisms. The more prisms we employ, of course the greater dispersion we get, the more is the light drawn out into its special varieties; and the greater also is the intensity of the light which it is necessary to employ in order to get the rays to pass through this greater number of prisms.

I will next show you a drawing of the actual arrangement used by Kirchhoff (Fig. 25). There you see the prisms employed, four in number, placed one behind another on a horizontal table of cast iron. The light passes through the slit at the end of this tube. Here (top of Fig. 25) is an enlarged representation of the slit, the breadth of which can be altered at pleasure by means of the screw; on this slit is placed a small reflecting prism to enable us to get two superposed spectra. The light passes through the fine vertical slit, the rays are rendered parallel by the lens fixed at the end of the tube (A); it then passes through these four prisms, and the rays thus split up into constituent parts fall on to the telescope (B), at the end of which the eye is placed. This, then, gives you the simplest, and at the same time the most delicate and complete, form of spectroscope.

We have here representations as truly painted as possible (see Frontispiece) of what is seen when we allow a light from such coloured flames as those which have been burning to fall on to the retina through a spectroscope properly arranged.

At the top of the diagram (No. 1) is a drawing showing a solar spectrum, and underneath we have the spectra of the alkalies and alkaline earths, potassium (No. 2), sodium (No. 7), and lithium (No. 8), calcium

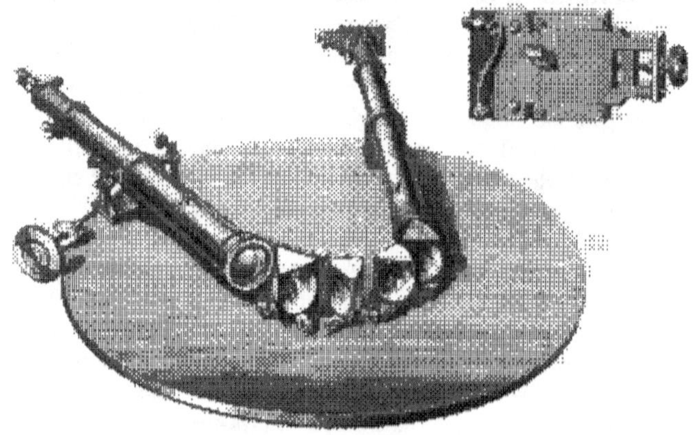

FIG. 25.

(No. 9), strontium (No. 10), and barium (No. 11), together with the two new metals rubidium and cæsium (Nos. 3 and 4), discovered by Bunsen, about which I shall speak in my next lecture; also the spectra of thallium and indium (Nos. 5 and 6), two other new metals, one of which was lately discovered by our countryman Mr. Crookes. You will perceive in the first place that each of these spectra is different from the rest, although they all possess the common characteristic of containing bright lines or bands, which occur in various portions of the spectrum, and indicate the peculiar kind of light which these various bodies, when brought into a state of glowing gas, emit. The sodium flame when observed by means of the spectroscope exhibits only one bright yellow line together with a faint continuous spectrum; in other words, this light is monochromatic, or nearly so: almost all the light which glowing sodium vapour gives off is light of one degree of refrangibility, and the spectrum is confined to one very narrow yellow band. The red light, which we saw was due to the presence of lithium, when seen through a prism gives this beautiful red line, together with this paler orange line. I need not describe the more complicated spectra of strontium, calcium, and barium: suffice it to say that they each yield peculiar bright bands, perfectly characteristic of the metal in question, as is seen at once by reference to the drawings.[1]

For the purpose of enabling any observer unacquainted with the spectra to identify with certainty the presence of any of the foregoing metals by means of their bright lines, and to lay down their positions in his own instrument, the following method of mapping the spectra has been devised by Bunsen. The millimetre scales (Fig. 26) represent the illuminated divisions seen with the scale of the spectroscope (*g*, Fig. 23): the exact position of the bright lines in any spectrum is shown by the black marks below the divisions; whilst their breadth, intensity, and gradation are indicated by the breadth, depth, and contour of these blackened surfaces. When the spectrum contains a continuous portion of light, this

[1] For the special description of these spectra see Appendix A, p. 72.

is shown by a continuous black band above the divisions. The positions of the fixed solar lines are given on the first horizontal scale, and those of the most prominent bands in several of the elements are placed as fiducial points at the bottom of the map.[1]

Having thus made ourselves acquainted with this new mode of chemical analysis, we may ask ourselves, "What improvement is this upon our ordinary chemical methods? What benefit is it to us that barium gives us these peculiar bands, that strontium yields certain different bands, that calcium produces others again? We know already that the chemical reactions of these bodies are very different, and we can detect these substances by ordinary chemical analysis." The answer to this is, that the new method is far more delicate than anything which we have hitherto employed, so delicate indeed as almost to pass belief, so that we have hereby obtained a means of examining the composition of terrestrial matter with a degree of exactitude hitherto unknown.

I will try to give you some idea of the delicacy of these spectrum reactions. I can show that the reaction for sodium is so sensitive that we can detect the presence of this element everywhere. There is not a speck of dust or a mote in the sunbeam which does not contain chloride of sodium. Sodium is a prevailing element in the atmosphere; we are constantly breathing in portions of the compound of this elementary substance together with the air which we inhale. Two-thirds of the earth's surface is covered with salt water, and the fine spray which is continually being carried up into the air by the dashing of the waves evaporates, leaving the

[1] For further information see Appendix C, p. 92.

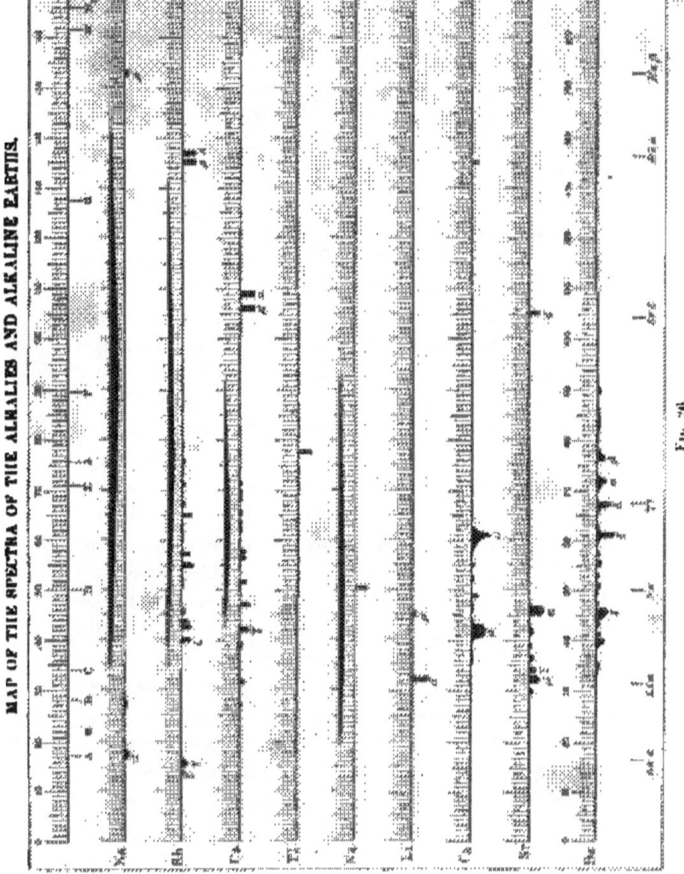
MAP OF THE SPECTRA OF THE ALKALIES AND ALKALINE EARTHS.
Na
Rb
Cs
Tl
K
Li
Ca
Sr
Ba

minute specks of salt which we see dancing as motes in the sunbeam. If I clap my hands, or if I shake my coat, or if I knock this dusty book, I think you will observe that this flame becomes yellow, and this not because it is the hand or coat of a chemist, but simply because the dust which everybody carries about with him is mixed with sodium compounds. When I place in the colourless flame this piece of platinum wire, which has been lying on the table for a few minutes since I heated it red-hot, you see there is sodium in it; there, we have for one moment a glimpse of a yellow flame. If I heat the wire in the flame, the sodium salts will all volatilize, and the yellow tinge will quite disappear; but if I now draw the wire once through my fingers, you observe the sodium flame will on heating the wire again appear. If I draw it through my mouth and heat it again, it will be evident that the saliva contains a very considerable quantity of sodium salts. Let me leave the wire exposed here, tied round this rod, so that the end does not touch anything, for ten minutes or a quarter of an hour; I shall then obtain the sodium reaction again, even if the wire be now perfectly clean. This is because sodium salts pervade the atmosphere, and some particles of sodium dust flying about in the air of the room settle on the wire, and show their presence in the flame.

I hope in the next lecture to consider the history of the subject, and to point out to you that this constant reaction of sodium puzzled the old observers very much. They thought this reaction must be due to the presence of water, for there was no other substance which was so commonly diffused; and it is only recently that this yellow reaction has been recognised as being due to this metal, sodium.

To refer for a moment to the distribution of lithium

compounds: we must remember that this substance, giving the beautiful red flame which you saw just now and the spectrum exhibiting the one bright red line, was until lately only known to exist in three or in four comparatively rare minerals. The moment, however, we come to examine substances by the method of spectrum analysis, we find that the brilliant red line, which is characteristic of the presence of lithium, occurs very frequently. And why, then, was not the red flame noticed before? Because when the light was examined by means of the eye alone, the red-coloured flame was masked by the presence of soda salts, and other substances affecting the flame, so that the red tint produced by the small quantity of lithium was unseen. But when we examine the flame with the prism, then all these lines range themselves into due order, no one interfering with the other. The presence of lithium may be thus easily detected, though it may be mixed with ten thousand times its bulk of sodium compounds, because, as you see by reference to this chart, the sodium line occurs in a different position to the lithium line, according to the differences in their refrangibilities. We now learn that this supposed rare substance is found to be most widely distributed,—not, it is true, in very large quantities, but still that it is one of the most widely diffused of the elementary bodies. Lithium not only occurs in very many minerals, but also in the juice of plants, in the ashes of the grape, in tea, coffee, and even in milk, in human blood, and in muscular tissue. And who knows what part this hitherto rare substance may not play even in the animal economy? It has been also found in meteoric stones, in the water of the Atlantic Ocean, as well as in that of most mineral springs and many rivers.

It is present in the ashes of tobacco, and, if we hold the end of a cigar in the colourless flame, we may always notice the red lithium line when the light is examined with a spectroscope. Dr. W. Allen Miller has lately found lithium in very large quantities in the water of a spring in the Wheal Clifford Mine in Cornwall.[1] This water contains 26 grains of lithium chloride in one gallon, and the spring flows at such a rate as to pour forth 800lbs. of this salt every twenty-four hours!

Here we have a table showing the great delicacy of the methods of spectrum analysis :—

1. *Sodium.* $\frac{1}{3000000}$ part of a milligramme, or $\frac{1}{180000000}$ part of a grain, of soda can easily be detected. Soda is always present in the air. All bodies exposed to the air show, when heated, the yellow soda line. If a book be dusted near the flame, the soda reaction will be seen.

2. *Lithium.* $\frac{1}{100000}$ part of a milligramme, or $\frac{1}{6000000}$ part of a grain, can be easily detected. Lithium was formerly only known to exist in four minerals : it is now found by spectrum analysis to be one of the most widely distributed elements. It exists in almost all rocks, in sea and river (Thames) water, in the ashes of most plants, in milk, human blood, and muscular tissue.

3. *Strontium.* $\frac{6}{100000}$ of a milligramme, or $\frac{1}{1000000}$ of a grain, of strontia is easily detected. Strontia has been shown to exist in very many limestones of different geological ages.

4. *Calcium.* $\frac{6}{100000}$ of a milligramme, or $\frac{1}{1000000}$ of a grain, of lime can be easily detected.

5. *Cæsium and Rubidium.* These new alkaline metals were discovered by Bunsen in the mineral waters of

[1] Chem. News, x. 181.

Baden and Dürkheim. Forty tons of mineral water yielded 200 grains of the salts of the new metals.

6. *Thallium.* A new metal discovered by Mr. Crookes in 1861, distinguished by the splendid green line which its spectrum exhibits. It is found in iron pyrites, and resembles lead in its properties.

7. *Indium.* Discovered in zinc blende by Professors Reich and Richter: found in very minute quantities. It is distinguished by the two indigo bands seen in its spectrum.

I will now endeavour to illustrate, by means of the electric lamp, the fact that all these bodies give off coloured lights, and that each of these coloured lights is of a peculiar kind; and I would wish first to show you that when we bring a small fraction of a grain of common salt, chloride of sodium, on to the lower carbon of the lamp, we obtain a distinct yellow band which was not seen before, for previously, you will remember, we had a perfectly continuous spectrum. This yellow band is due to the presence of sodium. You will probably see that there are other bands present as well as the sodium band, because it is impossible, owing to the delicacy of these reactions, to obtain any carbon which is perfectly free from other chemical salts, and the small impurities which exist in the carbon come out as evidence against us on the screen; yet I think you will see that we have the sodium line more distinctly visible than anything else.

No other metal but sodium gives this yellow band; still I must beg you to understand that this rough representation is not exactly that which you would see if you were to look at the yellow soda flame through a prism, by means of an accurate spectroscope. I would wish you to remember that this yellow line is in reality double when examined with a perfect optical arrangement, and that these lines are very fine, placed close together, and as thin as the finest spider's web. It is only because the arrangements I have to employ here for the purpose of exhibiting these lines on the screen are, optically considered, very crude and rough, that we get any appreciable breadth of this line.

Now allow me to show you the light which the body lithium gives off. For this purpose I will bring it on the same carbon, for by taking a new one we should not gain much, as all these poles are more or less impure. Here you observe the red line, which was not noticeable before. This splendid red band is due to the presence of lithium; and when we see it through an accurate instrument, it appears as fine as the finest slit of light which we can take. This bright red line is always found exactly in the same position; and the fixity of these lines is in fact the most important principle involved in our inquiry: they are unalterable in refrangibility.

I have next to direct your attention to the blue line which is now visible on my right. This is also caused by lithium, for when we heat up lithium vapour beyond a certain point, as high as I am now doing with the electric lamp, this blue band also becomes visible; but it is not visible when the temperature of the incandescent lithium vapour is lower. The blue ray may perhaps always be given off, even at lower temperatures; for if light requires to be of a certain intensity before it can affect the retina and become visible, and if, in order that the intensity of the light may thus increase, we must heat the vapour to a higher point, we have a complete explanation of the appearance of the blue line. It is important to notice that the positions of the red and of the orange lines seen at the lower temperature never shift or change in position the least when the temperature is changed. Hence the appearance of the red line is proof positive of the presence of lithium. In this lithium spectrum you will also notice the sodium line. We can never get rid of our friend sodium, he always remains stedfast to us; in fact, we should be sometimes glad to dispense with his presence, but it is not an easy matter to induce him to leave us.

I would next show you the spectra of metals of the alkaline earths. I will first bring a small quantity of strontium salt on to the pole, and we find that the strontium spectrum is characterised by a series of red lines, and also by a beautiful blue band almost identical, but not exactly so, with the blue line of lithium which I had the pleasure of showing you an instant ago. What a large number of bands we have here, especially in the red! Those red and blue bands are the ones to which I beg to draw your attention. These red lines now come out very distinctly; and we have here also the bright blue line flashing out brightly. This then is the strontium spectrum. In like manner I may show you the beautiful and characteristic spectra of barium, with its five green bands; and that of calcium, exhibiting special orange and green lines together with a purple band in the more refrangible part of the spectrum.

Now let us suppose that we have a mixture of compounds of all these substances which are capable of being volatilized, namely potassium, sodium, lithium, barium, strontium, and calcium, and let us expose this mixture to such a temperature that all the salts become volatilized, one after another, we shall see, in the first place, that the bands of those substances which are the most volatile appear first; that then, when these have burnt out, those next in order of volatility make their appearance; and that those which are the least volatile come out last. Thus we have the beautiful appearance of what may be called a natural dissolving view.

I place on the carbon poles a mixture containing a few grains of salts of all the above-mentioned metals. You see in the first place that the sodium line comes out at once, and afterwards the lines of the other metals gradually make their appearance. We have thus simply to place the smallest fraction of a grain of such a mixture as this before the slit of our spectroscope, and with the merest trace of substance we can in a moment obtain absolute and decisive evidence of the presence of all these substances, the lines coming out, as I said, like a dissolving view, one after another; and the minute quantity which we can thus detect is something almost marvellous. Here we have this splendid series of variegated bands, exhibiting the superposed spectra of all the substances I have mentioned. There you see the lithium red line; here the less refrangible red line of potassium; there the orange band of calcium and the red strontium bands; observe, if you please, the two blue bands, one due to strontium and the other to lithium. I shall have occasion to show you that there are some other very beautiful purple bands, characteristic of cæsium and rubidium, the new metals discovered by Bunsen, about the history of which I purpose speaking to you in the next lecture. Now the sodium is very nearly burnt out, and the lithium will soon disappear, whereas the green bands produced by the less volatile barium compounds will remain for a greater length of time.

In conclusion, gentlemen, I have to remind you that it is simply a question of temperature; it is only a matter of experimentation how, and in what way, we can best obtain the elementary bodies in the condition of glowing gas. Having done that, we can readily detect their presence by this very interesting and important property they possess, of each body emitting light of a peculiar and characteristic kind, light of various degrees of refrangibility; each giving what we term a discontinuous spectrum.

LECTURE II.—APPENDIX A.

DESCRIPTION OF THE SPECTRUM REACTION OF THE SALTS OF THE ALKALIES AND ALKALINE EARTHS.[1]

We now proceed to describe the peculiarities of the several spectra, the exact acquaintance with which is of practical importance, and to point out the advantages which this new method of chemical analysis possesses over the older processes.

SODIUM.

The spectrum reaction of sodium is the most delicate of all.

The yellow line Na *a* (see Chromolith. Table, No. 7), the only one which appears in the sodium spectrum, is coincident with Fraunhofer's dark line D, and is remarkable for its exactly defined form and for its extraordinary degree of brightness. If the temperature of the flame be very high, and the quantity of the substance employed very large, traces of a continuous spectrum are seen in the immediate neighbourhood of the line. In this case, too, the weaker lines produced by other bodies when near the sodium line are discerned with difficulty, and are often first seen when the sodium reaction has almost subsided.

The reaction is most visible in the sodium salts of oxygen, chlorine, iodine, bromine, sulphuric acid, and carbonic acid. But even in the silicates, borates, phosphates, and other non-volatile salts, the reaction is always evident. Swan[2] has already remarked upon the small quantity of sodium necessary to produce the yellow line.

The following experiment shows that the chemist possesses no reaction which in the slightest degree will bear comparison

[1] From Kirchhoff and Bunsen's first Memoir on Analysis by Spectrum Observations (Phil. Mag. vol. xx. 1860).

[2] Trans. Roy. Soc. Edin. vol. xxi. part iii. p. 411.

as regards delicacy with this spectrum-analytical determination of sodium. In a far corner of our experiment room, the capacity of which was about sixty cubic metres, we burnt a mixture of three milligrammes of chlorate of sodium with milk-sugar, whilst the non-luminous colourless flame of the lamp was observed through the slit of the telescope. Within a few minutes the flame, which gradually became pale yellow, gave a distinct sodium line, which, after lasting for ten minutes, entirely disappeared. From the weight of sodium salt burned and the capacity of the room, it is easy to calculate that in one part by weight of air there is suspended less than $\frac{1}{20000000}$ of a part of soda smoke. As the reaction can be observed with all possible comfort in one second, and as in this time the quantity of air which is heated to ignition by the flame is found, from the rate of issue and from the composition of the gases of the flame, to be only about 50 cub. cent. or 0·0647 grm. of air, containing less than $\frac{1}{20000000}$ of sodium salt, it follows that the eye is able to detect with the greatest ease quantities of sodium salt less than $\frac{1}{3000000}$ of a milligramme in weight. With a reaction so delicate, it is easy to understand why a sodium reaction is almost always noticed in ignited atmospheric air. More than two-thirds of the earth's surface is covered with a solution of chloride of sodium, fine particles of which are continually being carried into the air by the action of the waves. These particles of sea water cast thus into the atmosphere evaporate, leaving almost inconceivably small residues, which, floating about, are almost always present in the air, and are rendered evident to our eyesight in the sunbeam. These minute particles perhaps serve to supply the smaller organized bodies with the salts which larger animals and plants obtain from the ground. In another point of view, however, the presence of this chloride of sodium in the air is of interest. If, as is scarcely doubtful at the present time, the explanation of the spread of contagious disease is to be sought for in some peculiar contact-action, it is possible that the presence of so antiseptic a substance as chloride of sodium, even in almost infinitely small quantities, may not be without influence upon such occurrences in the atmosphere.

By means of daily and long-continued spectrum observations, it would be easy to discover whether the alterations of intensity in the line Na α produced by the sodium in the air have any connexion with the appearance and direction of march of an endemic disease.

The unexampled delicacy of the sodium reaction explains also the well-observed fact, that all bodies after a lengthened exposure to air show the sodium line when brought into a flame, and that it is only possible in a few salts to get rid of the line even after repeated crystallization from water which had only been in contact with platinum. A thin platinum wire, freed from every trace of sodium salt by ignition, shows the reaction most visibly on allowing it to stand for a few hours in the air: in the same way the dust which settles from the air in a room shows the bright line Na α. To render this evident it is only necessary to knock a dusty book, for instance, at a distance of some feet from the flame, when a wonderfully bright flash of yellow band is seen.

LITHIUM.

The luminous ignited vapour of the lithium compounds gives two sharply defined lines; the one a very weak yellow line, Li β, and the other a bright red line, Li α. This reaction exceeds in certainty and delicacy all methods hitherto known in analytical chemistry. It is, however, not quite so sensitive as the sodium reaction, only, perhaps, because the eye is more adapted to distinguish yellow than red rays. When nine milligrammes of carbonate of lithium mixed with excess of milk-sugar were burnt, the reaction was visible in a room of sixty cubic metres capacity. Hence, according to the method already explained, we find that the eye is capable of distinguishing with absolute certainty a quantity of carbonate of lithium less than $\frac{9}{1000000}$ of a milligramme in weight: 0·05 grm. of carbonate of lithium, burnt in the same room, was sufficient to enable the ignited air to show the red line Li α for an hour after the combustion had taken place.

The compounds of lithium with oxygen, iodine, bromine, and chlorine are the most suitable for the peculiar reaction; still the carbonate, sulphate, and even the phosphate, give almost as distinct a reaction. Minerals containing lithium, such as triphylline, triphane, petalite, lepidolite, require only to be held in the flame in order to obtain the bright line in the most satisfactory manner. In this way the presence of lithium in many felspars can be directly detected; as, for instance, in the orthoclase from Baveno. The line is only seen for a few moments, directly after the mineral is brought into the flame. In the same way the mica from Altenburg and Penig was found to contain lithium, whereas micas from Miask, Aschaffenburg, Modum, Bengal, Pennsylvania, &c., were found to be free from this metal. In natural silicates which contain only small traces of lithium this metal is not observed so readily. The examination is then best conducted as follows:—A small portion of the substance is digested and evaporated with hydrofluoric acid or fluoride of ammonium, the residue moistened with sulphuric acid and heated, the dry mass being dissolved in absolute alcohol. The alcoholic extract is then evaporated, the dry mass again dissolved in alcohol, and the extract allowed to evaporate on a shallow glass dish. The solid pellicle which remains is scraped off with a fine knife, and brought into the flame upon the thin platinum wire. For one experiment $\frac{1}{10}$ of a milligramme is in general quite a sufficient quantity. Other compounds besides the silicates, in which small traces of lithium require to be detected, are transformed into sulphates by evaporation with sulphuric acid or otherwise, and then treated in the manner described.

In this way we arrive at the unexpected conclusion that lithium is most widely distributed throughout nature, occurring in almost all bodies. Lithium was easily detected in forty cubic centimetres of the water of the Atlantic Ocean, collected in 41° 41′ N. latitude and 39° 14′ W. longitude. Ashes of marine plants (kelp), driven by the Gulf Stream on the Scotch coasts, contain evident traces of this metal. All the orthoclase and quartz from the granite of the Odenwald which we have examined contained lithium. A very pure spring water from the granite in Schleierbach, on the west side of the valley of the Neckar, was found to contain lithium, whereas the water from the red sandstone which supplies the Heidelberg laboratory was shown to contain none of this metal. Mineral waters, in a litre of which lithium could hardly be detected according to the ordinary methods of analysis, gave plainly the line Li *α* even if only a drop of the water on a platinum wire was brought into the flame.[1] All the ashes of plants growing in the Odenwald on a granite soil, as well as Russian and other potashes, contain lithium.

Even in the ashes of tobacco, in vine leaves, in the wood of the vine, and in grapes,[2] as well as in the ashes of the crops grown in the Rhine plain near Waghausel, Deidesheim, and Heidelberg, on a non-granitic soil, was lithium found. The milk of the animals fed upon these crops also contains this widely diffused metal.[3]

It is necessary to say that a mixture of volatile sodium and lithium salts gives the reaction of lithium alongside that of sodium with a precision and distinctness which are hardly perceptibly diminished. The red lines of the former substance are still plainly seen when the bead contains $\frac{1}{1000}$ part of lithium salt, and when to the naked eye the yellow soda flame appears untinged by the slightest trace of red. In consequence of the somewhat greater volatility of the lithium salt, the sodium reaction lasts longer than that of the other metal. In those cases, therefore, in which small quantities of lithium have to be detected in presence of large quantities of sodium salt, the bead must be brought into the flame whilst the observer is looking through the telescope. The lithium lines are often

[1] When liquids have to be brought into the flame, it is best to bend the end of the platinum wire, of the thickness of a horsehair, to a small ring, and to beat this ring flat. If a small drop of liquid be brought into this ring, enough adheres to the wire for one experiment.

[2] In the manufactories of tartaric the mother liquors contain so much lithium salts, that considerable quantities can thus be prepared.

[3] Dr. Folwarczny has been able to detect lithium in the ash of human blood and muscular tissue by the help of the line Li *α*.

only seen during a few moments amongst the first products of the volatilization.

In the production of lithium salts on the large scale, in the proper choice of a raw material, and in the arrangement of suitable methods of separation, this spectrum analysis affords most valuable aid. Thus it is only necessary to place a drop of mother liquor from any mineral spring in the flame and to observe the spectrum produced, in order to show that in many of these waste products a rich and hitherto unheeded source of lithium salts exists. In the same way, during the course of the preparation any loss of lithium in the collateral products and residues can be easily traced, and thus more convenient and economical methods of preparation may be found to replace those at present employed.[1]

POTASSIUM.

The volatile potassium compounds give, when placed in the flame, a widely-extended continuous spectrum, which contains only two characteristic lines, namely, one line, Ka *α*, in the outermost red, approaching the ultra-red rays, exactly coinciding with the dark line A of the solar spectrum, and a second line, Ka *β*, situated far in the violet rays towards the other end of the spectrum, and also identical with a particular dark line observed by Fraunhofer.

A very indistinct line, coinciding with Fraunhofer's line B, which, however, is only seen when the light is very intense, is not by any means so characteristic. The violet line is somewhat pale, but can be used almost as well as the red line for the detection of potassium. Owing to the position of these two lines, both situated near the limit at which our eyes cease to be sensitive to the rays, this reaction for potassium is not

[1] We obtain by such an improved method from two jars (about four litres) of a mother liquor from a mineral spring, which by evaporation with sulphuric acid gave 1·2 kil. of residue, half an ounce of carbonate of lithium of the purity of the commercial, the cost of which is about 140 florins the pound. A great number of other mineral-spring mother liquors which we examined showed a similar richness in compounds of lithium.

so delicate as the reaction for the two metals already mentioned. The reaction became visible in the air of our room when one gramme of chlorate of potassium mixed with milk-sugar was burnt. In this way, therefore, the eye requires the presence of $\frac{1}{1000}$ of a milligramme of chlorate of potassium in order to detect the presence of potassium.

Caustic potash, and all compounds of potassium with volatile acids, give the reaction without exception. Potash silicates, and other non-volatile salts, on the other hand, only produce the reaction when the metal is present in very large quantities. It is only necessary, however, to melt the substance with a bead of carbonate of sodium in order to detect potassium even when present in a very small quantity. The presence of the sodium does not in the least interfere with the reaction, and scarcely diminishes its delicacy. Orthoclase, sanidine, and adularia may in this way be easily distinguished from albite, oligoclase, Labradorite, and anorthite. In order to detect the smallest traces of potassium salt, the silicate requires only to be slightly ignited with a large excess of fluoride of ammonium on a platinum capsule, after which the residue is brought into the flame on a platinum wire. In this way it is found that almost every silicate contains potash. Salts of lithium diminish or influence the reaction as little as soda salts. Thus we need only to hold the end of a burnt cigar in the flame before the slit in order at once to see the yellow line of sodium and the two red lines of potassium and lithium, this latter metal being scarcely ever absent in tobacco ash.

STRONTIUM.

The spectra produced by the alkaline earths are by no means so simple as those produced by the alkalies. That of strontium is especially characterised by the absence of green bands. Eight lines in the strontium spectrum are remarkable, namely, six red, one orange, and one blue line. The orange line, Sr α, which appears close by the sodium line towards the red end of the spectrum, the two red lines, Sr β and, Sr λ, and, lastly, the blue line, Sr δ, are the most important strontium bands, both as regards their position and their intensity. For the purpose of examining the intensity of the reaction we quickly heated an aqueous solution of chloride of strontium, of a known degree of concentration, in a platinum dish over a large flame, until the water was evaporated and the basin became red-hot. The salt then began to decrepitate, and was thrown in microscopic particles out of the dish in the form of a white cloud carried up into the air. On weighing the residual quantity of salt, it was found that in this way 0·077 grm. of chloride of strontium had been mixed in the form of a fine dust with the air of the room, weighing 77,000 grms. As soon as the air in the room was perfectly mixed, by rapidly moving an umbrella, the characteristic lines of the strontium spectrum were beautifully seen. According to this experiment a quantity of strontium may be thus detected equal to the $\frac{1}{100000}$ part of a milligramme in weight. The chlorine and the other haloid salts of strontium give the action less vividly, the sulphate less distinctly, whilst the compounds of strontium with the non-volatile acids give either a very slight reaction or else none at all. Hence it is well first to bring the bead of substance alone into the flame, and then again after moistening with hydrochloric acid. If it be supposed that sulphuric acid is present in the bead, it must be held in the reducing part of the flame before it is moistened with hydrochloric acid, for the purpose of changing the sulphate into the sulphide, which is decomposed by hydrochloric acid. In order to detect strontium when combined with silicic, phosphoric, boracic, and other non-volatile acids, the following course of procedure gives the best results. Instead of fusing with carbonate of sodium in a platinum crucible, a conical spiral of platinum wire is employed: this spiral is heated to whiteness in the flame, and dipped whilst hot into finely powdered dried carbonate of sodium, which properly should contain so much water that a sufficient quantity adheres to the wire when it is once dipped into the salt. The fusion takes place in this spiral much more quickly than in a platinum crucible, as the mass of platinum requiring heating is small, and the flame comes into direct contact with the salt. As soon as the finely powdered mineral has been brought into the fused soda by means of a small platinum spatula, and the mass retained above the fusion point for a few minutes, the cooled mass has only to be turned upside down and knocked on the porcelain plate of the lamp in order to obtain the salt in one coherent bead. The fused mass is covered by a piece of writing-paper, and then broken by pressing it with the blade of a steel spatula until the whole is in the state of a fine powder. The powder is collected to one spot on the edge of the plate, and carefully covered with hot water, which is allowed to flow backwards and forwards over it, so that, after decanting and rewashing the powder several times, all the soluble salts are extracted without losing any of the residue. If a solution of chloride of sodium be employed instead of water, the operation may be conducted still more rapidly. The insoluble salt contains the strontium as carbonate, and one or two tenths of a milligramme of the substance brought on to the wire, and moistened with hydrochloric acid, is sufficient to produce the most intense reaction. It is thus possible, without help of platinum crucible, mortar, evaporating basin, or funnel and filter, to fuse, powder, digest, and wash out the substance in the space of a few minutes. The reactions of potassium and sodium are not influenced by the presence of strontium. Lithium also can be easily detected in presence of strontium, where the proportion of the former metal is not very small. The lithium line Sr α appears as an intensely red, sharply-defined band upon a less distinct red ground of the broad strontium band Sr β.

CALCIUM.

The spectrum produced by calcium is immediately distinguished from the four spectra already considered by the very characteristic bright green line Ca β. A second no less characteristic feature in the calcium spectrum is the intensely bright orange line, Ca α, lying considerably nearer to the red end of the spectrum than either the sodium line, Na α, or the orange band of strontium, $Sr\delta$. By burning a mixture consisting of chloride of calcium, chlorate of potassium, and milk-sugar, a white cloud is obtained which gives the reaction with as great a degree of delicacy as strontium salts do under similar circumstances. In this way we found that $\frac{6}{100000}$ of a milligramme in weight of chloride of calcium can be detected with certainty. Only the volatile compounds of calcium give this reaction; the more volatile the salt, the more distinct and delicate does the reaction become. The chloride, bromide, and iodide of calcium are in this respect the best compounds. Sulphate of calcium produces the spectrum, after it has become basic, very brightly and continuously. In the same way the reaction of the carbonate becomes more distinctly visible after the acid has been expelled. Compounds of calcium with the non-volatile acid remain inactive in the flame; but if they are attacked by hydrochloric acid, the reaction may be easily obtained as follows:—A few milligrammes of finely powdered substance are brought on to the moistened flat platinum ring in the moderately hot part of the flame, so that the powder is fritted but not melted on to the wire: if a drop of hydrochloric acid be now allowed to fall into the ring, so that the greater part of the acid remains hanging on to the wire, and if then the wire be brought into the hottest part of the flame, the drop evaporates in the spheroidal state without ebullition. The spectrum of the flame must be observed during this operation; and it will be noticed that at the moment when the last particles of liquid evaporate a bright calcium spectrum appears. If the quantities of the metal present are very small, the characteristic lines are only seen for a moment; if larger quantities are contained, the phenomenon lasts for a longer time.

Only in the silicates which are decomposed by hydrochloric acid can the calcium be thus found. In those minerals which are not attacked by that acid the following method may be best employed for the detection of calcium. A few milligrammes of the substance under examination, in a state of fine division, are brought upon a flat platinum lid, together with about a gramme of fluoride of ammonium, and the mixture is gently ignited until all the fluoride is volatilized. The slight crust of salt remaining is moistened with a few drops of sulphuric acid, and the excess of acid removed by heat. If about a milligramme of the residual sulphates be scraped together with a knife, and brought into the flame, the characteristic spectra of potassium, sodium, and lithium, supposing these three metals to be present, are first obtained, either simultaneously or consecutively. If calcium and strontium be also present, the corresponding spectra generally appear somewhat later, after the potassium, sodium, and lithium have been volatilized. When only traces of strontium and calcium are present, the reaction is not always seen: it becomes, however, immediately apparent on holding the bead for a few moments in the reducing flame, and, after moistening it with hydrochloric acid again, bringing it into the flame.

These easy experiments, such as either heating the specimen alone, or after moistening with hydrochloric acid, or after treating the powder with fluoride of ammonium either alone or in presence of sulphuric or by hydrochloric acid, provide the mineralogist and geologist with a series of most simple methods of recognising the components of the smallest fragment of many substances (such, for instance, as the double silicates containing lime) with a certainty which is attained in an ordinary analysis only by a large expenditure of time and material. The following examples will illustrate this statement.

1. A drop of sea water heated on the platinum wire shows at first a strong sodium reaction; and after volatilization of the chloride of sodium, a weak calcium spectrum is observed, which on moistening the wire with hydrochloric acid becomes at once very distinct. If a few decigrammes of the residual salts obtained by the evaporation of sea water be treated in the manner described under lithium with sulphuric acid and alcohol, the potassium and lithium reactions are obtained. The presence of strontium in sea water can be best detected in the boiler-crust from sea-going steamers. The filtered hydrochloric acid solution of such a crust leaves on evaporation and subsequent treatment with a small quantity of alcohol a residue slightly yellow-coloured from basic iron salt, which is deposited after some days, and

can then be collected on a small filter and washed with alcohol. The filter burnt on a fine platinum wire and held in the flame gives besides the calcium lines an intensely bright strontium spectrum.

2. Mineral waters often exhibit the reactions of potassium, sodium, lithium, calcium, and strontium by mere heating. If, for example, a drop of the Dürkheim or Kreuznach waters be brought into the flame, the lines Na α, Li α, Ca α, and Ca β are at once seen. If instead of using the water itself a drop of the mother liquor is taken, these bands appear most vividly. As soon as the chlorides of sodium and lithium have been to a certain extent volatilized, and the chloride of calcium has become more basic, the characteristic lines of the strontium spectrum begin to show themselves, and continue to increase in distinctness until at last they come out in all their true brightness. In this case, therefore, by the mere observation of a single drop undergoing vaporation, the complete analysis of a mixture containing five constituents is performed in a few seconds.

3. The ash of a cigar moistened with hydrochloric acid and held in the flame shows at once the bands Na α, Ka α, Li α, Ca α, Ca β.

4. A piece of hard potash-glass combustion tubing gave, both with and without hydrochloric acid, the lines Na α and Ka α: treated with fluoride of ammonium and sulphuric acid, the bands Ca α, Ca β, and traces of Li α were rendered visible.

5. Orthoclase from Baveno gives, either alone or when treated with hydrochloric acid only, the lines Na α and Ka α, with traces of Li α: with fluoride of ammonium and sulphuric acid, the bright lines Na α and Ka α, and a somewhat less distinct Li α, are seen. After volatilization of the bodies thus detected, the bead moistened with hydrochloric acid gives a scarcely distinguishable flash of the lines Ca α and Ca β. The residue on the platinum wire, when moistened with cobalt solution and heated, gives the blue colour so characteristic of alumina. If the well-known reaction of silicic acid be likewise employed, we may conclude from this examination made in the course

of a very few minutes that the orthoclase from Baveno contains silicic acid, alumina, potash, with traces of soda, lime, and lithia; and also that no trace of baryta or strontia is present.

6. Adularia from St. Gothard comported itself in a similar manner, with the exception that the calcium reaction was indistinctly seen, whilst that of lithium was altogether wanting.

7. Labradorite from St. Paul gives the sodium line Na α, but no calcium spectrum. On moistening the fragment with hydrochloric acid, the lines Ca α and Ca β appear very distinctly; with the fluoride of ammonium test a weak potassium reaction is obtained, and also faint indications of lithium.

8. Labradorite from the Corsican diorite gave similar reactions, except that no lithium was found.

9. Mosanderite from Brevig and Tscheffkinite from the Ilmengebirge showed when treated alone the sodium reaction: on the addition of hydrochloric acid the lines Ca α and Ca β appeared.

10. Melinophane from Lamoe gave the line Na α when placed in the flame; with hydrochloric acid the lines Ca α and Li α became visible.

11. Scheelite and sphene give, on treatment with hydrochloric acid, a very intense calcium reaction.

12. When small quantities of strontium are present together with calcium, the line Sr δ may be most conveniently employed for the detection of this metal. In this way the presence of small quantities of strontium can be easily detected in very many sedimentary limestones. The lines Na α, Li α, Ka α, and especially Li α, are observed as soon as the limestone is brought into the flame. Converted by hydrochloric acid into chlorides, and brought in this form into the flame, these minerals give the same bands; and not unfrequently the line Sr δ is also distinctly seen. This latter appears, however, only for a short time, and is in general best seen when the calcium spectrum begins to fade.

In this way the lines Na α, Li α, Ka α, Ca α, Ca β, and Sr δ were found in the spectra of the following limestones: limestone from the Silurian at Kugelbad near Prague, muschelkalk from Rohrbach near Heidelberg, limestone from the lias at Malsch in Baden, chalk from England. The following limestones gave the lines Na α, Li α, Ka α, Ca α, Ca β, but not the blue strontium band Sr δ:—marble from the granite near Auerbach,[1] limestone from the Devonian at Gerolstein in the Eifel, carboniferous limestone from Planitz in Saxony, dolomite from Nordhausen in the Hartz, Jura kalk from Streitberg in Franconia. From these few experiments it is evident that a more extended series of exact spectrum analyses, respecting the amount of strontium, lithium, sodium, and potassium which the various limestone formations contain, must prove of the greatest geological importance both as regards the order of their formation and their local distribution, and may possibly lead to the establishment of some unexpected conclusions respecting the nature of the oceans from which these limestones were originally deposited.

BARIUM.

The barium spectrum is the most complicated of the spectra of the alkalies and alkaline earths. It is at once distinguished from all the others by the green lines Ba α and Ba β (which are by far the most distinct), appearing the first, and continuing during the whole of the reaction. Ba γ is not quite so distinct, but is still a well-marked and peculiar line. As the barium spectrum is considerably more extended than those of the other metals, the reaction is not observed to so great a degree of delicacy: still 0·3 grm. of chlorate of barium burnt with milk-sugar gave a distinct band of Ba α which lasted for some time, when the air of the room was well mixed by moving an open umbrella about. Hence we may calculate, in the same manner as was done in the sodium experiment, that about $\frac{1}{1000}$ of a milligramme of barium salt may be detected with the greatest certainty.

[1] According to the method already described, a quantity of nitrate of strontium was obtained from 20 grms. of this marble such as to give a complete and vivid strontium spectrum. We have not examined the other limestones in the same way.

The chloride, bromide, iodide, and fluoride of barium, as also the hydrated oxide, the sulphate, and carbonate, show the reaction best. It may be obtained by simply heating any of these salts in the flame.

Silicates containing barium which are decomposed by hydrochloric acid also give the reaction if a drop of hydrochloric acid be added to them before they are brought into the flame. Baryta-harmotome, treated in this way, gives the lines Ca *α* and Ca *β*, together with the bands Ba *α* and Ba *β*. Compounds of barium with fixed acids, giving no reaction either when alone or after addition of hydrochloric acid, should be fused with carbonate of sodium as described under strontium, and the carbonate of barium thus obtained examined. If barium and strontium occur in small quantities together with large amounts of calcium, the carbonates obtained by fusion are dissolved in nitric acid, and the dried salt extracted with alcohol: the residue contains only barium and strontium, both of which can almost always be detected. When we wish to test for small traces of strontium or barium, the residual nitrates are converted into chlorides by ignition with sal-ammoniac, and the chloride of strontium is extracted by alcohol. Unless one or more of the bodies to be detected is present in very small quantities, the methods of separation just described are quite unnecessary, as is seen from the following experiment:—A mixture of the chlorides of potassium, sodium, lithium, calcium, strontium, and barium, containing at the most $\frac{1}{10}$ of a milligramme of each of these salts, was brought into the flame, and the spectra produced were observed. At first the bright yellow sodium line, Na *α*, appeared with a background formed by a nearly continuous pale spectrum: as soon as this line began to fade, the exactly defined bright red line of lithium, Li *α*, was seen, and still further removed from the sodium line the faint red potassium line, Ka *α*, was noticed; whilst the two barium lines, Ba *α*, Ba *β*, with their peculiar form, became visible in the proper position. As the potassium, sodium, lithium, and barium salts volatilized, their spectra became fainter and fainter, and their peculiar bands one after the other vanished, until, after the lapse of a few minutes, the lines Ca α, Ca β, Sr α, Sr β, Sr γ, and Sr δ became gradually visible, and, like a dissolving view, at last attained their characteristic distinctness, colouring, and position, and then, after some time, became pale, and disappeared entirely. The absence of any one or of several of these bodies is at once indicated by the non-appearance of the corresponding bright lines.

Those who become acquainted with the various spectra by repeated observation do not need to have before them an exact measurement of the single lines in order to be able to detect the presence of the various constituents; the colour, relative position, peculiar form, variety of shade and brightness, of the bands are quite characteristic enough to ensure exact results, even in the hands of persons unaccustomed to such work. These special distinctions may be compared with the differences of outward appearance presented by the various precipitates which we employ for detecting substances in the wet way. Just as it holds good as a character of a precipitate that it is gelatinous, pulverulent, flocculent, granular, or crystalline, so the lines of the spectrum exhibit their peculiar aspects, some appearing sharply defined at their edges, others blended off either at one or both sides, either similarly or dissimilarly; or some again appearing broader, others narrower: and just as in ordinary analysis we only make use of those precipitates which are produced with the smallest possible quantity of the substance supposed to be present, so in analysis with the spectrum we employ only those lines which are produced by the smallest possible quantity of substance, and require a moderately high temperature. In these respects both analytical methods stand on an equal footing; but analysis with the spectrum possesses a great advantage over all other methods, inasmuch as the characteristic differences of colour of the lines serve as the distinguishing feature of the system. Most of the precipitates which are valuable as reactions are colourless; and the tint of those which are coloured varies very considerably, according to the state of division and mechanical arrangement of the particles. The presence of even the smallest quantity of impurity is often sufficient entirely to destroy the characteristic colour of a precipitate; so that no reliance can be placed upon nice distinctions of colour as an ordinary chemical test. In spectrum analysis, on the contrary, the coloured bands are unaffected by such alteration of physical conditions, or by the presence of other bodies. The positions which the lines occupy in the spectrum give rise to chemical properties as unalterable as the combining weights themselves, and which can therefore be estimated with an almost astronomical precision. The fact, however, which gives to this method of spectrum analysis an extraordinary importance is, that the chemical reactions of matter thus reach a degree of delicacy which is almost inconceivable. By an application of this method to geological inquiries concerning the distribution and arrangements already mentioned, we are led to the unexpected conclusion, that not only potassium and sodium, but also lithium and strontium, must be added to the list of bodies occurring only indeed in small quantities, but most widely spread throughout the matter composing the solid body of our planet.

The method of spectrum analysis may also play a no less important part as a means of detecting new elementary substances; for if bodies should exist in nature so sparingly diffused that the analytical methods hitherto applicable have not succeeded in detecting or separating them, it is very possible that their presence may be revealed by a simple examination of the spectra produced by their flames. We have had opportunity of satisfying ourselves that in reality such unknown elements exist. We believe that, relying upon unmistakeable results of the spectrum analysis, we are already justified in positively stating that, besides potassium, sodium, and lithium, the group of the alkaline metals contains a fourth member, which gives a spectrum as simple and characteristic as that of lithium, a metal which in our apparatus gives only two lines, namely, a faint blue one, almost coincident with the strontium line Sr δ, and a second blue one lying a little further towards the violet end of the spectrum, and rivalling the lithium line in brightness and distinctness of outline.

APPENDIX B.

BUNSEN AND KIRCHHOFF ON THE MODE OF USING A SPECTROSCOPE.[1]

The apparatus is represented by Fig. 27. On the upper end of a cast-iron foot a brass plate is screwed, carrying the flint-glass prism (*a*), having a refracting angle of 60°. The tube *b* is also fastened to the brass plate: in the end of this tube nearest the prism is placed a lens, whilst the other end is closed by

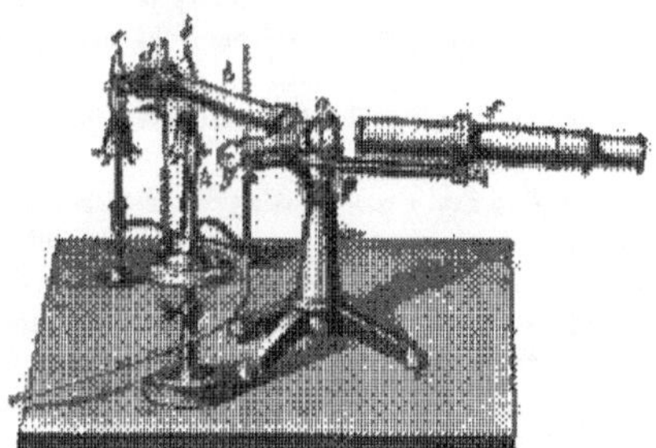

FIG. 27.

a plate in which a vertical slit has been made. Two arms are also fitted on to the cast-iron foot, so that they are moveable in a horizontal plane about the axis of the foot. One of these arms carries the telescope (*f*), having a magnifying power of 8, whilst the other carries the tube (*g*): a lens is placed in this tube at the

[1] Second Memoir on Spectrum Analysis, Phil. Mag. vol. xxii. 1861, pp. 334—498.

end nearest the prism, and at the other end is a scale which can be seen through the telescope by reflection from the front surface of the prism. This scale is a photographic copy of a millimetre scale, which has been produced in the camera of about $\frac{1}{15}$ the original dimensions.[1] The scale is covered with tinfoil, so that only the narrow strip upon which the divisions and the numbers are engraved can be seen. The upper half only of the slit is left free, as is seen by reference to Fig. 28; the lower half is covered by a small equilateral glass prism, which sends by total reflection the light of the lamp *d*, Fig. 27, through the slit, whilst the rays from the lamp *e* pass freely through the upper and uncovered half. A small screen placed above the prism prevents any light from *d* passing through the upper portion of the slit. By help of this arrangement the observer sees the spectra of the two sources of light immediately one under the other, and can easily determine whether the lines are coincident or not.[2]

FIG. 28.

We now proceed to describe the arrangement and mode of using the instrument.

The telescope *f* is first drawn out so far that a distant object is plainly seen, and screwed into the ring, in which it is held, care being taken to loosen the screws beforehand. The tube *b* is then brought into its place, and the axis of B brought into a straight line with that of *b*. The slit is then drawn out until it is distinctly seen on looking through the telescope, and this latter is then fixed by moving the screws, so that the middle of the slit is seen in about the middle of the field of view. After removing the small spring, the prism is next

[1] This millimetre scale was drawn on a strip of glass, covered with a thin coating of lampblack and wax dissolved in glycerine. The divisions and numbers, which by transmitted light showed bright on a dark ground, were represented in the photograph dark on a light ground. It would be still better to employ, for the spectrum apparatus, a scale in which the marks were light on a dark ground. Such scales are beautifully made by Salleron and Ferrier of Paris.

[2] This apparatus was made in the celebrated optical and astronomical *atelier* of C. A. Steinheil in Munich.

placed on the brass plate, and fastened in the position which is marked for it, and secured by screwing down the spring. If the axis of the tube *b* be now directed towards a bright surface, such as the flame of a candle, the spectrum of the flame is seen in the lower half of the field of the telescope on moving the latter through a certain angle round the axis of the foot. When the telescope has been placed in position, the tube *g* is fastened on to the arm belonging to it, and this is turned through an angle round the axis of the foot such that, when a light is allowed to fall on the divided scale, the image of the scale is seen through the telescope *f*, reflected from the nearer face of the prism. This image is brought exactly into focus by altering the position of the scale in the tube *g*; and by turning this tube on its axis it is easy to make the line in which one side of the divisions on the scale lie parallel with the line dividing the two spectra, and by means of the screw *δ* to bring these two lines to coincide.

In order to bring the two sources of light, *e* and *é*, into position, two methods may be employed. One of these depends upon the existence of bright lines in the inner cone of the colourless gas flame, which have been so carefully examined by Swan. If the lamp *e* be pushed past the slit, a point is easily found at which these lines become visible; the lamp must then be pushed still further to the left, until these lines nearly or entirely disappear; the right mantle of the flame is now before the slit, and into this the bead of substance under examination must be brought. In the same way the position of the source of light *é* may be ascertained.

The second method is as follows:—The telescope *f* is so placed that the brightest portion of the spectrum of the flame of a candle is seen in about the middle of the field of view; the flame is then placed before the ocular in the direction of the axis of the telescope, and the position before the slit determined in which the upper half of the slit appears to be the brightest; the lamp *e* is then placed so that the slit appears behind that portion of the flame from which the most light is given off after the introduction of the bead. In a similar way the position of the lamp *e* is determined by looking through the small prism and the lower half of the slit.

By means of the screw the breadth of the slit can be regulated in accordance with the intensity of the light, and the degree of purity of spectrum which is required. To cut off foreign light, a black cloth, having a circular opening to admit the tube *g*, is thrown over the prism *a* and the tubes *b* and *f*. The illumination of the scale is best effected by means of a luminous gas flame placed before it: the light can, if necessary, be lessened by placing a silver-paper screen close before the scale. The degree of illumination suited to the spectrum under examination can then be easily found by placing this flame at different distances.

APPENDIX C.

BUNSEN ON A METHOD OF MAPPING SPECTRA.[1]

For the purpose of facilitating the numerical comparison of the data of various spectrum observations we give in Fig. 29, p. 93, graphical representations of the observations which are taken from the guiding lines given in the chromolithograph drawings of the spectra published in our former memoirs, and in which the prism was placed at the angle of minimum deviation. The ordinates of the edges of the small blackened surfaces, referred to the divisions of the scale as abscissæ, represent the intensity of the several lines, with their characteristic gradations of shade. These drawings were made when the slit was so broad and the flame of such a temperature, that the fine bright line upon the broad Ca *a* band began to be distinctly visible. This breadth of the slit was equal to the fortieth part of the distance between the sodium line and the lithium line *a*. For the sake of perspicuity, the continuous spectra which some bodies exhibit

[1] Phil. Mag. Fourth Series, vol. xxvi. p. 247.

MAP OF THE SPECTRA OF THE ALKALIES AND ALKALINE EARTHS.

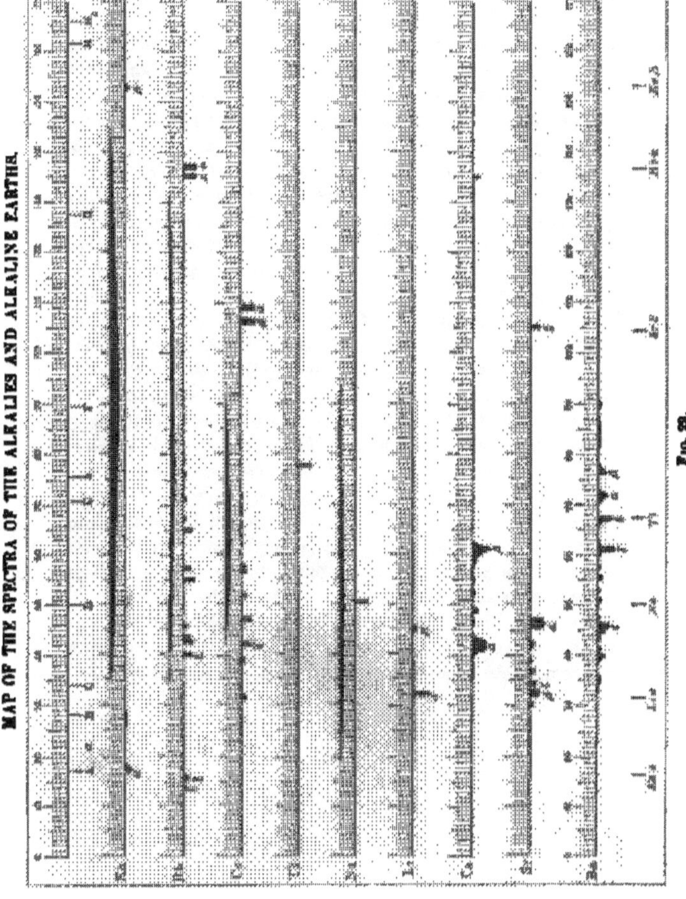

Fig. 29.

are specially represented on the upper edge of the scale, to the divisions of which they are referred as abscissæ. In order to render these drawings, which have reference to our instrument, applicable to observations upon the scale of any other apparatus, which we may call D, it is only necessary to prepare a reduced scale, which is laid upon the several drawings, and used in place of the divided scale given in the figure. The lines marked at the bottom of Fig. 27 serve for the preparation of this new scale: these lines denote the distances between the lines K α, Li α, Na, Tl, Sr δ, Rb α, and K β, measured according to the scale of our instrument. The position of each of these lines is determined by the edge of the line, which does not change its place on altering the breadth of the slit. The position of these same lines is read off on the scale of the instrument B, and the corresponding number written under each. A series of fixed points on the scale is thus obtained, and the complete divisions for the scale of instrument D are got by interpolating the values of the portions of the scale situated between the fixed points.

The sodium line is then inserted in this scale, which is pasted upon a straight-edge, and the divisions numbered in tens and fives. If this measure be now laid upon any one of the drawings, so that the sodium line on the measure coincides with the division 50 on the drawing, the scale on the measure will give the position of all the lines in the particular spectrum exactly as they are seen in the photographic scale of the instrument D. When the position of the line under observation has in this way been ascertained, it is easy to assure oneself of its exact identity by means of the small prism on the slit of the spectroscope.

LECTURE III.

Historical Sketch.—Talbot, Herschel, Bunsen, and Kirchhoff.—Discovery of New Elements by means of Spectrum Analysis.—Cæsium, Rubidium, Thallium, Indium.—Their History and Properties.—Spectra of the Heavy Metals.—Examination of the Light of the Electric Discharge.—Wheatstone.—Volatilization of Metals in the Electric Arc.—Kirchhoff, Ångström, Thalén, and Huggins.—Maps of the Metallic Lines.

Appendix A.—Spectrum Reactions of the Rubidium and Cæsium Compounds.

Appendix B.—Contributions towards the History of Spectrum Analysis. By G. Kirchhoff.

Appendix C.—On the Spectra of some of the Chemical Elements. By Wm. Huggins. With Maps and Tables.

I propose to point out to you to-day the properties of the new elementary bodies which have been discovered by means of spectrum analysis, the principles of which we considered in the last lecture. Before passing on to consider this point, I wish to direct your attention, for a few moments, to the history of the subject.

The experiments of which I gave you an account in the last lecture, and the results derived from these experiments, have been carried out chiefly by a German chemist and a German physicist, whose important discoveries have made the names of Bunsen and Kirchhoff celebrated throughout the scientific world.

But although these philosophers are the real discoverers of this method, because they carried it out with all due scientific accuracy and placed it on the sure foundation upon which it now rests, yet we must not suppose that the ground was before their time absolutely untrodden. No great discovery is made all at once. There are always stepping-stones by which such a position is reached, and it is right to know what has been previously done, and to give such credit as is their due to the older observers.

So long ago as 1752, Thomas Melville, while experimenting on certain coloured flames, observed the yellow soda flame, although he was unacquainted with its cause. In 1822 Brewster introduced his monochromatic lamp, in which the soda light is used; the first idea, however, being due to Melville. A simple experiment will prove to you the nature of this monochromatic soda light. I have here the means of producing a very intense soda flame, and I will throw the light on to this screen with painted letters. You will observe that no colour is noticeable in these letters. They appear in various degrees of shade or intensity, but no difference of colour is visible, because the light falling upon them is of a pure yellow colour. Now, if I throw a small quantity of magnesium powder into the flame, you will at once notice how brightly the various colours come out. We have here white light containing rays of every degree of refrangibility; hence the different colours appear, each letter being able to reflect its own peculiar rays.

Sir John Herschel, in the year 1822, investigated the spectra of many coloured flames, especially of the strontium and copper chlorides, and of boracic acid, and he writes in 1827 about this as follows: "The colours thus contributed by different objects to flame afford in many cases a ready and neat way of detecting extremely minute quantities of them."

Fox Talbot, whose name we know as being so intimately connected with the origin of the beautiful art of photography, makes the following suggestions respecting these spectra. Writing in 1826 he says: "The red fire of the theatres examined in the same way gave a most beautiful spectrum, with many light lines or maxima of light. In the red these lines were more numerous, and crowded with dark spaces between them" (these are the strontium lines which you see on the diagram), "besides an exterior ray greatly separated from the rest, and probably the effect of the nitre in the composition" (this is really the red potassium line caused by the nitre). "In the orange was one bright line, one in the yellow, three in the green, and several that were fainter." The blue line which he mentions is the blue strontium line which we saw so plainly. "The bright line in the yellow" (our friend sodium) "is caused without doubt by the combustion of sulphur." Talbot got wrong there, as did many of the early observers. They could not suppose that so minute a trace of sodium could produce that yellow light; and even Talbot says that the yellow line must be caused in certain cases by the presence of water. He continues: "If this opinion" (about the cause of formation of these lines) "should prove correct, and applicable to the other definite rays, a glance at the prismatic spectrum of a flame might show it to contain substances which it would otherwise require a laborious chemical analysis to detect." We cannot even now express the opinion entertained at the present moment more concisely than Talbot did in the year 1826. These early observers did not, however, determine the exact nature of the substance producing the colour, inasmuch as the extreme sensitiveness of this sodium reaction put them off the scent: they could not believe that sodium was present everywhere.

Both Herschel and Brewster found that the same yellow light was obtained by setting fire to spirits of wine diluted with water, and Talbot also mentions cases in which no soda was, as he thought, present, and yet this yellow line always made its appearance. Hence he says, "The only matter which these substances have in common is water," and he throws out the suggestion that this yellow line is produced by the presence of water. In February 1834 Talbot writes: "Lithia and strontia are two bodies characterised by the fine red tint which they communicate to the flame. The former of these is very rare, and I was indebted to my friend Mr. Faraday for the specimen which I subjected to the prismatic analysis. Now it is very difficult to distinguish the lithia red from the strontia red with the naked eye, but the prism betrays between them the most marked distinction which can be imagined. The strontia flame exhibits a great number of red rays well separated from each other by dark intervals, not to mention an orange and a very definite bright blue ray. The lithia exhibits one single red ray. Hence I hesitate not to say, that optical analysis can distinguish the minutest portions of these two substances from each other with as much certainty, if not more than, any known method." Still Talbot says further on, that "the mere presence of the substance, which suffers no diminution in consequence, causes the production of a red and green line to appear in the spectrum."

Professor William Allen Miller next made some interesting experiments in 1845 on the spectra of coloured flame produced by the metals of the alkaline earths, and came still nearer to the result which we now find Bunsen and Kirchhoff arrived at in 1861. Diagrams of these spectra accompany the memoir, but they are not characteristic enough to enable them to be used as distinctive tests for the metals, owing to the fact that a luminous flame was used. Hence the investigations of Miller in 1845 attracted less attention than they deserved.[1] The first person who pointed out this characteristic property of sodium was Professor Swan, in 1857, and it is to him that we owe the examination and the determination of the very great sensitiveness of this sodium reaction. So much then for the history of the method as applied to the detection of the alkalies and the alkaline earths.

We will now pass on to the consideration of the new elements which have been discovered by spectrum analysis. And, in the first place, I would direct your attention to the new alkaline metals discovered by Professor Bunsen in 1860. Shortly after he made his first experiments on the subject of spectrum analysis, Bunsen happened to be examining the alkalies left from the evaporation of a large quantity of mineral water from Dürkheim in the Palatinate. Having separated out all other bodies, he took some of these alkalies, and found, on examining by the spectroscope the flame which this particular salt or mixture of salts gave off, that some bright lines were visible which he had never observed

[1] See extract in Appendix B. from Kirchhoff's Contributions to the History of Spectrum Analysis, Phil. Mag. Fourth Series, vol. xxv. p. 250, 1863.

before, and which he knew were not produced either by potash or soda. So much reliance did he place in this new method of spectrum analysis that he at once set to work to evaporate so large a quantity as forty-four tons of this water in which these new metals, which he termed *cæsium* and *rubidium*, were contained in exceedingly minute quantities.

In short, he succeeded in detecting and separating the two new alkaline substances from all other bodies, and the complete examination of the properties of their compounds which he made with the very small quantity of material at his disposal remains a permanent monument of the skill of this great chemist. Both these metals occur in the water of the Dürkheim springs. I have here the numbers giving Bunsen's analysis, in thousand parts, of the mineral water of Dürkheim and of Baden-Baden.

The quantity of the new substance contained in the water from the Dürkheim springs is excessively small, amounting in one ton to about three grains of the chloride of cæsium and about four grains of the chloride of rubidium; whilst in the Baden-Baden spring we have only traces of the cæsium chloride, and a still smaller quantity than in the other spring of the rubidium chloride. From the forty-four tons of water which he evaporated down Bunsen obtained only about 200 grains of the mixed metals. You will easily appreciate the delicacy and accuracy of a method by which the presence of so minute a trace of the new metals as that contained in the water could be so readily detected.

Analysis of 1,000 parts of the Mineral Water in which the new Alkaline Metals, Cæsium and Rubidium, were discovered by Bunsen.

	Dürkheim.	Ungemach, at Baden-Baden.
Calcium Bicarbonate . . .	0·28350	1·475
Magnesium Bicarbonate . .	0·01460	0·712
Ferrous Bicarbonate . . .	0·00840	0·010
Manganous Bicarbonate . .	traces	traces
Calcium Sulphate	—	2·202
Calcium Chloride	3·03100	0·463
Magnesium Chloride . . .	0·39870	0·126
Strontium Chloride. . . .	0·00810	—
Strontium Sulphate . . .	0·01950	0·023
Barium Sulphate	—	traces
Sodium Chloride	12·71000	20·834
Potassium Chloride. . . .	0·09660	1·518
Potassium Bromide. . . .	0·02220	traces
Lithium Chloride	0·03910	0·451
Rubidium Chloride. . . .	0·00021	0·0013
Cæsium Chloride	0·00017	traces
Alumina	0·00020	—
Silica	0·00040	1·230
Free Carbonic Acid. . . .	1·64300	0·456
Nitrogen	0·00460	—
Sulphuretted Hydrogen . .	traces	—
Combined Nitric Acid. . .	—	0·030
Phosphates	traces	traces
Arsenic Acid.	—	traces
Ammoniacal Salts	traces	0·008
Oxide of Copper	—	traces
Organic Matter.	traces	traces
	18·28028	29·6393

Let me show you, in the first place, the colours produced by these two new metals when brought into a non-luminous gas flame. In the last lecture we noticed the beautiful violet tint which the potash flame exhibits. The tint yielded by these two metals is very similar indeed, and in fact, not only in the character of the light which they emit, but in all their chemical properties, the compounds of both the new bodies resemble potassium compounds very closely. Here I bring a small quantity of rubidium salt into the flame, and you observe the beautiful purple colour with which the flame is tinged. Now I throw in a little cæsium salt, and you notice we get a very similar kind of tint, rather more red, but still scarcely to be distinguished from the violet potash flame burning alongside.

If I next show you the spectra of cæsium and rubidium on the screen, and compare them with the spectrum of potassium (see Frontispiece, Nos. 2, 3, and 4), you will see that the spectra of these three metals exhibit (in accordance with their correspondence in other chemical properties) a striking analogy. Each of the three metals possesses a spectrum which is continuous in the middle, showing that under certain circumstances gases may emit light of every degree of refrangibility, and decreasing in intensity towards each end. In the case of potassium the continuous portion is most intense, in that of rubidium less intense, and in the cæsium spectrum this luminosity is least. In all three we observe the most intense and characteristic lines towards both the red and blue ends of the spectrum. The metal rubidium, as its name implies, is characterised by two splendid deep red lines (see Frontispiece, No. 3), both less refrangible than the potassium red line; but the two violet lines are even more characteristic, and serve as the most delicate test of the presence of the metal. No less than the 0·0002 part of a milligramme of rubidium can be detected by the spectrum reaction. The cæsium spectrum is chiefly characterised by the two blue lines from which it derives its name; they are remarkable for their brilliancy and sharpness of definition: while it is singular that cæsium exhibits no red lines whatever.

Since the discovery of these two bodies by Bunsen in 1860, chemists have been on the look-out for them, and have found both of them in very different situations; one of them, rubidium, being comparatively widely distributed. The celebrated French waters of Bourbonne-les-Bains contain 0·032 grm. of chloride of cæsium and 0·010 grm. of chloride of rubidium in one litre of water; whilst in the well-known mineral springs of Vichy, Gastein, Nauheim, Karlsbrunn, and many more, either one or both of the new metals has been discovered. And here the thought strikes one that the presence of these metals even in such minute quantities may possibly exert a not unimportant influence upon the medicinal qualities and effects of the waters. Rubidium has been found to be very widely diffused; it has been found in beetroot, in tobacco, in the ash of the oak (the *Quercus pubescens*), in coffee, in tea, and in cocoa: indeed of the new metals it is only rubidium which is found in vegetables and in vegetable products; whilst both new metals are found in tolerably large quantities in certain minerals, especially in lepidolite and petalite.

One very interesting example of the occurrence of the metal cæsium has been observed in a mineral termed *pollux*, which was analysed in the year 1846 by the well-known chemist, Plattner, and supposed to contain potassium. In calculating out the results of his analysis Plattner invariably found a considerable loss, the cause of which he was unable to account for. Spectrum analysis has now explained this anomaly, for since the discovery of the two new metals it has been found that it was not potassium, but the new metal cæsium, which was present, of the oxide of which no less than 34 per cent. is contained in this mineral. The want of agreement of the former analysis is therefore wholly attributable to the difference of the combining weights of these two bodies; that of potassium being only 39·1, whilst cæsium is 133; and if we use this last number in the calculation, we find that Plattner's analysis comes up exactly, as it ought to do, to 100 parts. So closely indeed are cæsium and potassium allied in their chemical characters, that it is only by the discriminating power of spectrum analysis that we have been able to ascertain even the existence of the new metal.

Having once proved the existence of these two new elementary bodies, Bunsen was of course easily able to find the means of separating them accurately one from the other, and from the well-known substance potassium, and at the present day the chemical history and characters of these two metals and their compounds are as well known as those of the commoner alkalies.

The reaction by which Bunsen separated the new metals from potassium can easily be rendered visible to you. I have here a small quantity of rubidium chloride in solution, and here again I have a solution of the double chloride of potassium and platinum.

The chloride of rubidium and platinum is much less soluble than the corresponding potassium compound, and hence, if I add the potassium double-chloride to this rubidium salt, I shall have a precipitation of the double chloride of rubidium and platinum; and this will indicate to you the mode by which Bunsen separated these two metals from each other.

Here you observe, by pouring in this solution, the

liquid at once becomes turbid, and we get a very considerable quantity of a heavy yellow granular precipitate of the new rubidium compound.

It is unnecessary now to enter into the analytical methods by which cæsium can be separated from rubidium; it is sufficient to state that the separation is based upon the different solubilities of the tartrates of the new metals, the acid tartrate of cæsium being much more soluble than the corresponding rubidium salt.

The isomorphous relations between the salts of rubidium and cæsium and those of potassium also point out the striking chemical analogy subsisting between these interesting bodies.

Shortly after the discovery of these two new alkaline metals the existence of a third new elementary substance was made known by our countryman Mr. Crookes. In the year 1861 he sent to the Exhibition a very small portion of a substance which he stated was a new element obtained from a certain seleniferous deposit from a sulphuric acid manufactory at Tilkerode in the Hartz. This body gives a most beautiful green tint to flame. If I bring a small quantity of this element into the flame, you see that it produces this exquisite green colour. And this was the reaction by which it was discovered. Mr. Crookes proved that this green light was due to some new elementary body; then he separated out the substance, and gave to it the name thallium, from *thallus*, a green twig.

Now the spectrum of thallium is very distinct and specific, consisting of one bright green line. I will show it to you with the electric lamp. Here you see this magnificent green band (Frontispiece, No. 5). The spark-spectrum of thallium is rather more complicated, as it exhibits five other lines in addition to the bright one in the green.[1] The line Tl α falls between 77·5 and 78 mm. on the photographic scale of the spectroscope, and appears to be partly coincident with one of the barium lines. This apparent coincidence is resolved when examined by a higher magnifying power. The chemical properties of this substance are also very remarkable. It stands about half-way between lead and the alkalies, resembling in many of its characters the metal lead, and it has been well described by Dumas as the "ornithorhynchus" amongst the metals. Thallium can, however, be perfectly separated from the alkalies and lead by means of the insolubility of its chloride and the solubility of its sulphate. The specific gravity of thallium is 11·8, and its combining weight is a very high one, 204.

The properties of thallium have been examined by a French chemist, M. Lamy, as well as by Mr. Crookes. It has been found to exist in very large quantities in certain varieties of iron pyrites, a substance from which we manufacture almost all our sulphuric acid.

The metal thallium can be easily obtained in the metallic state from its salts. This I can readily render evident to you all. We can here decompose a solution of the sulphate of thallium by a current of electricity, and then we shall observe the metallic thallium shooting out as a beautiful arborescent growth on the screen. Here you see the crystals of metallic thallium stretching out their long branches all over the screen.

The soluble salts of thallium act as a cumulative poison: they have been found in large quantities in animals which have been poisoned by this substance. The method for determining or detecting the presence of thallium in

[1] Miller, Proc. Roy. Soc. 1863, p. 167.

such a poisoned animal by means of spectrum analysis is extremely simple. If we had such a means of detecting some of the other metallic poisons as readily as that of thallium, the work of the toxicologist would be extremely easy; except that under these circumstances the very delicacy of the test becomes in itself a danger, as the most minute trace of the poisonous metals which might by chance be present would in this way be as easily detected as a larger quantity.

There is still one other elementary body of which I have to speak, namely, the metal indium.

Indium was discovered in 1864 by two German professors, Reich and Richter, of the celebrated Mining School of Freiberg. It also was detected by the peculiar spectrum, which consists simply of two indigo-coloured lines. These lines are best seen when a bead of an indium compound is held between two electrodes from which a spark passes. The lines In α and In β fall respectively upon divisions 107·5 and 140 mm. of the photographic scale of the spectroscope, when Na α = 50 and Sr δ = 100·5. (See Frontispiece, No. 6.) It was discovered in certain zinc ores, and has only been found in small quantities. Its chemical characters are still imperfectly understood; but in its properties it appears to stand about half-way between zinc and lead. Its combining weight appears to be 75·6, and its specific gravity is 7·277; and it forms definite compounds, which, however, have not yet been examined with a sufficient amount of attention to enable me to give you a detailed account of them.

I can here show you the indigo colour which indium compounds impart to the flame; and you now see on the screen that the spectrum of indium consists of two bright indigo-coloured lines, one situated in the blue and one in the ultra-blue, or indigo, portion of the spectrum. I need scarcely say that there is as yet no notion of any practical employment of any of these new substances, though chemists never can tell what important applications of their most recondite discoveries may not arise, even in the immediate future.

The subject to which I would wish in the next place to direct your attention is the mode by which we can determine by spectrum analysis the presence of metals

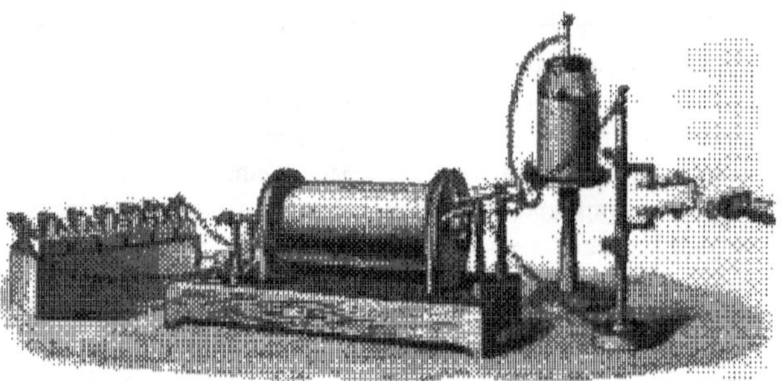

FIG. 30.

proper, or heavy metals. How, for instance, can we ascertain the presence of copper, or of gold, or of silver, or of zinc, or of iron? how can we volatilize these metals to make them give off the light which is peculiar to each one? I have here the means of doing this. We have again to employ our most valuable agent, electricity. By means of this battery and induction coil I can obtain an electric spark; and by means of the electric spark I can get what I require, namely, the volatilization of these metals. It is many years since the application of the electric spark to this particular branch of analysis was discovered. The first person who examined the nature of the electric spark was Wollaston, whose name I mentioned to you in my opening lecture as having first pointed out the existence of these very important dark lines in the solar spectrum. But it was Faraday who first declared that the electric spark consists solely of the material particles of the poles and the medium through which it passes. It was originally supposed that electricity had some existence apart from matter; but Faraday, by a most elaborate series of experiments, discovered that when the electric spark passes from one knob of the electric machine to your hand or knuckle, a quantity of matter passes too, partly consisting of the brass of the pole, and partly consisting of the air and moisture which exist between your knuckle and the brass knob. He speaks, in his experimental researches, of the electric spark as being produced by a current propagated along, and by, ponderable matter, and heated in the same manner, and according to the same laws, as a voltaic current heats and volatilizes a metallic wire. So that what we see and call the spark is really the ignition of the matter which exists in this arc; and when we take a spark from the electrical machine, the particles of the brass are actually carried over from the one pole to the other, in this case from the pole to the knuckle. If this is so, it is evident that, when we bring certain different metals in this arc, we must obtain different coloured sparks: thus, if I bring a small quantity of strontium salt into the spark, we shall have a very intensely red light, due to the ignition of the peculiar body strontium which is volatilized between the poles; and when I take some thallium, we have the green colour characteristic of this metal.

If we examine the light of such a spark with a spectroscope, we shall find that two superimposed spectra here present themselves; the one spectrum produced by the very bright points of light lying close to the poles, and the other by the less luminous portion of the arc lying further from the poles. The spectrum of the bright points is, as we shall see, that of the metal present, whilst the light from the less luminous portion in the centre exhibits the spectrum of the incandescent air, and shows the particular lines produced by the gases present in the atmosphere, viz. nitrogen, oxygen, and hydrogen (for in the atmosphere we have constantly the vapour of water present). Each gas gives us lines peculiar to itself; and in some cases, when the quantity of carbonic acid present in the air is considerable, we may even get the carbon lines.

It was Sir Charles Wheatstone, in the year 1835, who first pointed out that the spectra produced from the sparks of different metals were dissimilar; and he concluded that the electric spark resulted from the volatilization and not from the combustion of the matter of the poles themselves, for he observed the same phenomena *in vacuo* and in hydrogen, in which no combustion can occur; and in 1835 he writes as follows: "These differences are so obvious, that one metal may easily be distinguished from another by the appearance of its spark; and we have here a mode of discriminating metallic bodies more readily than that of chemical examination, and which may hereafter be employed for useful purposes."

You have here a copy of the diagram (Fig. 31) published in Wheatstone's paper, giving the lines which he saw in the metals. Subsequent research has shown that the number of the lines peculiar to each of these metals is very large, although on Wheatstone's diagram but a few of these are noticeable. On this drawing you see some of the bright lines of the metals mercury (Hg), zinc (Zn), cadmium (Cd), bismuth (Bi), tin (Sn), and

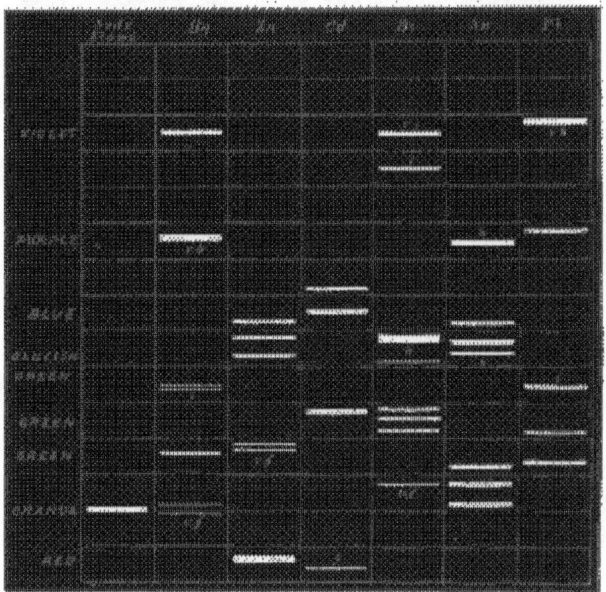

FIG. 31.

lead (Pb). The letters placed above and below each bright line indicate its degree of intensity: very bright, bright, faint, very faint.

It was, however, chiefly through the experiments of the Swedish philosopher Ångström, that we gained an intimate knowledge of the nature of the electric spark. In the year 1855 Ångström investigated the matter very thoroughly, and pointed out the important fact which I have explained, that the spark yields two superimposed spectra; one derived from the metal of the poles, and the other from the gas or air through which the spark passes.

Perhaps I had better show you, first of all, the beautiful spectra of some of these metals, as I can exhibit

FIG. 32.

them to you on a screen; and then explain to you the nature of the spectra which we see when we look at the spark through such a train of prisms as you have in the large spectroscope on the table. I cannot show you these lines on the screen with anything like the amount of accuracy or delicacy with which we can see them when we throw the image on the retina itself, for then we observe the true spectra. The lines then observed are excessively fine and extremely numerous, and each line possesses a fixed position, and does not interfere with the lines of any other metal. Still I can show you something very beautiful and interesting. I will endeavour by means of the electric lamp to throw the spectrum of metallic copper on the screen. I take a small piece of metallic copper and volatilize it between the incandescent carbon poles, and then you will see the green bands indicative of the presence of this metal. Here you observe these magnificent green bands, which are characteristic of copper; but when we examine the copper spark by throwing the image into the eye, we get a much more splendid effect, and see the way to a far more delicate method of detecting the presence of copper. In the next place I will take another carbon pole, and bring a small piece of zinc into it, and we shall see that zinc also gives its peculiar and beautiful lines perfectly characteristic of this special metal. If we examine this light by means of an accurate spectroscope, these broad bands are seen to consist of masses of bright lines, each one as fine as the most gauzy spider's web.

If I now take a mixture of zinc and copper, such as brass, we shall not only get the lines of the zinc, but we shall also see the bright copper lines. I have put a small piece of brass on the pole, and when I make the contact I shall volatilize this brass, and the result is a spectrum showing both the copper lines and the zinc lines. You will notice that what I have said with respect to the other metals, the alkaline earths, holds good with this,—that the most volatile of the metals burns out first. Now we can still see the less volatile copper, but the zinc lines have died away. In the same way I may show you that cadmium gives us a peculiar set of lines. If we volatilize some metallic cadmium, we shall have a series of lines somewhat resembling those of zinc, but not identical with them. There you observe three bands: these are the cadmium lines, something perfectly characteristic and distinct.

Fox Talbot observed these metallic lines in June 1834, by deflagrating thin sheets of metal by galvanism; he says: "Gold leaf and copper leaf each afforded a fine spectrum exhibiting peculiar definite rays. The effect of zinc was still more interesting: I observed in this instance a strong red ray, three blue rays, besides several more of other colours."

I will next show you the spectrum of silver. Here you observe those splendid green lines, and the beautiful purple lines in the distance: the latter are only visible when you look at the most refrangible end for some time. Nothing, surely, can be more magnificent than these spectra! These green lines are quite different in position and in character from the green copper lines. To exhibit that difference to you, I will put a bit of copper into this silver—which is chemically pure—and I think you will be able to see that we get the green copper lines distinctly arranged alongside of the green lines of silver. Thus, then, by means of the electric lamp, many of these lines can be rendered visible, although to see others distinctly we must employ a delicate spectroscope, and throw the light directly into the eye of the observer.

By the examination of the spark-spectrum chemists are now able to distinguish with the greatest ease between the rarest metals. We are able to detect the difference between erbium and yttrium, and didymium and lanthanum—metals which resemble one another in their properties so closely that it is extremely difficult to separate them from each other by ordinary chemical means. These substances all give distinct lines, and any one of these substances may be detected when mixed with the other: and we thus get a decisive answer as to their presence.

In order to examine with accuracy the spectra of the heavy metals, an arrangement, represented in Fig. 30, is necessary. This consists of a powerful induction coil used in conjunction with a delicate spectroscope, such as that used by Kirchhoff (see page 60). The light from the spark falls on to the slit, and is refracted by passing through the prisms.

For the purpose of intensifying the spark, the ends of the secondary coil are placed in contact with the coatings of a large Leyden jar. The electrodes, also of course connected with the poles of the secondary coil, consist of the metals under examination, either in the form of wire, or of irregular pieces held by forceps on a moveable stand. Many precautions must be taken, especially with two sets of electrodes, as it has been found that currents caused by the rapid passage of air between the poles are sufficient to carry over to a second set of electrodes, placed at a distance of a few inches, a very perceptible quantity of the materials undergoing volatilization.

We are indebted to the labours of Professors Kirchhoff, Ångström, and Thalén, and Mr. Huggins, for the most accurate sets of maps of the metallic lines which we possess. The positions of the metallic lines have been arranged by Kirchhoff with reference to the dark solar lines, whilst Huggins has used the bright air lines as a constant scale upon which to note the positions of the metal lines; but both experimenters use an arbitrary scale of divisions by which the lines are designated. Owing to the very large number of the lines of each of the metals, very great care is needed in the discrimination of these spectra: still, when the eye has been trained, the detection of the individual metal is perfectly certain. The spectra of the following elements were mapped by Kirchhoff:—

1. Sodium.	9. Strontium.	17. Antimony.	25. Aluminium.
2. Calcium.	10. Cadmium.	18. Arsenic.	26. Lead.
3. Barium.	11. Nickel.	19. Cerium.	27. Silver.
4. Magnesium.	12. Cobalt.	20. Lanthanum.	28. Gold.
5. Iron.	13. Potassium.	21. Didymium.	29. Ruthenium.
6. Copper.	14. Rubidium.	22. Mercury.	30. Iridium.
7. Zinc.	15. Lithium.	23. Silicon.	31. Platinum.
8. Chromium.	16. Tin.	24. Glucinum.	32. Palladium.

Copies of Kirchhoff's and Ångström's maps are found in Plates III., IV., and V., facing Lecture V.; and a copy of Huggins' maps is given on Plates I. and II., at the end of this lecture. The Tables of reference to the spectrum of each metal are found at the end of Appendix C.

The maps of two of the experimenters do not agree exactly with each other, because Kirchhoff altered the position of his prisms several times during the measurements, in order to bring the different rays as nearly as possible to the point of minimum deviation, whilst Huggins allowed the position of his prisms to remain unaltered. The spectra of the following metals have been drawn by Huggins:—

1. Sodium.	7. Thallium.	13. Antimony.	19. Lead.
2. Potassium.	8. Silver.	14. Gold.	20. Zinc.
3. Calcium.	9. Tellurium.	15. Bismuth.	21. Chromium.
4. Barium.	10. Tin.	16. Mercury.	22. Osmium.
5. Strontium.	11. Iron.	17. Cobalt.	23. Palladium.
6. Manganese.	12. Cadmium.	18. Arsenic.	24. Platinum.

A very interesting fact is noticed by all the observers, namely, that several of the bright lines of different metals seem to coincide. When, however, these cases of apparent coincidence are narrowly observed, most of the lines are found to show real though slight differences of refrangibility.

The following still remain as unresolved coincidences in Huggins' map, and future experiments with help of higher magnifying powers must decide whether these and similar coincidences are real, or only apparent; whether the lines in question really fall upon one another, or whether they only lie very close together:—

	DIVISION		DIVISION
Zinc and Arsenic	909	Tellurium and Nitrogen .	1366
Sodium and Lead	1000	Osmium and Arsenic . .	1737
Sodium and Barium . . .	1005	Chromium and Nitrogen .	2336

These six are then the only cases of coincidence observed by Huggins in examining many hundreds of bright lines of twenty-four elements, and even these may possibly disappear when investigated by a more powerful instrument.

Thalén has quite recently published[1] an interesting memoir upon the determination of the *wave-lengths* of the lines of the metal spectra. In Kirchhoff's and Huggins' maps the bright metal lines are arranged according to an arbitrary scale, but it is of great interest to know the absolute position of each line, and this can only be ascertained by the determination of its wave-length. The spectra of no less than forty-five metals are given in Thalén's tables, and amongst them are those of many rare metals not previously examined: thus we

[1] Annales de Chemie et de Physique, Oct. 1869, sér. [4], vol. xviii. p. 202.

find the lines of the spectra of glucinum, zirconium, erbium, yttrium, thorinum, uranium, titanium, tungsten, molybdenum, and vanadium, most accurately mapped. To give you a notion of the immense number of these lines, I should like to show you a drawing or two: for this purpose I will throw on the screen an enlarged image of one of Kirchhoff's maps (Plates III. and IV. preceding Lecture V.). Professor Kirchhoff was the first to examine the exact character of these metallic lines, and he drew an accurate map, not only of these metallic lines, but of the dark lines in the spectrum of the sun; and this is a copy of one of his diagrams, to which I would now briefly allude. The dark lines here represent the dark lines in the sun. With these I have at present nothing to do. We shall devote a subsequent lecture to a detailed discussion of this most remarkable subject. To-day I would simply draw your attention to the short lines at the lower part of the diagram, which indicate to us the positions of the bright metal lines with regard to the fixed dark solar lines, these latter being taken as a sort of inch-rule, by which the positions of the other lines are reckoned. The lines which you see joined by a horizontal line, and marked Fe (for Ferrum), are the iron lines; and I beg you to notice the very large number and the very beautifully fine nature of these iron lines. On Kirchhoff's map each line is accompanied by a letter, being the chemical symbol of the element to which this line belongs; here an aluminium line, here an antimony line, here a calcium line, here again a number of iron lines connected together; and so I might go through all these diagrams, showing the number of lines which Kirchhoff has mapped,—and this for only a small portion of the spectrum. The one end of this diagram is in the yellow, and the other end in the green; so that we have, on this map, only a very, very small portion of the metal lines which would be seen if we were looking at the whole length of the visible spectrum.

I would next illustrate this fact by showing you another beautiful drawing of these metal lines, made by our countryman, Mr. Huggins (see Plates I. and II. at the end of this lecture). This map, which is copied from Mr. Huggins' paper in the *Philosophical Transactions* for 1864, will also give you an idea of the very great number of these metal lines. We have here about twenty metals, and each pair of these horizontal lines includes the spectrum of a particular metal. For instance, if we take silver, here is one line, here three, here two, here a number of other lines: thus we go on through the whole spectrum, and have altogether a great number of silver lines. On the right we have the red end, on the left the blue end of the spectrum, and at the top, for the sake of comparison, are the chief lines of the solar spectrum and the air lines. From this table you may not only form an idea of the large number of metal lines existing, but you will see that the lines of any one metal do not coincide or interfere with those of any other.

These lines are by no means all the peculiar rays which such highly heated metallic vapours emit, for Professor Stokes has shown that the bright sparks from poles of iron, aluminium, and magnesium give off light of so high a degree of refrangibility, that distinct bands are situated at a distance beyond the last visible violet ray, ten times as great as the length of the whole visible spectrum from red to violet! These bands cannot of course be seen under ordinary circumstances, but when allowed to fall on a fluorescent body, such as paper moistened by quinine solution, they can easily be rendered visible; or we may photograph them, and make them leave their impression on the sensitive film. In order that these highly refrangible rays may be seen, no glass lenses or prisms must be used, as the rays of high refrangibility cannot pass through glass: quartz, on the other hand, permits them to pass; hence all the lenses and prisms must be made of quartz.

In new and interesting subjects like those which now occupy our attention, the mind is very apt to be led away into speculations, which, however engrossing they may prove, are foreign to the spirit of the exact scientific inquirer. Such speculations might in this case have special reference to the possibility or probability of arriving, by the help of the observations of the bright lines which bodies give us, at some more intimate knowledge of the composition of the so-called elements. We might speculate as to the connexion, for instance, between the wave-lengths of the various bright lines of the metal and the particular atomic weight of the substance; or we might ask, Can we find out any relation between the spectra of the members of some well-known chemical family, as of the alkaline metals, potassium, sodium, cæsium, and rubidium? Such questions as these naturally occur to every one. At present, however, this subject is in such an undeveloped state that these speculations are useless, because they are premature, and the data are insufficient; but doubtless a time will come when these matters will be fully explained, and a future Newton will place on record a mathematical theory of the bright lines of the spectrum as a striking monument of the achievements of exact science.

In the next lecture I purpose to show how this method of analysis can be applied to the detection of the non-metallic elements, whether they be solid, liquid, or gaseous at the ordinary temperature.

LECTURE III.—APPENDIX A.

SPECTRUM REACTIONS OF THE RUBIDIUM AND CÆSIUM COMPOUNDS.[1]

CÆSIUM and rubidium are not precipitated either by sulphuretted hydrogen or by carbonate of ammonium. Hence both metals must be placed in the group containing magnesium, lithium, potassium, and sodium. They are distinguished from magnesium, lithium, and sodium by their reaction with bichloride of platinum, which precipitates them like potassium. Neither rubidium nor cæsium can be distinguished from potassium by any of the usual reagents. All three substances are precipitated by tartaric acid as white crystalline powders; by hydrofluosilicic acid as transparent opalescent jellies; and by perchloric acid as granular crystals: all three, when not combined with a fixed acid, are easily volatilized on the platinum wire, and they all three tinge the flame violet. The violet colour appears indeed of a bluer tint in the case of potassium, whilst the flame of rubidium is of a redder shade, and that of cæsium still more red. These slight differences can, however, only be perceived when the three flames are ranged side by side, and when the salts undergoing volatilization are perfectly pure. In their reactions, then, with the common chemical tests, these new elements cannot be distinguished from potassium. The only method by means of which they can be recognised when they occur together is that of spectrum analysis.

The spectra of rubidium and cæsium are highly characteristic, and are remarkable for their great beauty (Frontispiece, Nos. 3 and 4). In examining and measuring these spectra we have

[1] Extract from Professors Kirchhoff and Bunsen's Second Memoir on Chemical Analysis by Spectrum Observations (Phil. Mag. vol. xxii. 1861).

employed an improved form of apparatus (Fig. 33), which in every respect is much to be preferred to that described in our first memoir. In addition to the advantages of being more manageable and producing more distinct and clearer images, it is so arranged that the spectra of two sources of light can be examined at the same time, and thus, with the greatest degree of precision, compared both with one another and with the numbers on a divided scale.

In order to obtain representations of the spectra of cæsium and rubidium corresponding to those of the other metals which we have given in our former paper, we have adopted the following course.

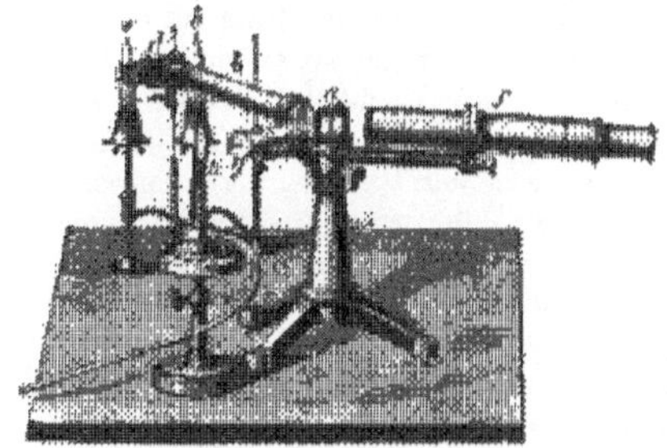

FIG. 33.

We have placed the tube *g* (Fig. 33) in such a position that a certain division of the scale, viz. No. 100, coincided with Fraunhofer's line D in the solar spectrum, and then observed the position of the dark solar lines A, B, C, D, E, F, G, H, on the scale: these several readings we called A, B, C, &c. An interpolation scale was then calculated and drawn, in which each division corresponded to a division on the scale of the instrument, and in which the points corresponding to the observations A, B, C, &c. were placed at the same distances apart as the same lines on our first drawings of the spectrum. By help of this scale, curves of the new spectra were drawn (Fig. 26, p. 63), in which the ordinates express the degrees of luminosity at the various points on the scale, as judged of by the eye. The lithographer then made the designs represented in the Frontispiece from these curves.

As in our first memoir, so here we have represented only those lines which, in respect to position, definition, and intensity, serve as the best means of recognition. We feel it necessary to repeat this statement, because it has not unfrequently happened that the presence of lines which are not represented in our drawings has been considered as indicative of the existence of new bodies.

We have likewise added a representation of the potassium spectrum to those of the new metals for the sake of comparison, so that the close analogy which the spectra of the new alkaline metals bear to the potassium spectrum may be at once seen. All three possess spectra which are continuous in the centre, and decreasing at each end in luminosity. In the case of potassium this continuous portion is most intense, in that of rubidium less intense, and in the cæsium spectrum the luminosity is least. In all three we observe the most intense and characteristic lines towards both the red and blue ends of the spectrum.

Amongst the rubidium lines, those splendid ones named Rb α and Rb β are extremely brilliant, and hence are most suited for the recognition of the metal. Less brilliant, but still very characteristic, are the lines Rb δ and Rb γ. From their position they are in a high degree remarkable, as they both fall beyond Fraunhofer's line A; and the outer one of them lies in an ultra-red portion of the solar spectrum, which can only be rendered visible by some special arrangement. The other lines, which are found on the continuous part of the spectrum, cannot so well be used as a means of detection, because they only appear when the substance is very pure, and when the luminosity is very great. Nitrate of rubidium, and the chloride, chlorate, and perchlorate of rubidium, on account of their easy volatility, show these lines most distinctly. Sulphate of rubidium and similar salts also give very beautiful spectra. Even silicate and phosphate of rubidium yield spectra in which all the details are plainly seen.

The spectrum of cæsium is especially characterised by the two blue lines Cs α and Cs β: these lines are situated close to the blue strontia line Sr δ, and are remarkable for their wonderful brilliancy and sharp definition. The line Cs δ, which cannot be so conveniently used, must also be mentioned. The yellow and green lines represented on the figure, which first appear when the luminosity is great, cannot so well be employed for the purpose of detecting small quantities of the cæsium compounds; but they may be made use of with advantage as a test of the purity of the cæsium salt under examination. They appear much more distinctly than do the yellow and green lines in the potassium spectrum, which, for this reason, we have not represented.

As regards *distinctness* of the reaction, the cæsium compounds resemble in every respect the corresponding rubidium salts: the chlorate, phosphate, and silicate gave the lines perfectly clearly. The *delicacy* of the reaction, however, in the case of the cæsium compounds, is somewhat greater than in that of the corresponding compounds of rubidium. In a drop of water weighing four milligrammes, and containing only 0·0002 milligramme of chloride of rubidium, the lines Rb α and Rb β can only just be distinguished; whilst 0·00005 milligramme of the chloride of cæsium can, under similar circumstances, easily be recognised by means of the lines Cs α and Cs β.

If other members of the group of alkaline metals occur together with cæsium and rubidium, the delicacy of the reaction is of course materially impaired, as is seen from the following experiments, in which the mixed chlorides contained in a drop of water, weighing about four milligrammes, were brought into the flame on a platinum wire.

When 0·003 milligramme of chloride of cæsium was mixed with from 300 to 400 times its weight of the chloride of potassium or sodium, it could be easily detected. Chloride of rubidium, on the other hand, could be detected with difficulty when the quantity of chloride of potassium or chloride of sodium amounted to from 100 to 150 times the weight of the chloride of rubidium employed.

0·001 milligramme of chloride of cæsium was easily recognised when it was mixed with 1,500 times its weight of chloride of lithium; whilst 0·001 milligramme of chloride of rubidium could not be recognised when the quantity of chloride of lithium added exceeded 600 times the weight of the rubidium salt.

APPENDIX B.

CONTRIBUTIONS TOWARDS THE HISTORY OF SPECTRUM ANALYSIS. BY G. KIRCHHOFF.[1]

In my "Researches on the Solar Spectrum and the Spectra of the Chemical Elements"[2] I made a few short historical remarks concerning earlier investigations upon the same subject. In these remarks I have passed over certain publications in silence —in some cases because I was unacquainted with them, in others because they appeared to me to possess no special interest in relation to the history of the discoveries in question. Having become aware of the existence of the *former* class, and seeing that more weight has been considered to attach to the *latter* class of publications by others than by myself, I will now endeavour to complete the historical survey.

Amongst those who have devoted themselves to the observation of the spectra of coloured flames, I must in the first place mention Herschel and Talbot. Their names need special notice, as they pointed out with distinctness the service which this mode of observation is capable of rendering to the chemist. For a knowledge of their researches I am mainly indebted to Professor W. Allen Miller, who gave an extract from them in a lecture republished in the number of the *Chemical News* for 19th April, 1862. It is there stated that in the volume of the *Transactions of the Royal Society of Edinburgh* for 1822, at page 455, Herschel shortly describes the spectra of chloride of strontium, chloride of potassium, chloride of copper, nitrate of

[1] Communicated to the Phil. Mag. Fourth Series, vol. xxv. p. 250, by Professor Roscoe.

[2] Published by Macmillan and Co. Cambridge and London, 1862.

copper, and boracic acid. The same observer says, in his article on Light in the "Encyclopædia Metropolitana," 1827, page 438: "Salts of soda give a copious and purely homogeneous yellow; of potash, a beautiful pale violet." He then describes the colours given by the salts of lime, strontia, lithia, baryta, copper, and iron, and continues: "Of all salts, the muriates succeed the best, from their volatility. The same colours are exhibited also when any of the salts in question are put (in powder) into the wick of a spirit-lamp. The colours thus communicated by the different bases to flame afford in many cases a ready and neat way of detecting extremely minute quantities of them. The pure earths when violently heated, as has recently been practised by Lieut. Drummond, by directing on small spheres of them the flames of several spirit-lamps, urged by oxygen gas, yield from their surfaces lights of extraordinary splendour, which, when examined by prismatic analysis, are found to possess the peculiar definite rays in excess which characterise the tints of flames coloured by them; so that there can be no doubt that these tints arise from the molecules of the colouring matter, reduced to vapour and in a state of violent ignition."

Talbot says:[1] "The flame of sulphur and nitre contains a red ray which appears to me of a remarkable nature. This red ray appears to possess a definite refrangibility, and to be characteristic of the salts of potash, as the yellow ray is of the salts of soda, although, from its feeble illuminating power, it is only to be detected with a prism. If this should be admitted, I would further suggest, that whenever the prism shows a *homogeneous* ray of any colour to exist in a flame, this ray indicates the formation or the presence of a *definite chemical compound*." Somewhat further on, in speaking of the spectrum of red fire and of the frequent occurrence of the yellow line, he says: "The other lines may be attributed to the antimony, strontia, &c. which enter into this composition. For instance, the orange ray may be the effect of the strontia, since Mr. Herschel found in the flame of muriate of strontia a ray of that colour. If this opinion should be correct, and applicable to the other definite rays, a glance at

[1] Brewster's Journal of Science, vol. v. 1826; Chemical News, April 27, 1861.

the prismatic spectrum of a flame may show it to contain substances which it would otherwise require a laborious chemical analysis to detect." In a subsequent communication[1] the same physicist, after a striking description of the spectra of lithium and strontium, continues: "Hence I hesitate not to say that optical analysis can distinguish the minutest portions of these two substances from each other with as much certainty as, if not more than, any other known method."

In these expressions the idea of "chemical analysis by spectrum observations" is most clearly put forward. Other statements, however, of the same observers, occurring in the same memoirs from which the foregoing quotations are taken (but not mentioned by Professor Miller in his abstract), flatly contradict the above conclusions, and place the foundations of this mode of analysis on most uncertain ground.

Herschel, in page 438 of his article on Light, almost immediately before the words quoted above, says: "In certain cases when the combustion is violent, as in the case of an oil-lamp urged by a blowpipe (according to Fraunhofer), or in the upper part of the flame of a spirit-lamp, or when sulphur is thrown into a white-hot crucible, a very large quantity of a definite and purely homogeneous yellow light is produced; and in the latter case forms nearly the whole of the light. Dr. Brewster has also found the same yellow light to be produced when spirit of wine, diluted with water, and heated, is set on fire."

Talbot states: "Hence the yellow rays may indicate the presence of soda; but they nevertheless frequently appear where no soda can be supposed to be present."[2] He then mentions that the yellow light of burning sulphur, discovered by Herschel, is identical with the light of the flame of a spirit-lamp with a salted wick, and states that he was inclined to believe that the yellow light which occurred when salt was strewed upon a platinum foil placed in a flame "was owing to the water of crystallization rather than to the soda: but then," he continues, "it is not easy to explain why the salts of potash, &c. should

[1] Phil. Mag. 1834, vol. iv. p. 114; Chemical News, April 27, 1861.
[2] Brewster's Journal, vol. v. 1826.

not produce it likewise. Wood, ivory, paper, &c., when placed in the gas flame, give off, besides their right flame, more or less of this yellow light, which I have always found the same in its characters. The only principle which these various bodies have in common with the salts of soda is *water*; yet I think that the formation or presence of water cannot be the origin of this yellow light, because ignited sulphur produces the *very same*, a substance with which water is supposed to have no analogy." "It may be worth remark," he adds in a note, "though probably accidental, that the specific gravity of sulphur is 1·99, or almost *exactly twice* that of water." "It is also remarkable," he continues in the text, "that alcohol burnt in an open vessel, or in a lamp with a metallic wick, gives but little of the yellow light; while, if the wick be of cotton, it gives a considerable quantity, and that for an unlimited time. (I have found other instances of a change of colour in flames, owing to *the mere presence* of the substance, which *suffers no diminution in consequence*. Thus a particle of muriate of lime on the wick of a spirit-lamp will produce a quantity of red and green rays for a whole evening without being itself sensibly diminished.)"[1]

In a later portion of the memoir he attributes the yellow line in one place to the presence of soda salts, in another to that of sulphur. Thus, in the above-mentioned statement concerning the spectrum of red fire, he says, "The bright line in the yellow is caused, without doubt, by the combustion of the sulphur."[2]

Hence we must admit that the conclusion that the aforesaid yellow line can be taken as a positive proof of the presence of sodium compounds in the flame can in no way be deduced from Herschel and Talbot's researches. On the contrary, the numerous modes in which the line is produced would rather point to the conclusion that it is dependent upon no chemical constituent of the flame, but arises by a process whose nature is unknown, which may occur, sometimes more easily, sometimes with diffi-

[1] Brewster's Journal, vol. v. 1826.

[2] A short statement of Herschel and Talbot's results, as here quoted, was made by me in a lecture at the Royal Institution on April 5, 1862, and reprinted in the *Chemical News* for May 10, 1862.—H. E. R.

culty, with the most different chemical elements. If we accept such an explanation concerning this yellow line, we must form a similar opinion respecting the other lines seen in the spectrum, which were far more imperfectly examined; and in this we should be strengthened by the statement of Talbot, that a piece of chloride of calcium by *its mere presence* in the wick of a flame, and *without suffering any diminution*, causes a red and a green line to appear in the spectrum.

The experiments of Wheatstone,[1] Masson, Ångström, Van der Willigen, and Plücker upon the spectra of the electric spark or electric light (to which I have already referred in my "Researches on the Solar Spectrum and Spectra of the Chemical Elements." Macmillan, London, 1862, p. 8), as well as those of Despretz,[2] from which the physicist concluded that the positions of the bright lines in the spectrum of the light from a galvanic battery were unaltered by variation of the intensity of the current, might serve to support the view that the bright lines in the spectrum of an incandescent gas are solely dependent upon the several chemical constituents of the gas; but they could not be considered as *proof* of such an opinion, as the conditions under which they were made were, for this purpose, too complicated, and the phenomena occurring in an electric spark too ill understood. The demonstrative power of the above experiments as regards the question at issue is rendered less cogent by the difference visible in the colour of the electric light in different parts of a Geissler's tube; by the circumstance noticed by Van der Willigen, who obtained different spectra by passing an electric spark from the same electrodes through gas of constant chemical composition, if the density of the gas was varied within sufficient limits; and lastly by an observation which Ångström cursorily mentions. This physicist says:[3] "Wheatstone has already noticed that when the poles consist of two different

[1] Wheatstone not merely experimented with the spark from an electrical machine, but likewise with the voltaic induction-spark. (Report of the British Association, 1835; Chemical News, March 23 and March 30, 1861.)

[2] Comptes Rendus, vol. xxxi. p. 419 (1850).

[3] Pogg. Ann. vol. xciv. p. 160 (translated in Phil. Mag. for May 1855).

metals the spectrum contains the lines of both metals. Hence it became of interest to see whether a compound of these metals, especially a chemical compound, also gives the lines of both metals, or whether the compound is distinguished by the occurrence of new lines. Experiment shows that the first supposition is correct. The sole difference noticed is, that certain lines were wanting, or appeared with less distinctness; but when they were observed, they always appeared in the position in which they occurred in the separate metals." In the following sentence, however, he states, "That in the case of zinc and tin the lines in the blue were somewhat displaced in the direction of the violet end, but the displacement was very inconsiderable." Had such a displacement, however small, *really* occurred, we must conclude either that the bright lines of the electric spark obey other laws than those of a glowing gas, or that these latter are *not* solely dependent on the separate chemical constituents of the gas.

The question at issue respecting the lines of incandescent gases could only be satisfactorily solved by experiments carried out under the most simple conditions—such, for instance, as the examination of the spectra of flames. Observations of this kind were made in the year 1845 by Professor W. Allen Miller, but they do not furnish any contribution towards a solution of the question. Dr. Miller has the merit of having first published diagrams of the spectra of flames;[1] but these diagrams are but slightly successful, although, in a republication in the *Chemical News*[2] of the paper accompanying these drawings, Mr. Crookes remarks: "We cannot, of course, give the coloured diagrams with which it was originally illustrated; but we can assure our readers that, after making allowance for the imperfect state of chromolithography sixteen years ago,[3] the diagrams of the spectra given by Professor Miller are *more accurate* in several respects than the coloured spectra figured in recent numbers of the scientific periodicals." In reply to this "assurance"

[1] Phil. Mag. for August 1845. [2] Chemical News, May 18, 1861.

[3] Prof. Miller's diagrams are not printed by chromolithography, but, as is seen on inspection, tinted by hand.—H. E. R.

of Mr. Crookes I only have to remark that, by way of experiment, I have laid Professor Miller's diagrams before several persons conversant with the special spectra, requesting them to point out the drawing intended to represent the spectrum of strontium, barium, and calcium respectively, and that in no instance have the right ones been selected.

Swan was the first who endeavoured experimentally to prove whether the almost invariably occurring yellow line may be solely caused by the presence of sodium compounds. In his classical research "On the Spectra of the Flames of the Hydro-carbons"[1] (referred to both in my "Researches" and in the paper published by Bunsen and myself) Swan shows how small the quantity of sodium is which produces this line distinctly; he finds that this quantity is minute beyond conception, and he concludes: "When indeed we consider the almost universal diffusion of the salts of sodium, and the remarkable energy with which they produce yellow light, it seems highly probable that the yellow line R, which appears in the spectra of almost all flames, is in every case due to the presence of minute quantities of sodium."

The strict subject-matter of Swan's investigation was the comparison of the spectra of flames of various hydrocarbons. "The result of his comparison has been, that in all the spectra produced by substance, either of the form C_rH_s or of the form $C_rH_sO_t$, the bright lines have been identical. In some cases, indeed, certain of the very faint lines which occur in the spectrum of the Bunsen lamp were not seen. The brightness of the lines varies with the proportion of carbon to hydrogen in the substance which is burned, being greatest where there is most carbon. . . . . . The absolute identity which is thus shown to exist between the spectra of dissimilar carbo-hydrogen compounds is not a little remarkable. It proves, 1st, that the position of the lines in the spectrum does not vary with the proportion of carbon and hydrogen in the burning body—as when we compare the spectra of light carburetted hydrogen, CH_2, olefiant gas, C_2H_2, and oil of turpentine, $C_{20}H_{16}$; and 2dly,

[1] Trans. Roy. Soc. of Edinburgh, vol. xxi. p. 414.

that the presence of oxygen does not alter the character of the spectrum: thus ether, C_4H_5O, and wood spirit, $C_2H_4O_2$, give spectra which are identical with those of paraffin, $C_{20}H_{20}$, and oil of turpentine, $C_{20}H_{16}$.

"In certain cases, at least, the mechanical admixture of other substances with the carbo-hydrogen compound does not affect the lines of the spectrum. Thus I have found that a mixture of alcohol and chloroform burns with a flame having a very luminous green envelope—an appearance characteristic of the presence of chlorine—and no lines are visible in the spectrum. When, however, the flame is urged by the blowpipe, the light of the envelope is diminished, and the ordinary lines of the hydrocarbon spectrum become visible."

In this research Swan has made a most valuable contribution towards the solution of the proposed question as to whether the bright lines of a glowing gas are solely dependent upon its chemical constituents; but he did not answer it positively, or in its most general form; he did not indeed enter upon this question, for he wished to confine his investigation to the spectra of the hydrocarbons, and was only led to the examination of this yellow line by its frequent occurrence in these spectra.

No one, it appears, had clearly propounded this question before Bunsen and myself; and the chief aim of our common investigation was to decide this point. Experiments which were greatly varied, and were for the most part new, led us to the conclusion upon which the foundations of the "chemical analysis by spectrum observations" now rest.

APPENDIX C.

ON THE SPECTRA OF SOME OF THE CHEMICAL ELEMENTS. BY WILLIAM HUGGINS, Esq. F.R.A.S.[1]

1. I have been engaged for some time, in association with Professor W. A. Miller, in observing the spectra of the fixed stars. For the purpose of accurately determining the position of the stellar lines, and their possible coincidence with some of the bright lines of the terrestrial elements, I constructed an apparatus in which the spectrum of a star can be observed directly with any desired spectrum. To carry out this comparison, we found no maps of the spectra of the chemical elements that were conveniently available. The minutely detailed and most accurate maps and tables of Kirchhoff were confined to a portion of the spectrum, and to some only of the elementary bodies; and in the maps of both the first and the second part of his investigations the elements which are described are not all given with equal completeness in different parts of the spectrum. But these maps were the less available for our purpose because, since the bright lines of the metals are laid down relatively to the dark lines of the solar spectrum, there is some uncertainty in determining their position at night, and also in circumstances when the solar spectrum cannot be conveniently compared simultaneously with them. Moreover, in consequence of the difference in the dispersive power of prisms, and the uncertainty of their being placed exactly at the same angle relatively to the incident rays, tables of numbers obtained with one instrument are not alone sufficient to determine lines from their position with any other instrument.

[1] Phil. Trans. 1864, p. 139.

PLATE I.—HUGGINS' MAPS OF THE METALLIC LINES.

Plate I.—Huggins' Maps of the Metallic Lines.

PLATE II.—HUGGINS' MAPS OF THE METALLIC LINES.

It appeared to me that a standard scale of comparison such as was required, and which, unlike the solar spectrum, would be always at hand, is to be found in the lines of the spectrum of common air. Since in this spectrum about a hundred lines are visible in the interval between A and H, they are sufficiently numerous to become the fiducial points of a standard scale to which the bright lines of the elements can be referred. The air spectrum has also the great advantage of being visible, together with the spectra of the bodies under observation, without any increased complication of apparatus.

2. The optical part of the apparatus employed in these observations consists of a spectroscope of six prisms of heavy glass. The prisms were purchased of Mr. Browning, optician, of the Minories, and are similar in size and in quality of glass to those furnished by him with the Gassiot spectroscope. They all have a refracting angle of 45°. They increase in size from the collimator; their faces vary from 1·7 inch by 1·7 inch to 1·7 inch by 2 inches.

The six dispersing prisms and one reflecting prism were carefully levelled, and the former adjusted at the position of minimum deviation for the sodium line D. The train of prisms was then enclosed in a case of mahogany, marked *a* in the diagram (Fig. 34), having two openings, one for the rays from the collimator *b*, and the other for their emergence after having been refracted by the prisms. These openings are closed with shutters when the apparatus is not in use. By this arrangement the prisms have not required cleansing from dust, and their adjustments are less liable to derangement. The collimator *b* has an achromatic object-glass by Ross of 1·75 inch diameter, and of 10·5 inches focal length. The object-glass of the telescope, which is of the same diameter, has a focal length of 16·5 inches. The telescope moves along a divided arc of brass, marked in the diagram *c*. The centre of motion of the telescope is nearly under the centre of the last face of the last prism. The eyepiece was removed from the telescope, and the centre of motion was so adjusted that the image of the illuminated lens of the collimator, seen through the train of prisms, remained approximatively concentric with the object-glass of the telescope whilst the latter was moved through an extent of arc equal to the visible spectrum. All the pencils emerging from the last prism, therefore, with the exception of those of the extreme refrangible portion of the spectrum, are received nearly centrically on the object-glass of the telescope. The total deviation of the light in passing through the train of prisms is, for the ray D, about 198°. The interval from A to H corresponds to about 21° 14′ of arc upon the brass scale.

3. The measuring part of the apparatus consists of an arc of

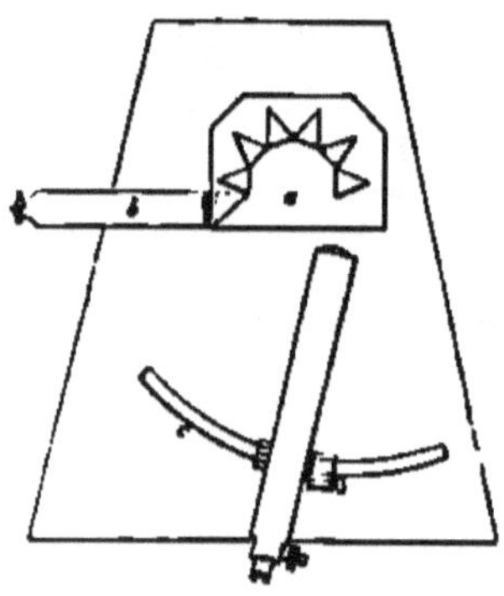

FIG. 34.

brass, marked *c* in the figure, divided to intervals of 15″. The distance traversed by the telescope in passing from one to the other of the components of the double sodium line D is measured by five divisions of 15″ each. These are read by a vernier.

Attached to the telescope is a wire micrometer by Dollond. This records sixty parts of one revolution of the screw for the interval of the double sodium line. Twelve of these divisions of the micrometer, therefore, are equal to one division of the scale upon the arc of brass. The micrometer has a cross of strong wires placed at an angle of 45° nearly with the lines of the spectrum. The point of intersection of these wires may be brought upon the line to be measured by the micrometer screw, or by a screw attached to the arm carrying the telescope. For the most part the observations were read off from the scale, and the micrometer has been only occasionally employed in the verification of the measures of small intervals. The sexagesimal readings of the scale, giving five divisions to the interval of the double line D, have been reduced to a decimal form, the units of which are intervals of 15″; and these are the numbers given in the Tables. An attempt was made to reduce the measures to the scale of Kirchhoff's Tables, but the spectra are not found to be superposable on his. This is due, in great part, probably to the prisms in his observations having been varied in their adjustment for different parts of the spectrum. The eyepieces are of the positive form of construction. One, giving the power of 15, is by Dollond; the other, of about 35, is by Cook.

4. The excellent performance of the apparatus is shown by the great distinctness and separation of the finer lines of the solar spectrum. All those mapped by Kirchhoff are easily seen, and many others in addition to these. The *whole* spectrum is very distinct. The numerous fine lines between *a* and A are well defined. So also are the groups of lines about and beyond G. H is seen, but with less distinctness.

As, with the exception of the double potassium line near A, no lines have been observed less refrangible than *a*, the maps and Tables commence with the line *a* of the solar spectrum and extend to H.

The observations are probably a little less accurate and complete near the most refrangible limit. Owing to the feebleness of the illumination of this part of the spectrum, the slit has to be widened; and, moreover, the cross-wires being seen with difficulty, the bisection of a line exactly is less certain.

5. For all the observations the spark of an induction coil has been employed. This coil has about fifteen miles of secondary wire, and was excited by a battery of Grove's construction, sometimes two, at others four cells having been employed.

Each of these cells has thirty-three square inches of acting surface of platinum. With two such cells the induction spark is three inches in length. A condenser is connected with the primary circuit, and in the secondary a battery of Leyden jars is introduced. Nine Leyden jars, the surface of each of which exposes 140 square inches of metallic coating, were employed. These are arranged in three batteries of three jars each, and the batteries are connected in polar series.

The metals were held in the usual way with forceps. The nearness of the electrodes to each other, their distance from the slit, and the breadth of the latter were varied, to obtain in each case the greatest distinctness. The amount of separation of the electrodes was always such that the metallic lines under observation extended across the spectrum. The two sets of discharging points were arranged in the circuit in series.

6. Some delay was occasioned by the want of accordance of the earlier measures, though the apparatus had remained in one place and could have suffered no derangement. These differences are supposed to arise from the effect of changes of temperature upon the prisms and other parts of the apparatus. This source of error could not be met by a correction applied to the zero point of measurement, as the discordances observed corresponded, for the most part, to an irregular shortening and elongation of the whole spectrum.

The principal air lines were measured at one time of observing, during which there was satisfactory evidence that the values of the measures had not sensibly altered; and these numbers have been preserved as the fiducial points of the scale of measures. The lines of the spectra of the metals have been referred to the nearest standard air line, so that only this comparatively small interval has been liable to be affected by differences of temperature. Upon these intervals the effect of such changes of temperature as the apparatus is liable to be subjected to is not, I believe, of sensible amount with the scale of measurement adopted. Ordinarily, for the brighter portion of the spectrum, the width of the slit seldom exceeded $\frac{1}{100}$ inch: when this width had to be increased in consequence of the feebler illumination towards the ends of the spectrum, the measure of the nearest air line as seen in the compound spectrum was again taken, and the places of the lines of the metal under observation were reckoned relatively to this known line.

By this method of frequent reference to the principal air lines the measures are not sensibly affected by the errors which might have been introduced from the shifting of the lines in absolute position in consequence of alterations either in the width of the slit, in the place and direction of the discharge before the slit, or in the apparatus from variations of temperature, flexure, or other causes.

The usual place of the electrodes was about ·7 inch from the slit, though occasionally they were brought nearer to the slit. When they are placed in such close proximity, the sparks charge the spectroscope by induction; but the inconvenience of sparks striking from the eyepiece to the observer may be prevented by placing the hand upon the apparatus, or putting the latter into metallic communication with the earth.

The spectrum of comparison was received by reflection from a prism placed in the usual manner over one-half of the slit. As the spectrum of the discharge between points of platinum, when these are not too close, is, with the exception of two or three easily recognised lines, a pure air spectrum, this was usually employed as a convenient spectrum of comparison for distinguishing those lines in the compound spectrum which were due to the particular metal employed as electrodes. The measures, however, of all the lines, including those of the air spectrum itself, were invariably taken from the light received into the instrument directly, and in no case has the position of a line been obtained by measures of it taken in the spectrum of the light reflected into the slit by the prism.

The measures of all the lines were taken more than once; and, when any discordance was observed between the different sets, the lines were again observed. The spectra of most of the metals were remeasured at different times of observing. In the measurement of the solar lines for their co-ordination with the standard air spectrum, the observations were repeated on several different occasions during the progress of the experiments. The line G of the solar table is the one so marked by Kirchhoff.[1] When no change in the instrument could be detected, the measures came out very closely accordant, for the most part identical. The discordances due to small alterations in the instrument itself were never greater than five or six of the units of measurement in the whole arc of 4,955 units. As the apparatus remained in one place free from all apparent derangement, these alterations are probably due to changes of temperature. The method employed to eliminate these discordances has been described.

Throughout the whole of the bright portion of the spectrum the probable error of the measures of the narrow and well-defined lines does not, I believe, exceed one unit of the scale.

* * * * * *

Notes to the Tables.

Upon a re-examination of the Tables I found that it frequently occurred that lines of two or more metals were denoted by the same number. It appeared probable that these lines having a common number were not coincident, but only approximated in position within the limits of one unit of the scale employed; and besides, there might be small errors of observation. I therefore selected about fifty of these groups of lines denoted by common numbers, and compared the lines of each group, the one with the other, by a simultaneous observation of the different metals to which they belong. Some of the lines were found to be too faint and ill-defined to admit of being more accurately determined in position relatively to each other.

The following lines appear with my instrument to be coincident:

Zn, As 909	Na, Ba 1005	Cd, As 1737
Na, Pb 1006	Te, N 1366	Cr, N 2336

Of a much larger number of groups, the lines were, by careful scrutiny, observed to differ in position by very small quantities,

[1] Untersuchungen ü. d. Sonnenspectrum, 2 Theil, Taf. iii. Berlin, 1863.

corresponding for the most part to fractional parts of the unit of measurement adopted in the Tables. These are :

Ba 450 Sb 458·8	Ba 621·5 Bi 621	Te 657 Cd 656	Bi 857·5 Sb 857	Cu 997 Sb 997·5	Sb 1081 Au 1081·5	Te 1185 Fe 1185·3	Zn 1707 Pd 1708
Ca 515 Au 516	Ca 627 Fe 623	Fe [illegible] Zn [illegible]	Cd [illegible] Sb [illegible]	Au 981 Sb 981·5	Ca 1236 Cu 1237	Tl 1365 Mn 1365·5	Tl 1851 Bi 1851·3
Te 545·5 Sb 545	Au 643 Ca 643 Fe 641·5	Sb 765 Te 765·5	Au 906·8 Mn 909	Ca 1032 Pd 1031·5 Ag 1031·3 Pb 1031·1	Fe 1276 Ag 1276·3	Pd 1541 Fe 1540·8	Sb 1900 Pb 1900·3 Ni 1900
Na 541 Bi 555·5	Ca 649 Sn 649	Na 818·3 Ca 818	Ca 921·3 Tl 921 Co 921 Sb 921·1	Te 1090·5	Fe 1438 Te 1438·3	Pb 1580·2 Fe 1580	

TABLE I.

Solar.	Air.	Na.	K.	Ca.	Ba.	Sr.	Mn.	Tl.	Ag.	Te.	Sn.	Fe.
[illegible]	[illegible]	[illegible]	[illegible]	[illegible]	[illegible]	[illegible]	[illegible]	[illegible]	[illegible]	[illegible]	[illegible]	[illegible]
Solar.	Air.	Na.	K.	Ca.	Ba.	Sr.	Mn.	Tl.	Ag.	Te.	Sn.	Fe.

The character of the lines is indicated as follows:—

A line sharply defined at the edges, and narrow when the slit is narrow, a.

A band of light, defined as a line, but remaining, even with a narrow slit, nebulous at the edges, b.

A band of light irresolvable into lines, c.

Nebulous band, too faint for measurement, d.

The comparative intensity of the lines is indicated by the smaller figures placed in the position of exponents against the numbers in the Tables. The scale extends from 1 to 10, and includes fractional parts of unity to represent the very faint lines.

From *a* to D. (See Plates I. and II.)

Cd.	[illegible]	Au.	Bi.	Hg.	Cu.	[illegible]	Pb.	Zn.	Cr.	[illegible]	Pd.	Pt.
[illegible]	[illegible]	[illegible]	[illegible]	[illegible]	[illegible]	[illegible]	[illegible]	[illegible]	[illegible]	[illegible]	[illegible]	[illegible]
Cd.	[illegible]	Au.	Bi.	Hg.	Cu.	[illegible]	Pb.	Zn.	Cr.	[illegible]	Pd.	Pt.

TABLE II.

Solar.	Air.	Na	K.	Ca.	Ba.	Sr.	Mn.	Tl.	Ag.	Te.	Sn.	Pb.
1st 1006												1511.8 s
				1031.4 s								
			1049.1 b		1034.6 s			1063.9 s	1031-2.0 s	1039.3.8		1040.8 s
			1055.5 b		1047.1 s	1061.0 s				1053.2 s		
			1073.1-6 b		1056.1 s							1066.8 s
	N 1100.8							1093.5 s			1076.1 s	
	N 1111.1					1108.1 s				1111.7 s		
	N 1125.9	1160.8 s			1119.1 b					1122.1 s		
	N 1150.8	1174.1 b										
	N 1171.1									1151.7 s		
	N 1177.9											
	N 1190.1											
	N 1147.7											1238.8 s
	N 1229.8									1200.8 b		1240.8 s
	N 1262.8					1203.1 s			1207.8 s			1247.8 s
	N 1263.5.2			1176.1 b							1213.0 s	1251.9 s
	N 1274.3.8			1178.8 s		1227.8 s,d			1223.8 s			1261.8 s
	N 1284.8			1232.8 s					1227.7 s	1230.1 s		1270.7 s
				1250.0 s					1240.8 s		1250.8 s	1278.8 s
	N 1285.6			1255.5.8 s		1260.9 s	1280.1 b		1262.7 s	1270.1 b		1286.7 s,d
				1260.4 s					1276.6.4 s			1363.8 s
	N 1306			1265.7 s								1391.7 s
			1328.1 s		1308.6 s				1346.57 s		1340.8 s	1400.7 s
	N 1355.9											1413.8 s
	N 1364.7				1387.1 s	1401.8 s						1439.7 s
				1365.7 b		1405.8 s	1420.4 b					1451.5.7 s
						1411.8 s						1454.8 s
						1424.8 s						1462.8 s
					1361.8 s							1468.7
						1364.8 s						1468.7 s
						1389.7 b						1462.7 s
						1385.8 s		1358.6 s		1357.1 s		1475.5.8 s
						1385.4 s	1379.5 s		1379.8 s	1366.1 s		1482.7 s
						1397.5.3 s	1413.7 s		1390.8 s	1386.8 s		1484.1 s,d
							1425.1 s		1421.8 s			1485.8.3 s
						1432.8 s	1436.7 s		1432.8 s	1435.24 s		1486.8 s
						1457.1.8 b	1443.7 s		1446 b		1480.1 s	1490.5.8 s
	N 1503.8			1540.1 s			1452.9 s					1496.8 s
	N 1516.8						1456.8 s					1510.8 s
							1478.5.8 s			1483.8 s		1517.7 s
	N 1547.4						1500.5.8 s	1505.8 s			1520.7 s	1541.8 s
							1515.7 s			1544.38 s	1550.8 s	1548.8 s
											1562.4 s	1560.8 s
							1556.1 / 1571 b				1580.1 s	1574.8 s
										1580.8 s		1593.7 s,d
B {1590 / 1600}	—			1589.54 s								1598.5 s
	—											1599.7 s
												1606.7 s
Solar.	Air.	Na.	K.	Ca.	Ba.	Sr.	Mn.	Tl.	Ag.	Te.	Sn.	Pb.

From D to E. (See Plates I. and II.)

Cd.	Sb.	Au.	Bi.	Hg.	Co.	As.	Pb.	Sn.	Cr.	Os.	Pd.	Pt.
[illegible]	[illegible]	[illegible]	[illegible]	[illegible]	[illegible]	[illegible]	[illegible]	[illegible]	[illegible]	[illegible]	[illegible]	[illegible]
Cd.	Sb.	Au.	Bi.	Hg.	Co.	As.	Pb.	Sn.	Cr.	Os.	Pd.	Pt.

TABLE III.

Scale.	Air.	Na.	K.	Cu.	[illegible]	Sr.	[illegible]	[illegible]	Ag.	[illegible]	[illegible]	Fe.
[illegible]	[illegible]	[illegible]	[illegible]	[illegible]	[illegible]	[illegible]	[illegible]	[illegible]	[illegible]	[illegible]	[illegible]	[illegible]
2210												
Scale.	Air.	Na.	K.	Cu.	[illegible]	Sr.	[illegible]	[illegible]	Ag.	[illegible]	[illegible]	Fe.

* When the induction spark is taken in oxygen, a faint line is seen nearly in the position of the nitrogen line [illegible]. Since the lines of oxygen have a diminished intensity when the spark passes in air, this line would be too faint to be distinctly observed in the air spectrum, in which it occurs in a position of close proximity to brighter lines of nitrogen.

From E to F. (See Plates I. and II.)

Cd.	Sb.	Au.	Bi.	Hg.	Co.	As.	Pb.	Zn.	Cr.	Os.	Pd.	Pt.
	...	...	...	...	[illegible]	...	...	...	[illegible]			
...	...	...	...	...	[illegible]	...	...	...	[illegible]	...	[illegible]	
...	[illegible]	...	...	...	[illegible]	[illegible]	...	...	[illegible]	...	[illegible]	
...	...	[illegible]	...	...	[illegible]	...	...	[illegible]	[illegible]	...	[illegible]	[illegible]
...	[illegible]	...	...	[illegible]	[illegible]	...	...	...	[illegible]	...	[illegible]	
...		...	[illegible]	...	[illegible]	...	...	[illegible]	...	[illegible]		
...		...	[illegible]	...	[illegible]	...	[illegible]	...	[illegible]			
...	[illegible]	...		...	[illegible]	...	...	...	[illegible]			
[illegible]	...	...	[illegible]	...	[illegible]	[illegible]	[illegible]	...	[illegible]	...	...	[illegible]
...	[illegible]	...	...	[illegible]	[illegible]	...	...	...	[illegible]			
...	...	...	[illegible]	...	[illegible]	...		[illegible]	...	...	[illegible]	
...	...	...	...	...	[illegible]	...	[illegible]	...	[illegible]			
...	[illegible]	...	...	...	[illegible]	...	...	[illegible]				
...	...	...	[illegible]	...	[illegible]	...	...	[illegible]	—	...	[illegible]	
[illegible]	...	...	...	...	[illegible]							
...	[illegible]	...	[illegible]	...	[illegible]	[illegible]	...	...	[illegible]	...	[illegible]	
...	...	[illegible]	...	...	...	...	...	[illegible]				
...	...	...	...	...	[illegible]	...	...	[illegible]		[illegible]		
...	[illegible]	...	...	...	[illegible]	...	[illegible]	[illegible]		...	[illegible]	[illegible]
...	...	...	...	...	[illegible]							
...	[illegible]											
...	...		...	...	[illegible]							
...	...		[illegible]									
—	...		[illegible]	...	[illegible]	[illegible]						
...	...		...	[illegible]	...	...	—	[illegible]				
—	[illegible]		...	...	...	...	—	[illegible]	[illegible]			
...	...		...	[illegible]	...	...		[illegible]				
—	...		[illegible]									
...	...		[illegible]									
—	—		...	...	...	[illegible]		—	[illegible]	—	[illegible]	
...	[illegible]	...	...	...	...	—	—	...	[illegible]			
...	...	—	...	...	[illegible]	—	—	...	[illegible]			
—	...	—	...	...	...	—	—	[illegible]	[illegible]			

Cd.	Sb.	Au.	Bi.	Hg.	Co.	As.	Pb.	Zn.	Cr.	Os.	Pd.	Pt.

TABLE IV.

From F. to H. (See Plates I. and II.)

Cd.	Sb.	Au.	Bi.	Hg.	Co.	As.	Pb.	Zn.	Cr.	Os.	Pd.	Pt.
[illegible]	[illegible]	[illegible]	[illegible]	[illegible]	[illegible]	[illegible]	[illegible]	[illegible]	[illegible]	[illegible]	[illegible]	[illegible]
Cd.	Sb.	Au.	Bi.	Hg.	Co.	As.	Pb.	Zn.	Cr.	Os.	Pd.	Pt.

LECTURE IV.

Mode of Obtaining the Spectra of Gases and other Non-metallic Bodies.—Plücker and Hittorf.—Huggins.—Influence of Change of Density and of Temperature.—Kirchhoff.—Frankland and Lockyer.—Variation of the Spectra of certain Metals with Temperature.—Spectra of Compounds.—Ångström's Conclusions.—Spectrum of the Bessemer Flame.—Selective Absorption.—Blood Bands.—Detection of Colouring Matters.—Phosphorescence.—Fluorescence.

APPENDIX A.—Description of the Spectra of the Gases and Non-metallic Elements.

APPENDIX B.—On the Effect of Increased Temperature upon the nature of the Light emitted by the Vapour of certain Metals or Metallic Compounds.

APPENDIX C.—Kirchhoff on the Variation of the Spectra of certain Elements.

APPENDIX D.—Ignited Gases under certain circumstances give continuous Spectra.—Combustion of Hydrogen in Oxygen under great pressure.

APPENDIX E.—On the Spectrum of the Bessemer Flame.

APPENDIX F.—On the Spectra of Erbium and Didymium Compounds.

APPENDIX G.—Description of the Micro-Spectroscope.

WE saw in the last lecture that the light of the electric spark consists of rays emitted from the incandescent materials, first of the poles from which the discharge passes, and in the second place of the air or gas surrounding those poles; and I would remind you that Ångström was the first who pointed out that the ordinary spark thus yields a double spectrum. If the density of the gas through which the discharge passes be diminished to a certain point within a few millimetres of a vacuum, the electricity is able to pass through a longer column of gas than it can do under the ordinary atmospheric pressure; and we thus may obtain this beautiful phenomenon of the discharge *in vacuo*, for we see this tube ten feet long, on being rendered nearly vacuous by the air-pump, becomes filled with purple light; though if the whole of the gas be withdrawn no electrical discharge can pass. The character of the light thus emitted

FIG. 35.

depends then upon the nature of the gas through which the electric spark or the silent discharge passes.

If we seal up a quantity of hydrogen gas, of carbonic acid gas, and of nitrogen gas, in separate tubes, and allow an electric spark to pass through these tubes (see Fig. 35), the spark which passes through the hydrogen has a red colour, and that which passes through the nitrogen has a yellow colour, while that which passes through the carbonic acid gas has a blue colour: and these differences of colour are due simply to the effect of the gas enclosed in the tube. I can vary the experiment by taking Geissler's tubes (Fig. 36) containing these gases only in very minute quantities, so that the electric discharge can pass through a longer capillary column of gas: we then find that the small quantity of gas in the exhausted tubes becomes heated up to incandescence, and gives off its peculiar rays in a line of brilliantly coloured light.

I have here a hydrogen vacuum tube, next a tube containing a carbonic acid vacuum, then one containing nitrogen, then one containing chlorine, then one containing iodine. I have only to connect these with the induction coil, and the discharge will pass through the whole of these tubes; and at once you see the variety

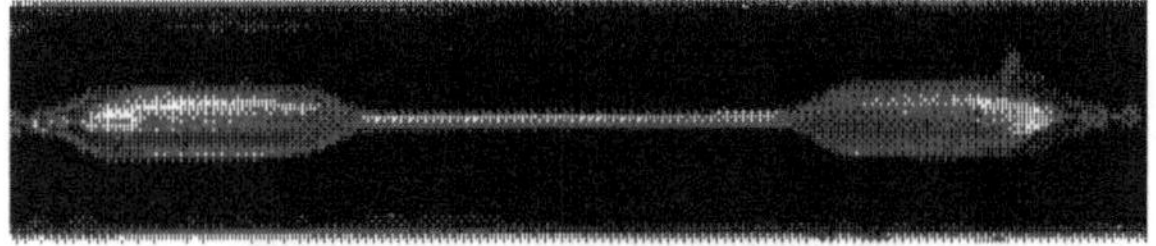

FIG. 36.

of bright colours obtained, entirely due to the small traces of the various gases which are here present in the tubes. If we examine the character of these lights by means of the spectroscope, we shall obtain the peculiar and characteristic spectra of each of these gases.

Here are some large tubes, in which we can see the same effects of the ignition of the small quantities of these various gases by means of the electric spark (Figs. 37, 38); and you observe the beautiful striated appearance which the light exhibits—a phenomenon which physicists are at present quite unable to explain.

I regret that it is impossible to exhibit the spectra of these luminous gases on the screen, owing to the slight intensity of the light which they emit. I must ask you to be content with my references to diagrams to explain to you the exact character of the light which these gases give off.

Thus, when we examine the peculiar red colour which hydrogen exhibits, we find that the spectrum consists of three distinct bright lines; one bright red line so intense as almost to overpower the others, one bright greenish-blue line, and one dark blue or indigo line. These are exhibited to you in the diagram. (See fig. of hydrogen spectrum, No. 8 on the chromolith. plate facing Lecture VI.) The bright red hydrogen line is always seen when an electric spark is passed through moist air: this is due

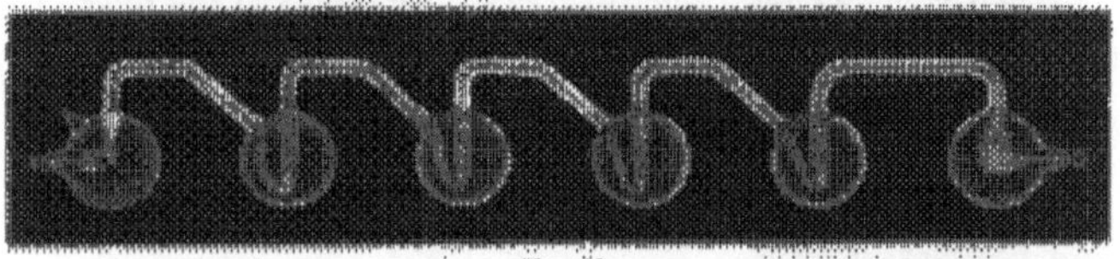

FIG. 37.

to the decomposition of the aqueous vapour which the air contains. If the air be carefully dried by passing it over hygroscopic substances, the red line disappears. Hence the spectroscope can be made a means of testing the presence of moisture.

A very remarkable fact, and one to which I shall have frequently to refer in the subsequent lectures, is that these three lines of hydrogen are found to be coincident with three well-known dark lines in the sun, of which I spoke to you in the first lecture. This red hydrogen line possesses exactly the same degree of refrangibility as the dark line C in the solar spectrum; the green hydrogen line corresponds to the well-known solar line F; whilst the blue hydrogen line is identical in position with a dark line near G in the sun's spectrum.[1] We shall see in a subsequent lecture how such coincidences point out to us the existence of hydrogen and other elements in the solar atmosphere.

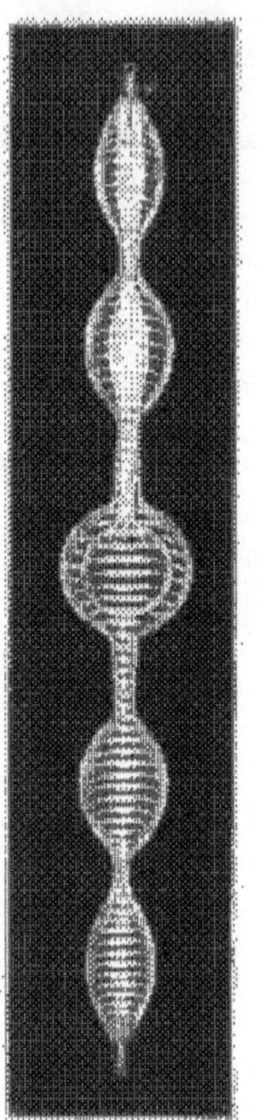

FIG. 38.

When a spark is passed through the air, the lines of both nitrogen and oxygen are seen. This air spectrum has been carefully mapped by Mr. Huggins, who employed it as a scale to which to refer the metal lines in his drawings. He observed the lines simultaneously given off from two sets of poles, one set being of gold and the other set of platinum (in order to eliminate any confusion arising from the presence of metal lines); and he took those lines which were common to both these spectra as being those due to the components of the air. The spectrum thus obtained remains perfectly constant with reference to the position and relative characteristics of its lines when other metals are employed as electrodes. It is, however, found that the air spectrum varies *as a whole* in distinctness according to the volatility of the

[1] Ångström maps a dark line in the violet portion of the solar spectrum, and termed by him (*h*), as coincident with a fourth hydrogen line, which is not seen unless the gas be heated to a very high temperature.

metal used as poles, the air being more or less replaced by metallic vapours in the neighbourhood of the electrodes. The bright hydrogen lines due to aqueous vapour are also seen in moist air, whilst the spectrum of lightning has been examined by Grandeau and Kundt, and found to exhibit in addition the nitrogen and hydrogen spectra, also the bright yellow sodium line.

The nitrogen spectrum is more complicated than that of hydrogen, but still perfectly definite and characteristic. (See No. 9 on the chromolith. plate in Lecture VI.)

Some very singular observations have been made by Plücker and Hittorf[1] upon certain changes which the spectra of gases undergo when enclosed in these tubes. They find that the spectrum of highly rarefied nitrogen undergoes a change when the intensity of the electric discharge varies; and they explain this by supposing that the nitrogen exists in various allotropic conditions, resembling for instance oxygen and ozone, their idea being that the changes in the intensity of the electric discharge may cause changes in the allotropic conditions of the nitrogen, and that these give rise to a variation in the appearance of the spectrum. These variations, however, it is important to observe, are not noticed in nitrogen gas when under the pressure of the atmosphere, however much we may increase the intensity of the spark. Plücker has even found that under certain conditions of increased electrical tension the fine lines of hydrogen are seen to become broader and broader, until at last the hydrogen gas may be made to emit light of every degree of refrangibility, so that its spectrum becomes continuous.

These variations which the hydrogen spectrum, as obtained in Geissler's tubes, undergoes when the density

[1] Plücker and Hittorf (Phil. Trans. 1865, p. 1).

of the gas on the one hand and the intensity of the spark on the other are altered, have been carefully examined by Huggins, as well as by Lockyer and Frankland, and by Wüllner; and these observations become of special interest, when we learn that the fact of this broadening of the hydrogen lines has been shown to possess important bearings upon the conclusions which spectrum analysis enables us to draw concerning the physical condition of the sun and fixed stars. To this point I will, however, direct your attention on a future occasion.

In the same way each of the non-metallic elements yields a characteristic spectrum when its vapour is heated to incandescence; but in the case of some of the elements, such as silicon, the difficulty of obtaining the spectrum is very great.

In his original memoir on the spectra of the chemical elements Kirchhoff plainly points out that under varying conditions of density, temperature, and thickness of the layer of incandescent gas the spectrum of the same body must vary in its appearance, some of the lines coming out more brightly under certain circumstances than others, and thus giving a different character to the spectrum.[1] That the change, in a gaseous spectrum, may go so far as to produce a continuous spectrum is also certain. The flames of many gases, such as this blue one of carbonic oxide, burning in the air to form gaseous products of combustion, give continuous spectra; indeed, we may see the beginning of such a continuous spectrum in every soda flame; and Dr. Frankland has lately observed that when oxygen and hydrogen gases are inflamed under great pressure, they emit white light and show an unbroken spectrum. From these facts

[1] See Appendix C. to this lecture.

there is no doubt that, when intensely heated and under certain circumstances, gaseous bodies can be made to yield continuous spectra. This, however, in no way interferes with the fixity of position of the bright lines, nor can it influence the deductions derived from this fact.

Several interesting observations have been made with respect to the changes produced in the spectra of some of the metals by increase of temperature. Let me, in the first place, show you that new lines may make their

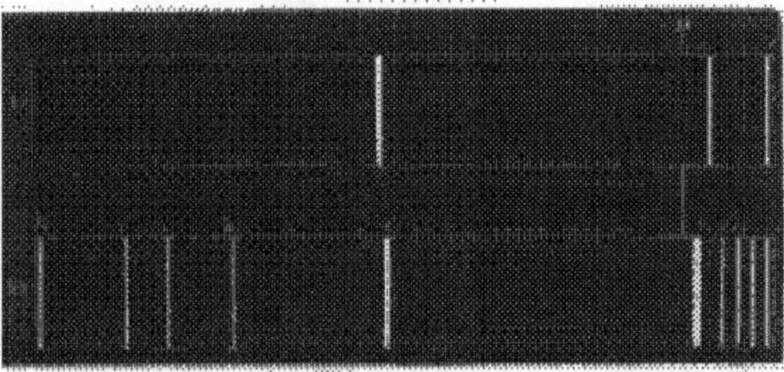

FIG. 39.

appearance in the spectra of certain elements when the temperature is raised. Thus, for instance, if we heat lithium, either the metal or its salts, in the electric arc, we obtain a splendid blue band (see Fig. 39), in addition to the red and orange rays α and β seen in the flame spectrum, showing that the undulations in this particular set of vibrations have become more intense. The same phenomenon is observed in the case of the strontium spectrum, where no less than four new lines (ϵ, η, κ, and λ, Fig. 39) make their appearance on increasing the temperature of the incandescent vapour of the metal. The analogy between the production of these more highly refrangible rays and that of the overtones or harmonics of a vibrating string will occur to all.

By reducing the temperature, and therefore the intensity of the spark, only the most prominent lines of a metallic spectrum may be seen. Thus Lockyer and Frankland have shown that the magnesium (*b*) lines vary in length and intensity when the electrodes are separated, so that in a certain position one of the four well-known magnesium lines disappears. We shall see the application of this observation in a subsequent lecture.

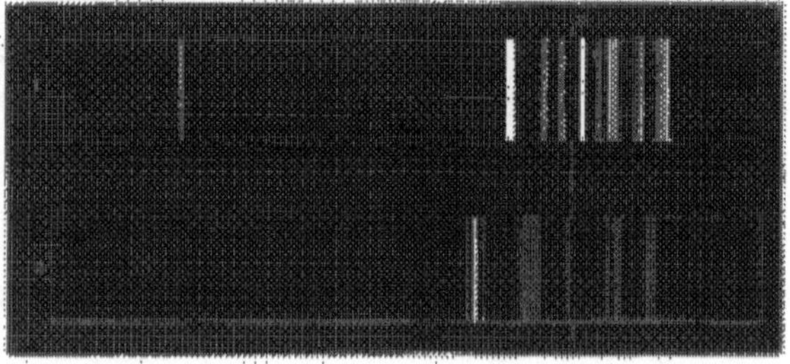

FIG. 40

The second set of facts with regard to the effect of increased heat has reference to the changes which the spectra of *compound* bodies undergo when the temperature is increased. This change is clearly seen in the following experiments. Let us first put a piece of fused chloride of calcium, a common lime salt, into the colourless gas flame: we observe a peculiar spectrum, which is represented roughly on this diagram, in which the red is supposed to be on the right and the blue on the left hand (Fig. 40, No. 1, and Frontispiece, No. 9). If, however, we now pass an electric spark over some pieces of chloride of calcium, and then look at the coloured spark, we find that the spectrum thus obtained is not the same as that observed in the flame. Here you notice the difference between these two spectra: the lower drawing gives you the spark spectrum and the upper one what we may call the flame spectrum. This difference can be readily explained. It is a well-known fact that certain chemical compounds, when they are heated up above a given temperature, decompose into their constituent elements; but that, below that temperature, these compounds are capable of existing in a permanent state. When we once get the spark spectrum, we find that no alteration in the intensity of the spark can then alter the position of those lines. The position of the red lithium line never varies, although the blue line comes out. It naturally strikes every observer that these bands seen in the flame spectrum are produced by a compound of calcium (say the oxide or chloride), which remains undecomposed at the temperature of the flame. When we increase the temperature, as in the spark spectrum, we get the true spectrum of the metal. The position of these true metallic lines never alters at all, although, owing to increased intensity in the electric spark, new lines may sometimes make their appearance. Hence we can fully rely upon the spectrum test as a proof of the presence of the particular metal.

No such change in the character of the spectra is noticed in the case of those metals whose compounds are easily decomposed: thus we do not see any such phenomenon in the alkaline metals, although it is observed in the case of barium, strontium, and calcium. Another fact which bears out the truth of this explanation has been observed by Plücker, that, in the case of bodies whose spectra change from bands to lines on increase of temperature, a recombination of the elements occurs on cooling, and the band spectrum of the compound reappears. Many other observations crowd upon us to convince us that compound substances capable of existing in the state of glowing gas yield spectra different from those of their constituent elements. Thus the spectrum of terchloride of phosphorus exhibits lines differing from those of either phosphorus or chlorine, and the chloride and iodide of copper each yields a distinct set of bands bearing no resemblance to the bright lines of the metal.

It is, here, important to learn that a distinguished spectroscopist, Professor Ångström,[1] does not indorse Plücker's conclusions respecting the existence of several spectra for one element, inasmuch as the spectra observed in the Geissler's tubes with low intensity are, according to Ångström, those of *compound bodies*, and it is only when the discharge becomes disruptive that the constant spectrum of the element appears. Ångström further states, the results of his experiments in no way bear out Plücker's conclusions, for by successively augmenting the temperature he finds that, although the intensity of the rays varies in a most complicated manner, and even new bright lines may appear, still independently of all these changes the spectrum of each substance always preserves its individual character. Thus we find ourselves in the midst of conflicting evidence, and we must endeavour to hold our minds unbiassed until the results of further research render it possible for us to come to a decision on this important subject.

[1] See Appendix A. Lect. V.

The examination of the spectrum of carbon is a subject of much interest. The character of the lines which this blue flame of coal gas and air emits was first described in the year 1857 by Professor Swan. Since that time the various spectra of the carbon compounds have been care-

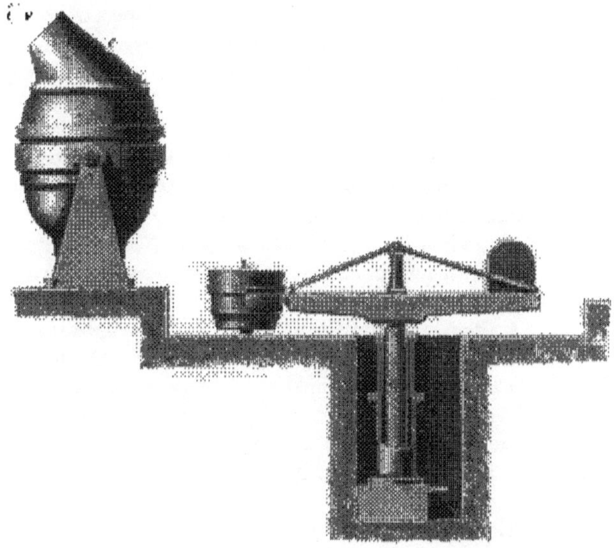

FIG. 41.

fully examined by Dr. Attfield, Dr. W. M. Watts, and others, and it has been found that the different compounds of this element, when brought into the condition of luminous gases, either by combustion or when heated up by the electric spark, give somewhat different spectra.[1]

[1] See Appendix E. on the Spectrum of the Bessemer Flame.

Thus this beautiful purple flame of cyanogen gas exhibits a great number of very peculiar lines, which differ in position and in intensity from the lines observed in this flame of the coal gas burning mixed with air. (See fig. of carbon spectra, Nos. 10, 11, on the chromolith. Plate facing Lecture VI.)

I may mention, in connexion with these different carbon spectra, the application of spectrum analysis to the important branch of steel manufacture which has been introduced and is well known under the name of the Bessemer process. In this process five tons of cast iron are in twenty minutes converted into cast steel. Steel differs from cast iron in containing less carbon, and by the Bessemer process the carbon is actually burnt out of the molten white-hot cast iron by a blast of atmospheric air. The arrangement employed for this purpose is shown on this diagram (Fig. 41).

The molten cast iron is run into a large wrought-iron vessel termed the converter (C), lined with refractory clay. The converter is capable of being turned round on a pivot (A), through which pivot passes a tube in connexion with a powerful blowing apparatus, by means of which air can be thrown into the bottom of the vessel, through a sort of tuyère or blowhole into the molten iron. The oxygen of the air burns out the carbon and silicon which the cast iron contains, and the heated gases issue in the form of a flame (F) from the mouth of the converter during the time that the molten iron is being burned. This flame varies in appearance, and it is of the utmost importance that the operation should be stopped instantly when the proper moment has arrived. If the blast be continued for ten seconds after the proper point has been attained, or if it be discontinued ten seconds before that point is reached, the charge becomes either so viscid that it cannot be poured from the converter into the ladle (L), from which it has to be transferred to the moulds, or it contains so much carbon as to crumble up like cast iron under the hammer.

Those who are accustomed to work this process are able by the simple inspection of the flame to tell with more or less exactitude when the air has to be turned off. To those who are uninitiated in this peculiar appearance of the flame no difference at all can be detected at the point in which it is necessary to stop, but by the help of the spectroscope this point can be at once ascertained beyond shadow of doubt, and that which previously depended upon the quickness of vision of a skilled eye has become a simple matter of exact scientific observation. The light which is given off by the Bessemer flame is most intense,—indeed, a more magnificent example of combustion in oxygen cannot be imagined. A cursory examination of the flame spectrum in its various phases reveals complicated masses of dark absorption bands and bright lines, showing that a variety of substances are present in the flame in the state of glowing gas.

By a simultaneous comparison of the lines in the Bessemer spectrum with those of well-known substances I was able in the year 1863 to detect the following substances in the Bessemer flame: sodium, potassium, lithium, iron, carbon, hydrogen, and nitrogen. At a certain stage of the operation I found that all at once the carbon lines disappeared, and we got a continuous spectrum. The workman by experience has learned that this is the moment at which the air must be shut off;

but it is only by means of the spectroscope that this point can be exactly determined.

No. 2, Fig. 42, represents the *general appearance* of the Bessemer spectrum towards the close of the "blow," drawn according to the plan proposed by Bunsen (see page 63). The striking analogy between the flame spectrum and that of carbon (No. 1, Fig. 42) renders it evident that the principal lines of the Bessemer spectrum are due to carbon in some form. When the spiegeleisen is brought into the converter, a very bright flame issues from the mouth of the vessel, and this flame exhibits a spectrum (No. 3) which really contains the same lines as that of the Bessemer flame, although the general

FIG. 42.

appearance of the spectrum is completely changed by the alteration of the relative brightness of the lines.

Those who are practically engaged in working this process would like spectrum analysis to do a great deal more; they would like to be told whether there is any sulphur, phosphorus, or silicon in their steel: questions which unfortunately at present spectrum analysis cannot answer, for this very good reason, that these substances do not appear at all as gases in the flame, but that they either remain unvolatilized in the molten metal, or swim on its surface in the slag of the ore; and consequently the lines of these bodies are not seen in the spectrum of the flame.

The next point to which I would direct your attention is one of a slightly different kind. We find that certain substances—not only gases, but liquids, and even solid bodies—exert at the ordinary temperature of the air as elective absorption power upon white light when it passes through them. In the following lecture I shall have occasion to show you, in various ways, the absorptive effect which glowing sodium vapour exerts upon the particular kind of yellow light which sodium itself gives off; but I would now consider some cases of selective absorption occurring at the ordinary temperature, and just indicate to you a most interesting and important branch of this subject which has been, to a certain extent, worked out, but in which a rich harvest of investigation still remains open. I refer to the absorption spectra obtained by the examination of various coloured gases and liquids, especially of blood and other animal fluids. In the first place, then, it has long been known that certain bodies have at the ordinary temperature the power of selecting a kind of light and absorbing it. In Fig. 43 we have a representation of the selective absorption exhibited by two coloured gases; No. 1 shows the dark bands seen when white light passes through the violet vapours of iodine, whilst No. 2 gives the bands first observed by Brewster in red nitrous fumes. Some coloured gases, such as chlorine, do not give any dark absorption bands; whilst, on the other hand, certain colourless gases, such as air and aqueous vapour, exert a remarkable power of selective absorption, and exhibit spectra filled with dark lines. I shall return to this subject of the atmospheric lines in a subsequent lecture. Perhaps the most striking instance of the formation of these absorption lines in the case of liquids is the one which I will now show you of this colourless solution of a salt of the rare metal didymium. Now all the didymium salts possess the power of absorbing from white light certain definite rays, so that if I place the solution in the path of our continuous spectrum we get the broad absorption bands by which, as Dr. Gladstone has shown, the presence of didymium can be recognised, when present even in very minute quantities. It is very remarkable that, although these didymium absorption lines are so black, and serve as such a reliable test of the presence of this metal, yet the fraction of the total light

FIG. 43.

which is absorbed is so small that the solution appears colourless. From the recent experiments of Bunsen on this subject we learn that the various didymium compounds do not exhibit exactly the same absorption lines, and that if light is allowed to fall upon a crystal, the dark bands also differ according to the direction in which the light passes through (see Appendix F).

"The differences thus observed," says Bunsen, "cannot as yet be connected with other phenomena. They remind one of the gradual alteration in pitch which the notes from an elastic rod undergo when the rod is weighted."

Remembering the changes seen by Bunsen in the absorption spectra of didymium, we must accept with great caution any conclusions as to chemical composition derived solely from variation in the absorption spectra.[1] Indeed, it has been shown that even experienced observers may be led to false conclusions by relying too implicitly on the complicated absorption bands which certain mixtures may yield.

The solutions of many other coloured metallic salts possess a similar property of yielding definite absorp-

FIG. 44.

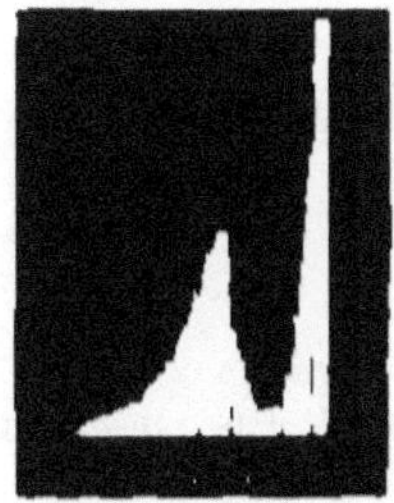

FIG. 45.

tion lines, and Dr. Gladstone finds that with very few exceptions all the compounds of the same base, or acid, have the same effect on the rays of light: thus the chromium salts (both green and purple) exhibit the same form of absorption spectrum (Fig. 44). Fig. 45 shows the bands produced by potassium permanganate solution, contained in a wedge-shaped vessel. The right hand

[1] It has been found that the absorption bands attributed to a new metal contained in zircons are due to a mixture of salts of zirconium and uranium. This mixture appears to afford a most delicate means of detecting the presence of uranium, for bands appear in the mixture which are not seen when the two metals are examined separately in much larger quantities (Sorby).

corresponds to the red end of the spectrum, and the letters refer to the position of Fraunhofer's lines. The absorptive action of the solution is most powerful at the upper part of each drawing, which represents the spectrum seen where the layer of solution was thickest, and diminishing towards the lower part of the figure.

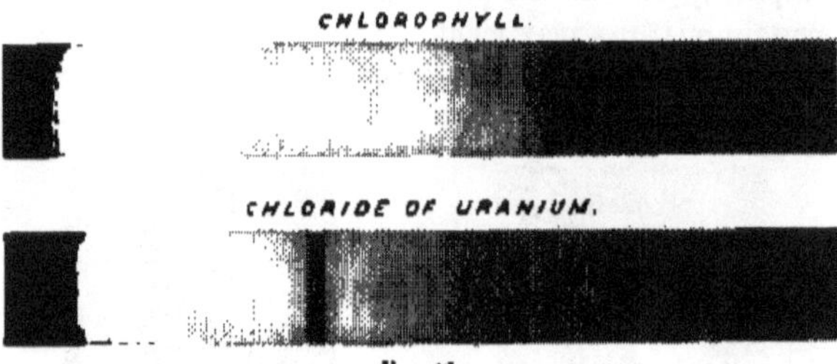

FIG. 46.

There are a variety of other substances which have this selective power: thus here is the absorption spectrum of chlorophyll, the green colouring matter of leaves, and here that of chloride of uranium.

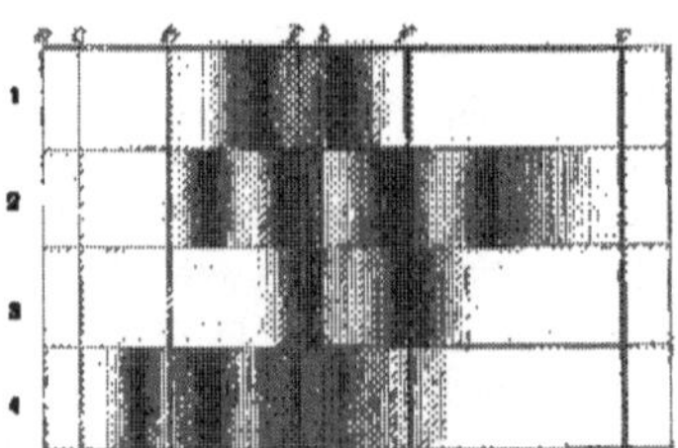

FIG. 47.

Another interesting case of the selective absorption of liquids is exhibited to you in these purple solutions containing the two colouring matters of madder—alizarine and purpurine. Spectrum No. 4 (Fig. 47) shows the peculiar dark bands of the purple solution of alkaline alizarine, a substance which has lately been artificially prepared from a hydrocarbon (anthracene) contained in coal-tar, by two German chemists, Messrs. Graebe and Liebermann; whilst No. 1 gives that of the alkaline solution of purpurine. The bands of artificial alizarine are found to correspond exactly to those of the natural vegetable colouring matter obtained from the *Rubia tinctorum*. The marked difference between the spectrum of purpurine dissolved in carbon disulphide (No. 2, Fig. 47)

DARK BANDS IN BLOOD.

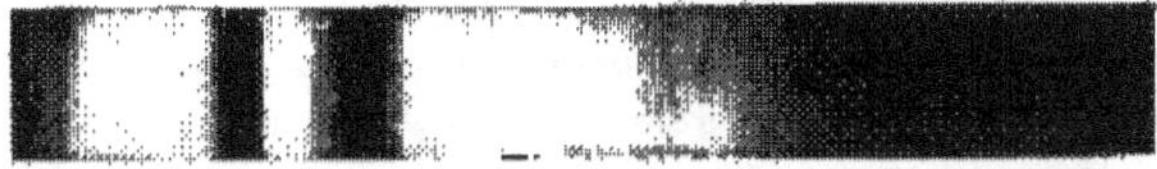

Fig. 48.

and that of an etherial solution (No. 3, Fig. 47) of the same body reminds one of the difference in colour which iodine exhibits when dissolved in chloroform and in alcohol. Professor Stokes, who has examined the spectra of these colouring matters, says:—"The characters of these subjects are so marked that I do not know any substance with which either of them could be confounded, even if we restricted ourselves to *any one* of the solutions yielding the peculiar spectra. Not only so, but these properties enable us to detect small quantities, in the case of purpurine the merest trace, of the substance present in the midst of a quantity of impurities."[1]

If I take a solution of blood, and place the cell containing it before the slit, we get these distinct dark absorption bands, due to the presence of the blood (Fig. 48). This is the red blood: deoxidized blood gives a different appearance. Here you see the two bands due to the oxyhæmoglobin; whilst this portion of deoxidized blood gives only one black band, somewhat similar, but not identical in position with the dark band in magenta which I now throw upon the screen. This subject was first examined by Professor Stokes, who published a paper on the subject in the *Proceedings of the Royal Society for* 1864. From this we learn that "the colouring matter of blood, like that of indigo, is capable of existing in two states of oxidation, distinguishable by a difference of colour and a fundamental difference in the action on the spectrum." These two forms may be made to pass one into the other by suitable oxidizing and reducing agents, and they have been termed red and purple cruorine.

I have here a drawing of Mr. Stokes's diagram of the blood bands. At the top (Fig. 49, No. 1) you see the position of the two bands of the scarlet cruorine, or oxyhæmoglobin. The deoxidized blood is seen in No. 2 to have only one dark band. By the action of an acid on blood the cruorine is converted into hæmatin, yielding a different absorption spectrum; and this hæmatin is capable of reduction and oxidation like cruorine. The absorption bands of hæmatin are represented in Nos. 3 and 4.

One very interesting point to which I must refer is the fact that the blood, when it contains very small

[1] Journal of the Chemical Society, vol. xii. p. 21.

quantities of carbonic oxide gas in solution, exhibits a very peculiar set of bands. And the poisoning by carbonic oxide—for, as is well known, the poison of burning charcoal is due to this gas—can be readily detected by the peculiar bands which the blood containing carbonic oxide in solution exhibits; and hence we have these absorption lines coming out as a most valuable aid in toxological research.

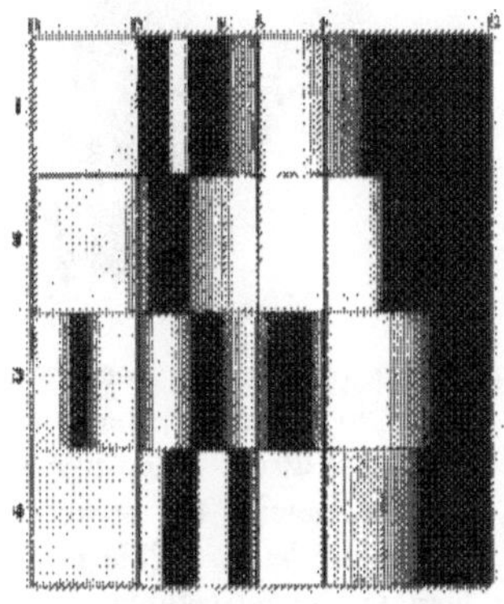

FIG. 49.

A valuable suggestion as to the mode of accurately measuring the position of these various absorption bands of the blood colouring matters has recently been made by Mr. Ray Lankester,[1] in referring to the numerous and well-marked absorption bands of the red nitrous fumes N_2O_4 (No. 2, Fig. 46) as a fixed scale upon which the places of the blood bands can be easily marked.

The instrument by which all these beautiful absorption phenomena can be observed with delicacy and accuracy

[1] Journal of Anatomy and Physiology, vol. iv. p. 119.

is simply a spectroscope placed in connexion with a microscope (Fig. 50). Here we have the instrument.[1]

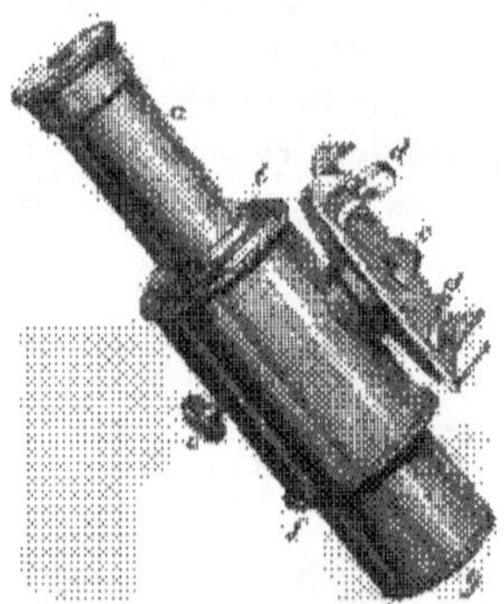

FIG. 50.

The eyepiece contains prisms, so placed as to enable the refracted ray to pass in a straight line to the eye. Such spectroscopes are termed direct-vision instruments. This (Fig. 51) is a diagram showing the structure of the eye-piece which I hold in my hand. This is the first lens of the eyepiece; here is the adjustable slit, for we must have a line of light in order to get a pure spectrum. When the light passes through the second lens, the rays are rendered parallel; and then they pass through this compound prism, consisting of three crown-glass prisms placed in one direction and two flint-glass prisms placed in the opposite direction, so that we see the spectrum by looking straight at the source of light, or have a direct-vision spectroscope. In this way, then, the absorption bands can be very beautifully seen; and,

[1] W. Huggins, "On the Prismatic Examination of Microscopic Objects" (Trans. Microscopical Society, May 10, 1865).

what is important, we can, by means of this little moveable mirror, send a ray of light (shown in the dotted line) through the slit at one side of the instrument, which being reflected upwards passes through the prisms along with the other light which comes from the object under the microscope, and so observe the two spectra one above the other: and thus it is that we can detect, for instance, the presence of blood. Supposing we wish to

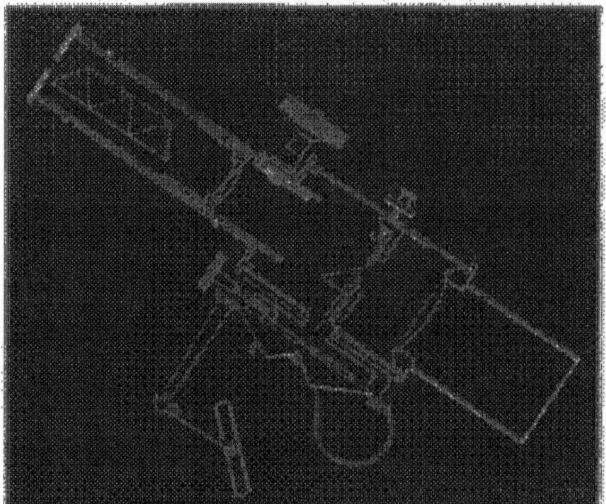

FIG. 51.

know whether a substance is blood which we have in solution: nothing is easier than to place a small quantity of the liquid supposed to be blood on the table of the microscope, and to bring a small quantity of blood in a tube, placed before the slit in the side of the instrument, so as to compare the spectrum obtained from the body under examination with that of the body which we know is really blood. This instrument, which in the hands of Mr. Sorby has taught us how to detect $\frac{1}{1000}$th part of a grain of the red colouring matter in a blood stain, and by means of which I have seen the characteristic bands in the blood circulating in a frog's foot, is a most beautiful one, and the method of microscopic spectrum analysis must every year become a more and more trusted and valuable means of research in medico-legal investigations.

Before concluding my remarks upon the application of spectrum analysis to the examination of terrestrial matter, I would wish to direct your attention for a moment to an interesting, though as yet but imperfectly understood, source of luminosity—namely, the phenomenon of phosphorescence. We have doubtless most of us observed the beautiful appearance of light in the ocean at night, when the motion of the ship stirs up myriads of brightly shining particles, and the crest of each wave is lighted up with liquid fire. I may produce this phosphorescence artificially, for if I write upon this board in the darkened room with a piece of phosphorus, we see the letters shining with a pale light. In this case certainly, and probably also in that of the infusoria, glowworms, and other animals and plants which phosphoresce, the evolution of light is accompanied by an act of oxidation or slow combustion. But we have here evidence of a case of phosphorescence which depends upon the power of substances to absorb light and to give it out again after lapse of some time. If I expose the powders contained in these tubes to any source of intense light, such as the sunlight or the rays obtained from this burning magnesium wire we shall find that on again darkening the room the tubes, or rather the substances which they contain, remain visible; each glowing with a different colour—red, green, or blue—and exhibiting this phosphorescence for several hours after their exposure to the light.

By means of his ingenious phosphoroscope, M. Ed. Becquerel[1] has shown that a large number of bodies

FIG. 52.

phosphoresce for a very short space of time after they are withdrawn from the action of the light. In order to show the phosphorescence of this small crystal of fluorspar, I put it in the small stirrup (*a*, Fig. 52) placed between the two rotating screens of the phosphoroscope,

[1] La Lumière, vol. i. p. 207

and set our powerful electric light to shine upon the back of the closed cylindrical box (Fig. 52, A B). From the peculiar construction of the instrument, you cannot see the light when by means of the handle I rotate the perforated screens which move inside the fixed box; for these screens (P P and M M) are so arranged that when the front slit (o) in the box is open that at the back is closed, and *vice versâ*. If I now cause the perforated screens to rotate, the crystal will be exposed to the light placed behind the box for a series of very short spaces of time, and will also be seen by you through the front hole for a series of equally short moments. If I turn the handle slowly, after darkening the room, you do not see the fluor-spar, but if I increase the speed of rotation so that the times of illumination and of observation do not exceed the $\frac{1}{1000}$th part of a second, you will observe that the crystal glows with a very perceptible amount of light, or in other words it becomes phosphorescent. This crystal of nitrate of uranium produces a more brilliant appearance. But many substances, such as sulphur, quartz, the metals, and liquids, cannot be made thus to phosphoresce.

Becquerel has carefully examined the spectra of these phosphorescent bodies, and he has found that the light emitted by many of them is of a peculiar kind; that they, in fact, give broken spectra, or bands of differently coloured rays. Thus on Fig. 53 we have a representation of the phosphorescent spectra of several substances: alumina, when phosphorescent, emits a red light, and its spectrum (No. 1) exhibits four bands between the lines C and B in the solar spectrum; diamond (No. 3) emits, when phosphorescent, light of many degrees of refrangibility, giving an almost continuous spectrum stretching from B in the red to beyond G in the indigo; aragonite (4) also gives a continuous spectrum; whilst native phosphate of lime (5), fluor-spar (6), and nitrate of uranium (7), each

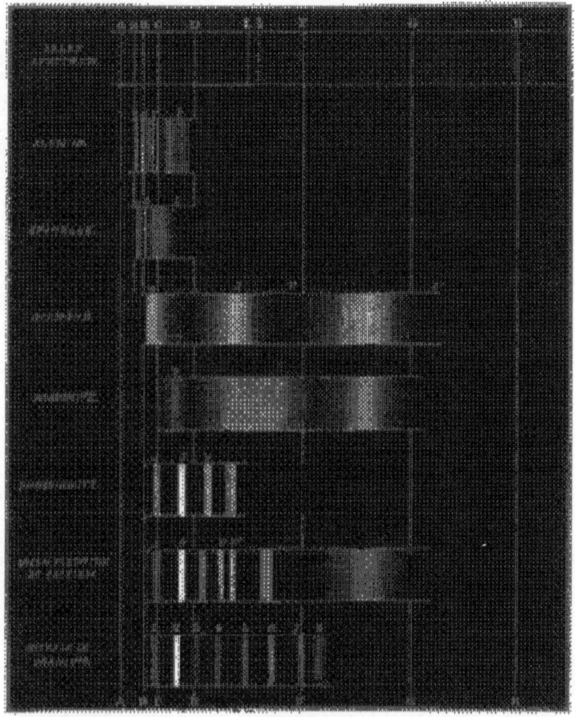

FIG. 53.

phosphoresces with the emission of a peculiar light, as is seen in the varying character of the above spectra.

The different rays of the solar spectrum possess a very

different power of producing phosphorescence. By far the most powerful in this respect are the more refrangible rays. Phosphorescent bodies exposed to the chemically active portion of the spectrum emit light, which, as we have seen, varies from red to violet, and as a rule depends only upon the nature of the substance.

Another interesting property which certain substances possess when exposed especially to these blue rays is that of fluorescence. This piece of uranium glass appears self-luminous when held in the scarcely visible violet rays of our electric lamp; and this is produced by a change of refrangibility of the light, the emitted rays being always of lower refrangibility than the exciting rays. This phenomenon of fluorescence may be used, thanks to the researches of Stokes, for the purpose of identifying certain substances, such as quinine, or the very interesting substance resembling quinine lately discovered in animal fluids and tissues by Dr. Bence Jones; but the spectra which fluorescent bodies emit are generally continuous, and in this direction it does not therefore seem likely that spectrum analysis will give us much help.

In the next lecture I hope to bring before you the simple facts upon which Professor Kirchhoff founded his discovery of the chemical composition of the solar atmosphere.

LECTURE IV.—APPENDIX A.

DESCRIPTION OF THE SPECTRA OF THE GASES AND NON-METALLIC ELEMENTS.

The spectra of the gases are obtained (1) by passing the electric spark from poles of certain metals whose lines are known, through the gas under the ordinary atmospheric pressure; or (2) by observing the electric discharge passed through a capillary tube (Geissler's) containing the gas in a rarefied state. Kirchhoff and Huggins have adopted the first, Plücker and Hittorf (Phil. Trans. 1865, p. 1) the second method.

The Air Spectrum.—"The lines given in this spectrum are present with all the electrodes when the spark is taken in air at the common pressure. The lines thus obtained between one set of electrodes of platinum and the other of gold were observed simultaneously. The lines common to both these spectra were measured as those due to the components of the air. The spectrum thus obtained remains invariably constant, with reference to the position and relative characteristics of its lines, with all the metals which have been employed. The air spectrum varies *as a whole*, however, in distinctness according to the metal employed as electrodes, owing to the difference in the volatility of the metals, the air in and around the electrodes being more or less replaced by the metallic vapours." The air spectrum is made up of the spectra of the following components —nitrogen, oxygen, and hydrogen. Grandeau[1] and Kundt[2] have observed the spectrum of lightning; and, in addition to the the nitrogen and hydrogen spectra, have seen the bright yellow sodium line.

[1] Chemical News, ix. 66. [2] Pogg. Ann. cxxxv. p. 315.

Huggins has employed the air lines (seen on Plates and in the Tables, Appendix C, Lecture III.) as a scale of reference for recognising the bright lines of the metals.

Hydrogen.—The spectrum of hydrogen seen under the ordinary pressure consists of three bright lines (see Chromolith. No. 8, facing Lecture VI.).

H α coincident with Fraunhofer's C in the red.
H β " " F " bluish green.
H γ " " G " violet.

The lines H α, H β, and H γ are seen fine and very bright when the gas is rarefied; but if the reduction of pressure be continued, the red line H α gradually disappears, whilst H β, though fainter, remains well defined. Plücker finds that when the intensity of the spark is increased the bands H β and H γ begin to broaden; and when the tension of the gas is increased to 360 mm. and a Leyden jar introduced into the circuit to raise the temperature of the discharge, the bright lines are found to give way to a continuous spectrum. This change from lines to a continuous spectrum is not observed under the ordinary atmospheric pressure. Wüllner has recently shown[1] that by intensifying the discharge through a Geissler's tube containing hydrogen, the tube and the abraded particles of the glass become highly heated, so that first the sodium line and afterwards the calcium lines make their appearance, whilst at last the spectrum becomes continuous, and the sodium line is reversed, giving a dark absorption line.

Nitrogen.—In the spectrum of the electric spark when taken in a current of pure nitrogen, under the ordinary pressure, a few of the lines of common air are wanting, but no new lines appear. The lines of the air-spectrum which remain in nitrogen preserve their relative brightness and their distinctive character. In the Tables these lines are distinguished by the letter N (pp. 142—149). Plücker and Hittorf have observed some remarkable changes which the nitrogen spectrum undergoes when the current is intensified. Nitrogen, like other gases, does not allow the induction current to pass when it is in an extreme state of rarefaction; but when its tension is only a fraction of a

[1] Pogg. Ann. cxxxv. p. 174.

millimetre the current passes, and the gas becomes luminous. At a comparatively low temperature nitrogen thus ignited emits a golden coloured light, giving a series of bands (see Chromolith. No. 9, facing Lecture VI.); above this point the colour becomes bluish, and a new spectrum of bands appears. If a Leyden jar be enclosed in the circuit, the temperature again rises, and a brilliant white light is emitted, the spectrum again changing to one of bright lines on a dark ground. These lines do not change their position with alteration of temperature, though the brilliancy of all does not increase in the same ratio. Plücker designates the spectra consisting of broad bands "spectra of the first order;" whereas those composed of fine bright lines on a dark background are termed "spectra of the second order." The nitrogen spectrum of the second order is doubtless that of the air-spectrum. The differences thus observed are attributed by Plücker to the existence of allotropic conditions of nitrogen which decompose at high temperatures (for analogous phenomena, see Appendix B, Lecture V.). According to Kundt, the spectrum of lightning varies with the nature of the discharge, the difference being due to the appearance of the two nitrogen spectra; one of these (viz. the second spectrum of the first kind) is also seen when the discharge of electricity from a point is observed. The discharge of forked lightning gives a spectrum consisting of bright lines, being the nitrogen spectrum of the second order.

Oxygen.—The lines given by this gas are given in Huggins' Tables, and designated by the letter O. The same experimenter found that some few lines appeared in the spectra of both nitrogen and oxygen. On further examination he finds that the phenomenon is produced by the superposition in the air-spectrum of lines of oxygen and nitrogen. Plücker, operating as with nitrogen, obtained only one "secondary" spectrum of oxygen, but the lines appeared to expand so as to form a continuous spectrum at a higher temperature.

Sulphur.—When sulphur burns in the air, or when carbon disulphide burns in nitric oxide, a continuous spectrum is observed. If a little sulphur be introduced into a narrow

Geissler's tube, and the air withdrawn, a band spectrum of the first order is seen upon warming the tube and passing the spark through. On continuing to heat the tube, these bands change to bright lines. A figure of these two spectra is given in Plücker and Hittorf's memoir.

Selenium likewise yields a characteristic spectrum.

Phosphorus yields a spectrum of the second order when treated like sulphur. The characteristic lines are three bright bands in the green, having the positions 58, 70, and 74 to 75, on the scale of the spectroscope when Na = 50. The green line Pβ appears with one prism to be coincident with the green barium line Baδ. The green bands may be seen by observing the spectrum of the green spot which makes its appearance in the interior of a hydrogen flame when the slightest trace of a phosphorus compound is placed in contact with the dissolving zinc (Cristofle and Beilstein, *Annales de Chimie et de Physique*, 4 Sér. iii. 280).

Chlorine, Bromine, and Iodine.—When enclosed in Geissler's tubes, each gives a peculiar spectrum of bright lines, which expand, and ultimately form continuous spectra when the temperature is increased. Figures of these spectra are given in the memoir above referred to.

Carbon.—The complicated question of the carbon spectra has been carefully investigated by W. M. Watts (Phil. Mag. Oct. 1869). He finds that there are four distinct modifications of the spectrum of carbon, or, at any rate, of the spectrum obtained from carbon compounds.

1. The carbon spectrum No. 1 is obtained when olefiant gas and oxygen are burnt together in an oxyhydrogen jet. This spectrum was first described by Swan, and afterwards by Attfield. It can be obtained from each of the following carbon compounds: olefiant gas, cyanogen,[1] carbonic oxide, carbon disulphide, carbon tetrachloride, amyl alcohol, marsh gas, and naphthalin (No. 10 of the Chromolith. facing Lecture VI.), and must, therefore, be produced by carbon vapour.

[1] The cyanogen spectrum varies according to its mode of production. No. 11 Chromolith. Plate facing Lecture VI. shows the spectrum of the flame of cyanogen burning in air.

2. Carbon spectrum No. 2 is obtained when carbonic oxide or olefiant gases are heated in a Geissler's vacuum-tube, when the pressure on the gas does not exceed 12 mm. of mercury.

3. Carbon spectrum No. 3 is seen in the Bessemer-flame, observed not only in the flame during the process of conversion, but also in the "Spiegel-flame," and in the coke-flame of the converter and of other furnaces. This is not identical with the carbonic oxide spectrum.

4. The fourth modification of the carbon spectrum is obtained from the induced spark, either from carbonic acid or carbonic oxide, when a Leyden jar is introduced into the circuit. The spectrum thus obtained consists of sharply-defined lines, and not of bands as seen in the former modifications.

For the exact description and maps of these carbon spectra the memoir above mentioned must be consulted.

APPENDIX B.

ON THE EFFECT OF INCREASED TEMPERATURE UPON THE NATURE OF THE LIGHT EMITTED BY THE VAPOUR OF CERTAIN METALS OR METALLIC COMPOUNDS.

BY H. E. ROSCOE AND R. B. CLIFTON.[1]

In a letter communicated to the *Philosophical Magazine* for January last we stated that, in examining, with Steinheil's form of Kirchhoff and Bunsen's apparatus, the spectra produced by passing the induction spark over beads of the chlorides and carbonates of lithium and strontium, we had observed an apparent coincidence between the blue lithium line, which is seen only when the vapour of this metal is intensely heated, and the common blue strontium line called Sr δ. We further stated that on investigating the subject more narrowly by the application of several prisms and a magnifying power of 40, we came to the conclusion that the lithium blue line was some-

[1] Proc. Lit. Phil. Soc. Manchester, read April 1, 1862.

what more refrangible than the strontium δ, but that two other more refrangible lines were observed to be coincident in both spectra. Having constructed a much more perfect instrument than we at that time possessed, we are now able to express a definite opinion on the subject, and beg to lay a short notice of our observations before the Society. Our instrument is in all essential respects similar to the magnificent apparatus employed by Kirchhoff in his recent investigations on the solar spectrum and the spectra of the chemical elements. It consists of a horizontal plane cast-iron plate, upon which three of Steinheil's Munich prisms, each having a refracting angle of 60°, are placed; and of two tubes fixed into the plate, one being a telescope having a magnifying power of 40, moveable with a slow-motion screw about a vertical axis placed in the centre of the plate, and the other being a tube carrying at one end the slit, furnished with micrometer screw, through which the beam of light passed, and at the other end an object-glass for the purpose of rendering the rays parallel. The luminous vapours of the metals under examination were obtained by placing a bead of the chloride or other salt of the metal on a platinum wire, between two platinum electrodes, from which the spark of a powerful induction coil could be passed. In order to obtain a more intense, and therefore a hotter, spark than can be got from the coil alone, the coatings of a Leyden jar were placed in connexion with electrodes of the secondary current respectively. When this arrangement was carefully adjusted, the two yellow sodium lines were observed to be separated by an apparent interval of two millimetres, as seen at the least distance of distinct vision.

The position of the blue line, or rather blue band, of lithium was then determined with reference to the fixed reflecting scale of Steinheil's instrument, by volatilizing the carbonate of lithium in the first place on a platinum wire between platinum electrodes, and secondly on a copper wire between copper electrodes. A bead of pure chloride of strontium was then placed on new platinum and copper wires between two new platinum and copper electrodes, and the position of the blue line Sr δ read off upon the same fixed scale: a difference of one division on the scale was seen to exist between the positions of the two lines, the lithium line being the more refrangible. The salts of the two metals were then placed between the poles at the same time, and both the blue lines were simultaneously seen, separated by a space about equal to that separating the two sodium lines. When experimenting with this complete instrument, we were unable to observe any other blue lines in the pure lithium spectrum than the one above referred to: we have, however, noticed the formation of four new violet lines in the intense strontium spectrum, and we now believe that the other two lithium lines mentioned in our letter to the *Philosophical Magazine* are caused by the presence of the most minute trace of strontium floating in the atmosphere, and derived from a previous experiment. We have convinced ourselves by numerous observations that the currents of air caused by the rapid passage of the electric spark between the electrodes are sufficient to carry over to a second set of electrodes placed at the distance of a few inches a very perceptible quantity of the materials undergoing volatilization. The greatest precautions must hence be taken when the spectra of two metals have to be compared; and no separate observations of the two spectra can be relied upon, unless one is made a considerable space of time after the other, and unless all the electrodes which have been once used are exchanged for new ones.

Kirchhoff, in his interesting Memoir on the Solar Spectrum and the Spectra of the Chemical Elements,[1] noticed in the case of the calcium spectrum that bright lines which were invisible at the temperature of the coal-gas flame became visible when the temperature of the incandescent vapour reached that of the intense electric spark.

We have confirmed this observation of Kirchhoff's, and have extended it, inasmuch as we, in the first place, have noticed that a similar change occurs in the spectra of strontium and barium; and, in the second place, that not only new lines appear at

[1] Kirchhoff on the Solar Spectrum, &c. Translated by H. E. Roscoe. (Cambridge, Macmillan, 1862.)

the high temperature of the intense spark, but that the broad bands, characteristic of the metal or metallic compound at the low temperature of the flame or weak spark, totally disappear at the higher temperature. The new bright lines which supply the part of the broad bands are generally not coincident with any part of the band, sometimes being less and sometimes more refrangible. Thus the broad band in the flame spectrum of calcium named Ca β is replaced in the spectrum of the intense calcium spark by five fine green lines, all of which are less refrangible than any part of the band Ca β; whilst, in the place of the red or orange Ca α, three more refrangible red or orange lines are seen (see Fig. 39). The total disappearance in the spark of a well-defined yellow band seen in the calcium spectrum at the lower temperature was strikingly evident. We have assured ourselves, by repeated observations, that, in like manner, the broad bands produced in the flame spectra of strontium and barium compounds, and especially Sr α, Sr β, Sr γ, Ba α, Ba β, Ba γ, Ba δ, Ba ϵ, Ba η, disappear entirely in the spectra of the intense spark, and that new bright non-coincident lines appear. The blue Sr δ line does not alter either in intensity or in position with alterations of temperature thus effected, but, as has already been stated, four new violet lines appear in the spectrum of strontium at the higher temperature.

If, in the present incomplete condition of this most interesting branch of inquiry, we may be allowed to express an opinion as to the possible cause of the phenomenon of the disappearance of the broad bands and the production of the bright lines, we would suggest that, at the lower temperature of the flame or weak spark, the spectrum observed is produced by the glowing vapour of some compound, probably the oxide, of the difficultly reducible metal; whereas at the enormously high temperature of the intense electric spark these compounds are split up, and thus the true spectrum of the metal is obtained.

In conclusion, we may add that in none of the spectra of the more reducible alkaline metals (potassium, sodium, lithium) can any deviation or disappearance of the maxima of light be noticed on change of temperature.

APPENDIX C.

KIRCHHOFF ON THE VARIATION OF THE SPECTRA OF CERTAIN ELEMENTS: 1862.

"I close this section with the following remarks: The position of the bright lines, or, to speak more precisely, the maxima of light in the spectrum of an incandescent vapour, is not dependent upon the temperature, upon the presence of other vapours, or upon any other conditions except the chemical constitution of the vapour. Of the validity of this conclusion Bunsen and I have assured ourselves by experiments made for that special object, and I have confirmed it by many observations made with the extraordinary delicate instrument just described. The appearance of the spectrum of the same vapour may, nevertheless, be very different under different circumstances. Even the alteration of the mass of the incandescent gas is sufficient to effect a change in the character of the spectrum. If the thickness of the film of vapour, whose light is being examined, be increased, the luminous intensities of all the lines increase, but in different ratios. By virtue of a theorem which will be considered in the next section, the intensity of the bright lines increases more slowly than that of the less visible lines. The impression which a line produces on the eye depends upon its breadth as well as upon its brightness. Hence it may happen that one line being less bright, although broader, than a second, is less visible when the mass of incandescent gas is small, but becomes more distinctly seen than the second line when the thickness of the vapour is increased. Indeed, if the luminosity of the whole spectrum be so lowered that only the most striking of the lines are seen, it may happen that the spectrum appears to be totally changed when the mass of the vapour is altered. Change of temperature appears to produce an effect similar to this alteration in the mass of the incandescent vapour. If the temperature be raised, no deviation of the maxima of light is observed, but the intensities of the lines increase so differently, that those which are most plainly seen at a high temperature are not the most visible at a low temperature."

APPENDIX D.

IGNITED GASES UNDER CERTAIN CIRCUMSTANCES GIVE CONTINUOUS SPECTRA.—COMBUSTION OF HYDROGEN IN OXYGEN UNDER GREAT PRESSURE.[1]

It has long been known that the flames of several gases, such as carbonic oxide, burning in the air to form gaseous products of combustion, give continuous spectra. Dibbits[2] in 1864 pointed out that, when oxygen and hydrogen are burnt in exactly the proportions to form water, a faint continuous spectrum alone is seen, neither the hydrogen nor the oxygen lines being visible. He states that the following gaseous products of combustion, viz. water, hydrochloric acid, sulphur dioxide, and carbon dioxide, exhibit continuous spectra when they are heated to incandescence. Frankland has recently shown that, when hydrogen gas is burnt in oxygen gas under a pressure gradually increasing up to twenty atmospheres, the feeble luminosity of the flame becomes gradually augmented, until at a pressure of ten atmospheres the light emitted by a jet about one inch long is amply sufficient to enable an observer to read a newspaper at a distance of two feet from the flame. Examined by the spectroscope, the spectrum of this flame is bright and perfectly continuous from red to violet. A similar increase of luminosity was observed in the case of carbonic oxide gas burning in oxygen under pressure; and with this gas the spectrum, both when burning under the pressure of the atmosphere and higher pressures, is a continuous one. This

[1] Frankland, Proc. Roy. Soc. xvi. p. 419.
[2] Pogg. Ann. cxxii. 497.

has also long been known to be the case with the combustion of carbon disulphide in oxygen or nitric oxide, and with that of arsenic and phosphorus in oxygen. In these combustions Dr. Frankland believes it to be impossible that the continuous spectrum is due to glowing *solid* matter, as the temperature at which these products of combustion are volatilized is much below the point at which bodies become luminous, and he expresses the opinion ("Lectures on Coal Gas," see *Journal of Gas Lighting*, March 1867) that the luminosity of a candle or coal-gas flame is not due to the incandescence of the particles of solid carbon separated out and heated in the flame, according to the generally received explanation of Davy, but that it is produced by the ignition of highly condensed gaseous hydrocarbons; and he considers himself supported in this view by the fact that the luminosity of a candle flame diminishes proportionally to the diminution of the atmospheric pressure under which it burns.[1]

However extensively future research may modify the proposition that gases give discontinuous spectra, it is well to remember that the *theory of exchanges*, upon which the science of Spectrum Analysis is based, does not give us any information as to whether a gas yields a continuous or a broken spectrum. This theory states that a gas—or any other body—which when incandescent is perfectly transparent to a certain class of rays, cannot emit these rays; but that it *must* emit any rays to which it is not perfectly transparent.

If a glowing gas under great pressure absorbs some of each kind of the rays which fall upon it, it *must* emit a continuous spectrum. Even under diminished pressure many gases exhibit traces of a continuous spectrum: this is seen clearly in a flame coloured by sodium or potassium salt. Kirchhoff has shown that when the temperature or density of a glowing gas is increased and the luminosity of the spectrum becomes more intense, the dark portions of the spectrum *must* increase in luminosity more rapidly than the bright portions. Hence it does not appear surprising that by increase of temperature and pressure the

[1] Frankland, Phil. Trans. vol. cli. p. 629, for 1861.

spectrum originally consisting of bright lines or bands upon a scarcely visible continuous background should gradually change into a spectrum exhibiting all the colours with an equal degree of intensity.[1]

APPENDIX E.

ON THE SPECTRUM OF THE BESSEMER FLAME.

BY W. M. WATTS, D.Sc.[2]

The October number of the *Philosophical Magazine* contains translations of two papers by Professor Lielegg, giving the results of his observations on the spectrum of the Bessemer flame. As these results are published as entirely new, and no mention is made of any prior observations, it is only right that attention should be called to the fact that as long ago as 1862 the same results had been obtained by Professor Roscoe, and were published in the form of a short preliminary notice in the "Proceedings of the Manchester Literary and Philosophical Society" for February 24th, 1863. As the note is extremely short, I venture to transcribe it in full:—

"Professor Roscoe stated that he had been for some little time, and is still, engaged in an interesting examination of the spectrum produced by the flame evolved in the manufacture of cast steel by the Bessemer process, on the works of Messrs. John Brown and Co. of Sheffield. The spectrum of this highly luminous and peculiar flame exhibits during a certain phase of its existence a complicated but most characteristic series of bright lines and dark absorption bands. Amongst the former the sodium, lithium, and potassium lines are most conspicuous; but these are accompanied by a number of other, and as yet undeter-

[1] H. St. Claire Deville (Phil. Mag. Fourth Series, vol. xxxvii. p. 111) explains the increase of luminosity in gases burnt under pressure by the consequent increase of the temperature of the flame, and does not endorse Frankland's views with reference to the source of light in a candle flame. This is in fact the same explanation of the phenomena as that given by Kirchhoff.

[2] Phil. Mag. (4) xxxiv. 437.

mined, bright lines: whilst among the absorption bands those formed by sodium vapour and carbonic oxide can be readily distinguished. Professor Roscoe expressed his belief that this first practical application of the spectrum analysis will prove of the highest importance in the manufacture of cast steel by the Bessemer process, and he hoped on a future occasion to be in a position to bring the subject before the Society in a more extended form than he was at present able to do."

In a lecture delivered before the Royal Institution (May 6, 1864) a year later than the communication quoted above, Dr. Roscoe described the Bessemer spectrum more fully, and pointed out the existence of lines produced by carbon, iron, sodium, lithium, potassium, hydrogen, and nitrogen.

An important practical result of the observations on which these communications were based was the discovery that the exact point of decarbonization could be determined by means of the spectroscope with much greater exactitude than from the appearance of the flame itself, the change in which indicating the completion of the process is minute, and requires a lengthened experience to detect with certainty. This method of determining the point at which it is necessary to stop the blast was indeed at that time (1863) in constant use at Messrs. Brown's works at Sheffield, and has since been introduced with equal success by Mr. Ramsbottom (at the suggestion of Dr. Roscoe) at the London and North-Western Railway Company's steel-works at Crewe.

I was at that time acting as assistant to Professor Roscoe, and in that capacity conducted a lengthened examination of the Bessemer spectrum at the works at Crewe. The results of that investigation were not published at the time, on account of their incompleteness; and I have since then continued in Glasgow the same research, which has now extended itself into an inquiry into the nature of the various spectra produced by the carbon compounds. These experiments are still incomplete; but, under the circumstances of the publication of Professor Lielegg's papers, I have put together a few of the more important results obtained in the examination of the Bessemer spectrum.

The changes which take place in the spectrum from the commencement of the "blow" to its termination are extremely interesting. When the blast is first turned on, nothing is seen but a continuous spectrum. In three or four minutes the sodium line appears flashing through the spectrum, and then becoming continually visible; and gradually an immense number of lines become visible, some as fine bright lines, others as intensely dark bands; and these increase in intensity until the conclusion of the operation. The cessation of the removal of carbon from the iron is strikingly evidenced by the disappearance of nearly all the dark lines and most of the bright ones.

The spectrum is remarkable from the total absence of lines in the more refrangible portion; it extends scarcely beyond the solar line *b*.

No. 2, Fig. 54, represents the *general appearance* of the Bessemer spectrum towards the close of the "blow," drawn according to the plan proposed by Bunsen (see pages 59, 88). It must be remarked, however, that at the period of greatest intensity

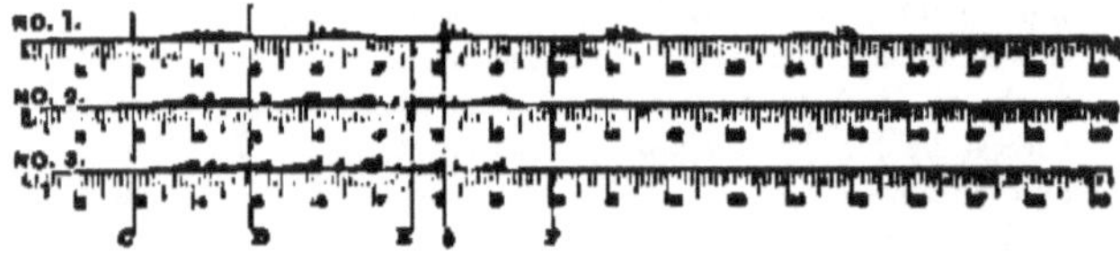

FIG. 54.

almost every bright band is seen to be composed of a great number of very fine lines.

The occurrence of *absorption* lines in the Bessemer spectrum is in itself extremely probable; and that this is the case appears almost proved by the great intensity of some of the dark lines of the spectrum. It was with this view that the investigation was commenced, with the expectation that the spectrum would prove to be a compound one, in which the lines of iron, carbon, or carbonic oxide, &c. would be found, some as bright lines, others reversed as dark absorption bands. To a certain extent this anticipation has been verified; but the great mass of the lines, including the brightest in the whole spectrum, have not as yet been identified.

In dealing with a complicated spectrum like that of the Bessemer flame, it is indispensable that the spectrum should be actually compared with each separate spectrum of the elements sought. This was the plan actually pursued; the spectroscope was so arranged that the spectrum of the Bessemer flame was seen in the upper half of the field of view, and the spectrum with which it was to be compared was seen immediately below. In no other way can any satisfactory conclusion be obtained as to the coincidence or non-coincidence of the lines with those of known spectra.

The spectrum of the Bessemer flame was thus compared with the following spectra:—

(1) Spectrum of electric discharge in a carbonic oxide vacuum.
(2) Spectrum of strong spark between silver poles in air.
(3) " " iron "
(4) " " iron poles in hydrogen.
(5) Solar spectrum.
(6) Carbon spectrum—oxyhydrogen blowpipe supplied with olefiant gas and oxygen.

The coincidences observed were, however, but very few, and totally failed to explain the nature of the Bessemer spectrum. The lines of the well-known carbon spectrum (given in No. 1) do not occur at all, either as bright lines or as absorption bands; nor was any coincidence observed between the lines of the Bessemer spectrum and those of the carbonic oxide vacuum tube.

The lines of lithium, sodium, and potassium are always seen, and are unmistakeable.

The three fine bright lines, 73·7, 76·8, and 82, are due to *iron*. The red band of hydrogen (C) is seen as a black band, more prominent in wet weather.

After the charge of iron has been blown, it is run into the ladle, and a certain quantity of the highly carbonized *spiegel-*

eisen is run into it. The effect of the addition of the spiegeleisen is the production of a flame which is larger and stronger when the blow has been carried rather far. This flame occasionally gives the same spectrum as the ordinary Bessemer flame; but more commonly a quite different spectrum (No. 3) is seen, which reminds one at first of the ordinary carbon spectrum, but differs from it very remarkably.

In the carbon spectrum, which is drawn in No. 1, each group of lines has its strongest member on the left (*i.e.* less refrangible), and fades gradually away towards the right hand: in the spectrum of the spiegel flame the reverse is the case; each group has its brightest line most refrangible, and fades away into darkness on the least refracted side. A comparison of the drawing of the spectrum of the spiegel flame (No. 3) with that of the Bessemer flame (No. 2) will show that they really contain the same lines; but the general appearance of the spectrum is completely changed by alteration of the relative brightness of the lines. This was shown by direct comparison of the actual spectra.

There can be no doubt that the principal lines of the Bessemer spectrum are due to carbon in some form or other. My own belief is that they are due to incandescent carbon vapour. The experiments in which I am at present engaged have already shown the existence of *two* totally different spectra, each capable of considerable modification (consisting in the addition of new lines), corresponding to alterations in the temperature or mode of producing the spectrum, and each due to incandescent carbon. It is possible that the Bessemer spectrum may prove to be a third spectrum of carbon, produced under different circumstances from those under which the ordinary carbon spectrum is obtained; and the intensity of the dark bands is more probably due to contrast with the extreme brilliancy of the bright lines than to their actual formation by absorption.

APPENDIX F.

ON THE SPECTRA OF ERBIUM AND DIDYMIUM AND THEIR COMPOUNDS.

Bunsen[1] has shown that the rare earth erbia is distinguished from all other known substances by a peculiar optical reaction of the greatest interest. This *solid* substance when strongly heated in the non-luminous gas flame gives a spectrum containing *bright lines*, which are so intense as to serve for detecting this substance. This singular phenomenon does not, however, constitute any exception to the law of exchanges; for Bunsen has shown that the bands of maximum intensity in the emission spectrum of erbia coincide exactly in position with the bands of greatest darkness in the absorption spectrum. A similar inversion of the didymium absorption bands has also been observed by Bunsen.[2]

Some very interesting observations have also been made by Bunsen upon the absorption spectrum of didymium,[3] from which we learn that the didymium spectrum, and also that of erbium, undergo changes if examined by polarized light according as the ordinary or extraordinary ray be allowed to pass through the crystal. These changes only become visible, however, when a powerful battery of prisms and a telescope of high magnifying power are employed. According to the direction in which the ray of polarized light is allowed to traverse the crystal of didymium sulphate is the position of the dark absorption bands found to vary; whilst the bands produced by the solution of the salt in water are again different. Very remarkable are the small alterations in the position of the dark bands of the didymium salts, dependent upon the nature of the

[1] Ann. Ch. Pharm. cxxxvii. p. 1.

[2] Ibid. cxxxi. p. 255; Phil. Mag. vol. xxviii. p. 246.

[3] Phil. Mag. vol. xxxii. 1866, p. 177.

compound in which the metal occurs. These changes are too minute to be seen with a small spectroscope, but are distinctly visible in the larger instrument. "The differences thus observed in the absorption spectra of different didymium compounds cannot in our complete ignorance of any general theory for the absorption of light *in media* be connected with other phenomena. They remind one of the slight gradual alteration in pitch which the notes from a vibrating elastic rod undergo when the rod is weighted, or of the change of tone which an organ-pipe exhibits when the tube is lengthened."

APPENDIX G.

DESCRIPTION OF THE SORBY-BROWNING MICRO-SPECTROSCOPE.

The construction of this instrument is represented in Figs. 51 (page 173) and 55. The prism is contained in a small tube (*a*),

FIG. 55.

which can be removed at pleasure, and which is shown in section in Fig. 51. Below the prism is an achromatic eyepiece, having an adjustable slit between the two lenses; the upper lens being furnished with a screw motion to focus the slit. A side slit capable of adjustment admits when required a second beam of light from any object whose spectrum it is desired to compare with that of the object placed on the stage of the microscope. This second beam of light strikes against a very small prism, suitably placed inside the apparatus, and is reflected up through the compound prism, forming a spectrum in the same field with that obtained from the object on the stage.

a is a brass tube carrying the compound direct-vision prism.

b, a milled head, with screw motion to adjust the focus of the achromatic eye-lens.

c, milled head, with screw motion to open or shut the slit vertically. Another screw at right angles to *c*, and which from its position could not be shown in the cut, regulates the slit horizontally. This screw has a larger head, and when once recognised cannot be mistaken for the other.

d d, an apparatus for holding a small tube, in order that the spectrum given by its contents may be compared with that from any other object placed on the stage.

e, a square-headed screw opening and shutting a slit to admit the quantity of light required to form the second spectrum. Light entering the round hole near *e* strikes against the right-angled prism which we have mentioned as being placed inside the apparatus, and is reflected up through the slit belonging to the compound prism. If any incandescent object is placed in a suitable position with reference to the round hole, its spectrum will be obtained, and will be seen on looking through it.

f shows the position of the field-lens of the eyepiece.

g is a tube made to fit the microscope to which the instrument is applied. To use this instrument, insert *g* like an eyepiece in the microscope tube, taking care that the slit at the top of the eyepiece is in the same direction as the slit below the prism. Screw on to the microscope the object-glass required, and place the object whose spectrum is to be viewed on the stage. Illuminate with stage mirror if transparent, with mirror and Lieberkühn and darken well if opaque, or by side-reflector bull's eye, &c.

Remove *a*, and open the slit by means of the milled head, not shown in cut, but which is at right angles to *d d*. When the slit is sufficiently open, the rest of the apparatus acts like an ordinary eyepiece, and any object can be focussed in the usual way. Having focussed the object, replace *a*, and gradually close the slit till a good spectrum is obtained. The spectrum will be much improved by throwing the object a little out of focus.

Every part of the spectrum differs a little from adjacent parts in refrangibility, and delicate bands or lines can only be brought out by accurately focussing their own parts of the spectrum. This can be done by the milled head *b*. Disappointment will occur in any attempt at delicate investigation if this direction is not carefully attended to.

When the spectra of very small objects are to be viewed, powers of from $\frac{1}{4}$ inch to $\frac{1}{16}$th or higher may be employed. The prismatic eyepiece is shown in section in Fig. 51.

Blood, madder, aniline red, permanganate of potash solution, are convenient substances to begin experiments with. Solutions that are too strong are apt to give dark clouds instead of delicate absorption bands.

PLATE III. (*See*

KIRCHHOFF'S MAPS OF THE SOLAR SPECTRUM

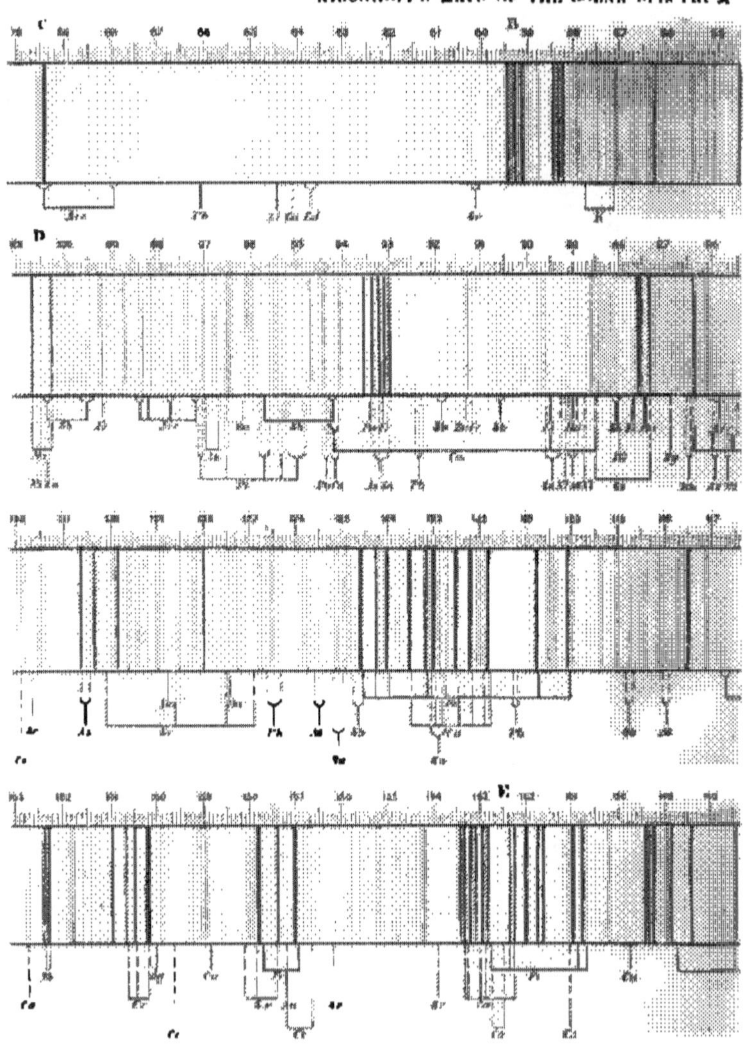

…ables at end of book.)

…OM A TO E. (*The portion from A to D mapped by K. Hofmann.*)

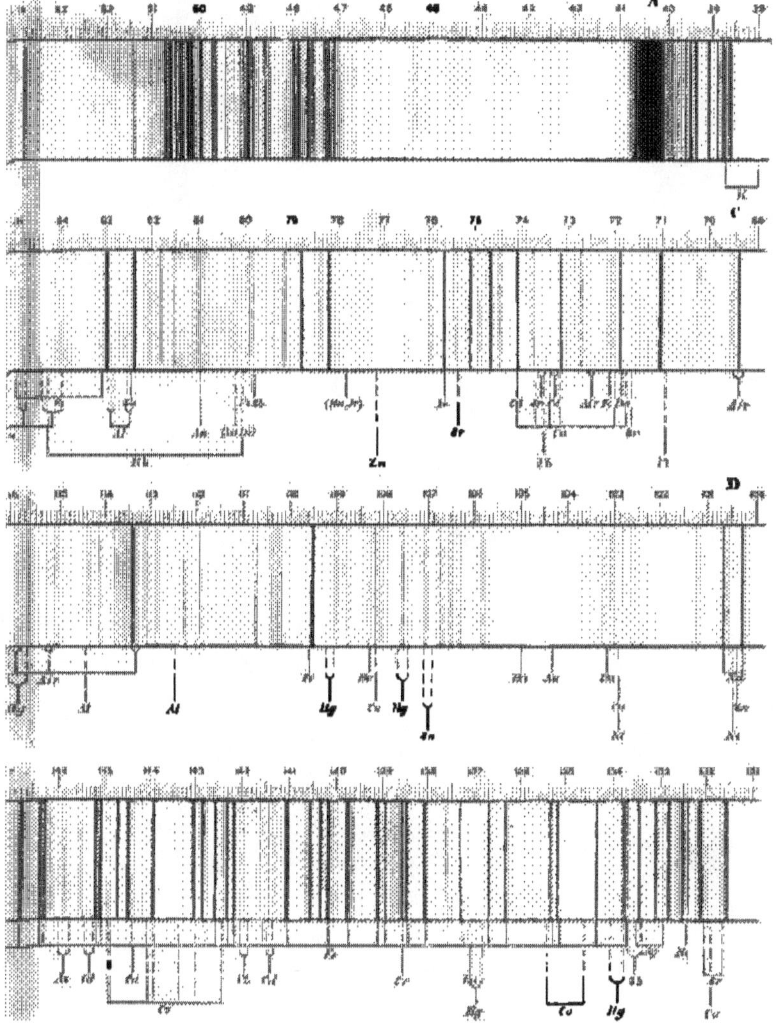

PLATE IV. (*See*

KIRCHHOFF'S MAPS OF THE SOLAR SPECTRUM,

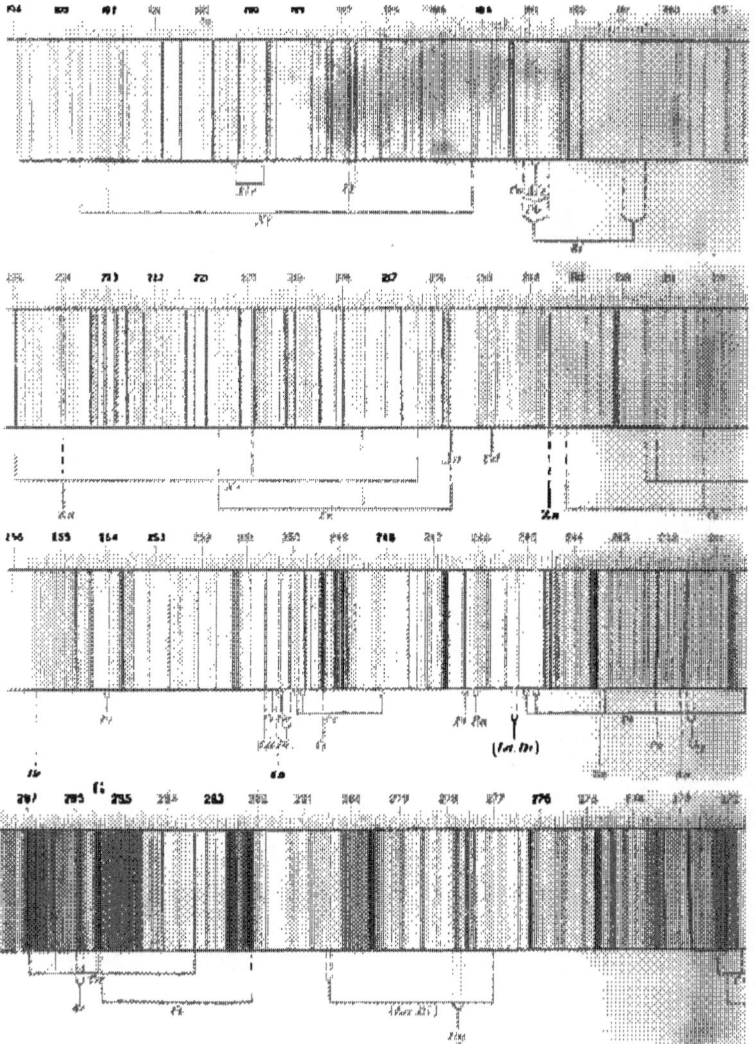

ables at end of book.)

...RUM *b* TO G. (*The latter half mapped by K. Hofmann.*)

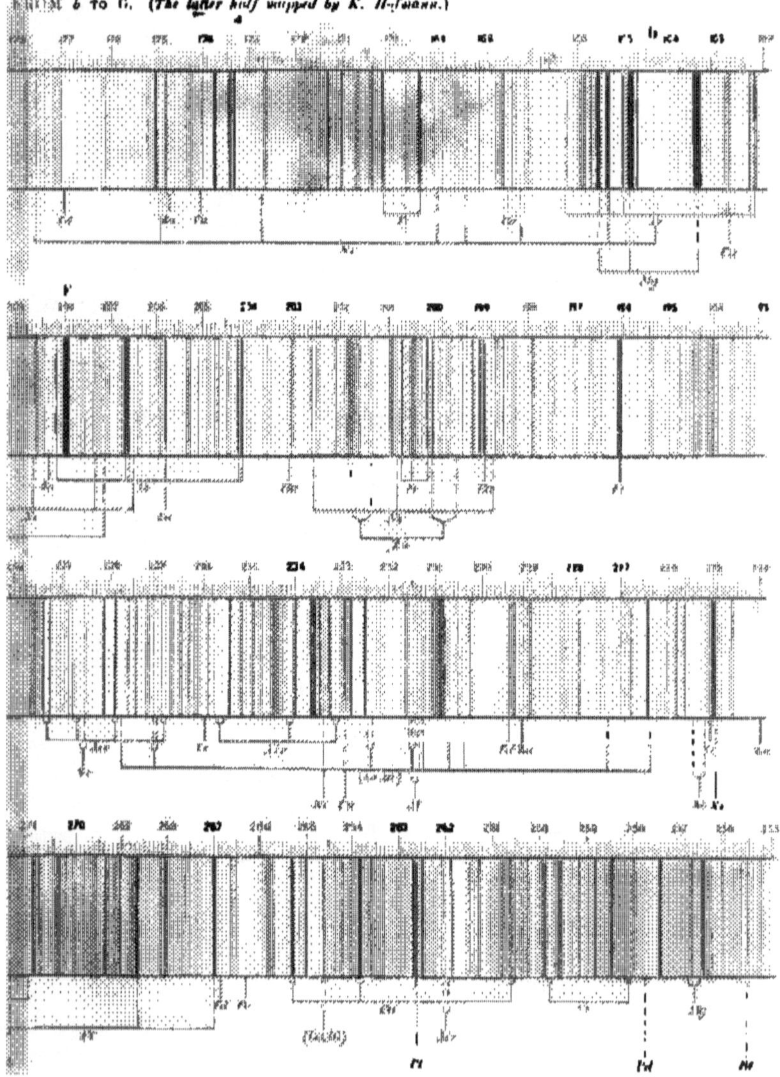

PLAT

ÅNGSTRÖM'S AND THALÉN'S MAPS OF THE SOLAR SPECTRU

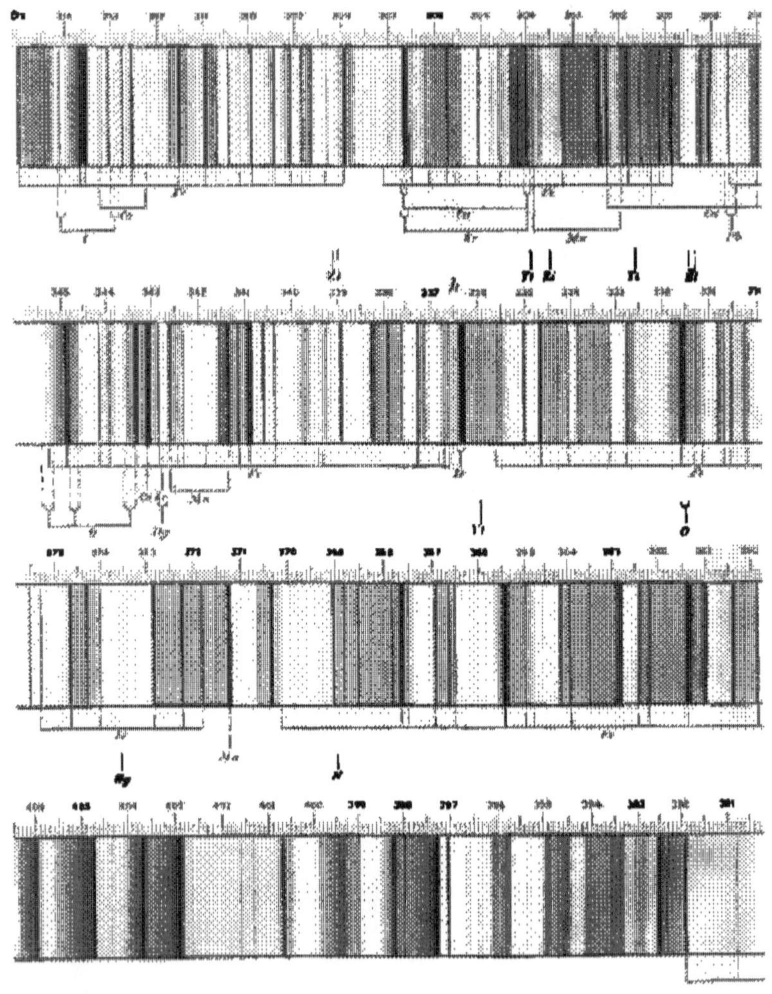

7.

FROM G TO H_2. (*Drawn to the same scale as Kirchhoff's maps.*)

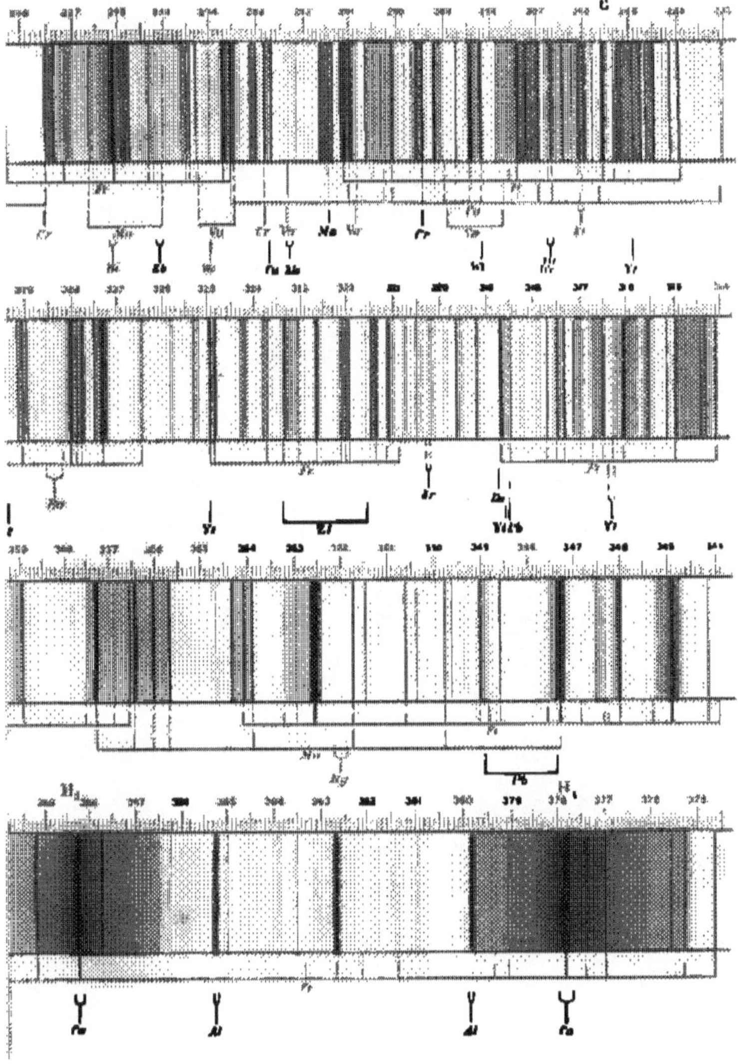

LECTURE V.

Foundation of Solar and Stellar Chemistry.—Examination of the Solar Spectrum.—Fraunhofer, 1814.—Kirchhoff, 1861.—Coincidence of Dark Solar Lines with Bright Metallic Lines.—Reversion of the Bright Sodium Lines.—Kirchhoff's Explanation.—Constituents of the Solar Atmosphere.—Lockyer and Janssen's Researches.—Eclipses of 1860 and 1869.—Constitution of the Red Solar Prominences—their gaseous Nature and rapid Motion.—The Chromosphere.—Physical Constitution of the Sun.—Sun-spots and Faculæ.

APPENDIX A.—"Recherches sur le Spectre normal du Soleil," by A. J. Ångström.

APPENDIX B.—The Indian Total Solar Eclipse of August 18, 1868.—Extracts from Report of the Council of the Royal Astronomical Society.

APPENDIX C.—Spectroscopic Observations of the Sun, being abstracts of the various Papers on this subject by Lockyer, Janssen, Huggins, and Zöllner.

WE have in this lecture the somewhat formidable task set before us of endeavouring to explain the grounds upon which Professor Kirchhoff concludes with certainty, that in the solar atmosphere, at a distance of about 91 millions of miles, substances such as iron, sodium, magnesium, and hydrogen, which we know well on this earth, are present in a state of luminous gas.

In beginning to consider this matter, we shall, however, do well to remember that the subject is still in its infancy; that it is only within the last few years that we have been at all acquainted with the chemistry of these distant bodies. We must not be surprised to find that some of our questions cannot be satisfactorily answered, and we may expect in several instances to meet with facts to which an explanation is still wanting.

In the first lecture I pointed out to you that sunlight differs from the light given off by solid and liquid substances, as well as from the light given off by gaseous bodies. If we were experimenting with sunlight now, and if I could throw the solar spectrum on to the screen, instead of this continuous spectrum of the incandescent carbon poles we should find that this bright band was cut up by a series of dark lines or shadows.

These lines I mentioned to you were first discovered in 1814 by Fraunhofer—at least they were first carefully observed by him—and have since gone by the name of Fraunhofer's lines.

Fraunhofer measured the distances (see Fig. 56) between these fixed lines, and he found that the distance from D to E, and from E to F, remained perfectly constant in the sunlight, that they are fixed lines which always appear in sunlight; and, moreover, as I think I mentioned to you on a previous occasion, he examined the light from the moon and from the planet Venus, and observed the same lines occur in moonlight and in planet-light, which is simply reflected sunlight, and he found that the relative distances between these lines were the same in light from these three sources. He then examined the light from some fixed stars, from Sirius and others, and he noticed that, although in some of these fixed stars certain lines exist which occur in sunlight, yet that other lines, always present in sunlight, are absent from the light of the stars: thus in Procyon and Capella he saw two solar lines D, but other well-known solar lines were wanting.

So long ago as 1814, Fraunhofer concluded that these

lines were caused by some absorptive power exerted in the star or in the sun.

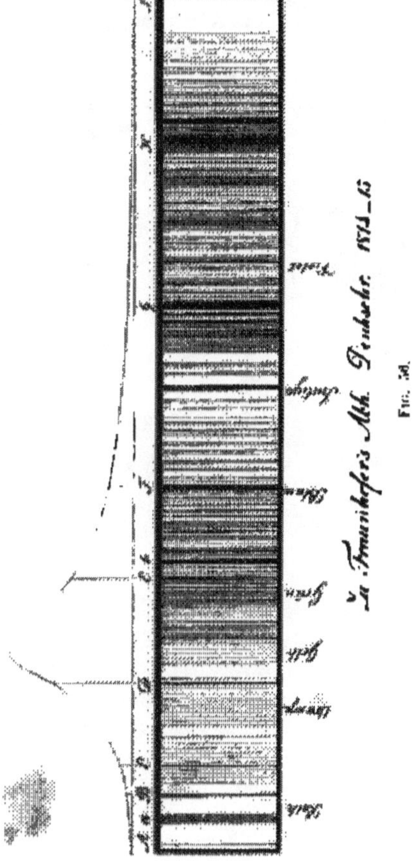

Fig. 38.

The exact mapping of these lines becomes a matter of very great importance, and, since the time of Fraunhofer, the best maps which have been made of these solar lines are those of Kirchhoff and Ångström. Facsimile drawings of these maps are, with the permission of the authors, given in Plates III., IV., and V.

I will now project, by means of the oxyhydrogen light, a photograph of one of the diagrams of Professor

FIG. 57.

Kirchhoff upon the screen, and show you the great number of lines existing in the solar spectrum (see Plate III. facing this Lecture). This is the line D in the yellow, which was noticed by Fraunhofer, and observed by him to be double. Thanks to the kindness of Mr. Browning, I have on the other end of the table a very beautiful instrument, which is so arranged that it enables me to show these double D lines. Reverting again to the map, we see a great number of lines varying in intensity, in depth of shade, as well as in breadth: here we come to E in the blue. I might in the same way show you that throughout the whole length of the spectrum similar groups of dark lines occur. From these diagrams you will, however, form an idea of the enormous number of these lines which exist in the solar spectrum.

On Plate IV. you see the lines existing at the blue end of the spectrum, going up as far as G in the blue. Here you observe these dark lines, to which Fraunhofer

FIG. 58.

gave the term G; and between these we have a very large number of lines mapped out with a very great degree of accuracy and care by Professor Kirchhoff by means of his delicate spectroscope (Fig. 59, p. 204).

In Fig. 57 we have a representation of a still larger spectroscope made by Mr. Browning, for Mr. Gassiot, in which there are nine prisms, and in which the light is actually bent round, so that the incident and emergent beams cross each other's path, as is seen in Fig. 58, giving a plan of the instrument and showing the path of the light through the prisms. With this we can see the D lines very beautifully doubled. To both these large instruments means of accurately measuring the distances between any lines are attached. In Kirchhoff's spectroscope a circular divided scale was used, fixed to the head of the micrometer screw by which the telescope was moved. The eyepiece was placed so that the cross wires made angles of 45° with the dark lines: the point of intersection

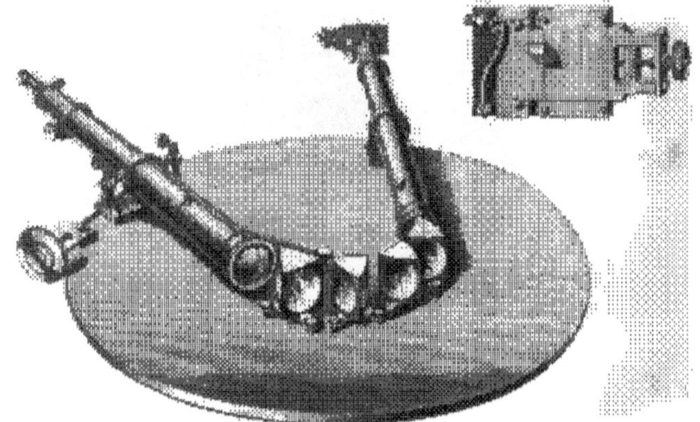

FIG. 59.

was then brought by means of the micrometer screw to coincide with each of these lines, and the divisions read off. A somewhat similar arrangement is seen in the instrument shown in Fig. 57.

Professor Kirchhoff did not draw the whole spectrum; he only got as far as G. Since his time, some very beautiful drawings have, however, been made by Ångström, whose name I had to mention in the third lecture as having given us the first notion respecting the true constitution of the electric spark. In Plate V. you have a copy of Ångström's drawings made in Upsala, which extend from G to H in the violet. I hope you will understand that these dark lines, betokening the absence of certain kinds of rays in the sunlight, not only exist in the visible portions of the spectrum, but also occur in the portions which contain the invisible heating- and chemically-active rays. I cannot show you any of the lines which are found in the ultra-red portion of the

FIG. 60.

spectrum, but I can show you those in the ultra-violet. Thanks to the beautiful researches of Professor Stokes on fluorescence, these lines have become perfectly well known (see Fig. 60).

The diagram shows the effect produced on a film of sensitized collodion which was exposed to the action of these ultra-violet rays passing through quartz prisms. The shaded spaces indicate the positions in which the intensity of the rays is small; they are the Fraunhofer's lines in the ultra-violet sunlight. You see that the lines stretch out a long way beyond the visible portion of the spectrum, that to which the eye is ordinarily sensitive, ending somewhere near the line H.

In order to point out to you the accuracy with which Professor Kirchhoff has drawn these very difficult maps of the solar lines, I will show you a copy of a very interesting photograph made by Mr. Rutherfurd of New York, who, as many present will be aware, has devoted himself with great success to astronomical photography. Mr. Rutherfurd has photographed those portions of the solar spectrum which are capable of producing a

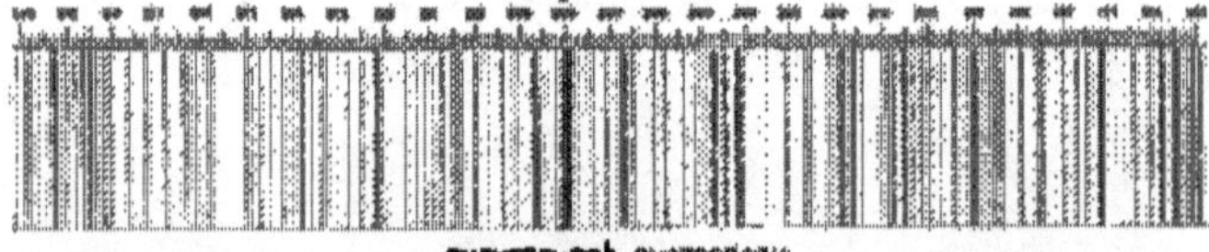

FIG. 61.

photographic image, for you will remember that it is only the blue and ultra-blue rays which are capable of thus acting chemically. You see here (Fig. 61) a copy of one of these photographs compared with Kirchhoff's drawing: at the bottom is Rutherfurd's photograph, and above is Kirchhoff's drawing. Let us compare the two. In the photograph there is this line F, for instance, and you will see that for every line Nature has drawn by means of the light itself there is a corresponding line in Kirchhoff's map: this will give you an idea of this philosopher's extreme accuracy.

When I first saw the photographs which Mr. Rutherfurd was good enough to send me, I really had some difficulty in believing that they had been photographed from the sun itself, so beautifully are they done, and so marvellously do they correspond with Kirchhoff's drawing: but on a careful scrutiny you will find some slight differences between them, especially in the relative intensities of the two sets of lines. This is readily understood if we remember that the map represents the variations of light and shade as affecting Kirchhoff's retina, whereas the photograph gives us the variations of the chemically active rays, indicated by decomposition of silver salt and subsequent development of the image.

Having fully mastered the facts concerning the composition of sunlight, I must now ask you to pass on to the examination of the first of Kirchhoff's discoveries by which the cause of these singular dark solar lines is explained. So long ago as 1814 Fraunhofer discovered that the dark lines D in the sunlight were coincident with the bright sodium lines. The fact of the coincidence of these lines is easily rendered visible if the solar spectrum is allowed to fall into the upper half of the field of our telescope, while the sodium spectrum occupies the lower half. The bright lines produced by the metal, as fine as the finest spider's web, are then seen to be exact prolongations, as it were, of the corresponding solar lines.

These facts, however, remained altogether barren of consequences, so far as regards the explanation of the phenomena, except to the bold minds of Ångström, Stokes, and William Thomson; the last two of whom, combining the facts with an ill-understood experiment of Foucault's made in 1849, foresaw the conclusion to which they must lead, and expressed an opinion which subsequent investigations have fully borne out. Clear light was, however, thrown upon the subject by Kirchhoff, in the autumn of 1859.[1] Wishing to test the accuracy of this asserted coincidence of the bright sodium line and the dark solar lines with his very delicate instrument, Professor Kirchhoff made the following very remarkable experiment, which is memorable as giving the key to the solution of the problem concerning the presence of sodium and other metals in the sun. "In order," says Kirchhoff, for I will now give his own words, "to test in the most direct manner possible the frequently asserted fact of the coincidence of the sodium lines with the lines D, I obtained a tolerably bright solar spectrum, and brought a flame coloured by sodium vapour in front of the slit. I then saw the dark lines D change into bright ones. The flame of a Bunsen's lamp threw the bright sodium lines upon the solar spectrum with unexpected brilliancy. In order to find out the extent to which the intensity of the solar spectrum could be increased without impairing the distinctness of the sodium lines, I allowed the full sunlight to shine through the sodium flame, and to my astonishment I saw that the dark lines D appeared with an extraordinary degree of clearness.

"I then exchanged the sunlight for the Drummond's or oxyhydrogen lime-light, which, like that of all incandescent solid or liquid bodies, gives a spectrum containing no dark lines.

[1] Berlin Acad. Bericht. 1859, 662; Phil. Mag. Fourth Series, xix. 193, xx. 1.

"When this light was allowed to fall through a suitable flame coloured by common salt, *dark* lines were seen in the spectrum in the position of the sodium lines.

"The same phenomenon was observed if, instead of the incandescent lime, a platinum wire was used, which being heated in a flame was brought to a temperature near its melting point by passing an electric current through it. The phenomenon in question is easily explained upon the supposition that the sodium flame absorbs rays of the same degree of refrangibility as those it emits, whilst it is perfectly transparent for all other rays."

FIG. 62.

Kirchhoff had in fact, as far as he had gone, produced artificial sunlight, because he had obtained the two double dark lines in his continuous spectrum. I will try to show the formation of the dark lines of the sodium: for this purpose we will again employ our electric lamp, and I will throw the continuous spectrum of the carbon points on to the screen, and then I will bring into the lower carbon, which is shaped like a cup, a small quantity of metallic sodium; and we shall thus see that the vapour of the sodium has the power of absorbing the particular kind of light which it emits, and that in place of the bright sodium line we shall have a dark line. There you observe the dark sodium line. As a further illustration I have here a diagram (Fig. 62) representing what is seen when we look at the spectrum of burning sodium with an instrument such as that which Kirchhoff used. At the bottom (No. 2) we have a drawing of the ordinary sodium spectrum, giving us these

FIG. 63.

bright double lines on a dark background, and above (No. 1) we see a drawing of the spectrum of burning sodium. Instead of two *bright* yellow lines, we here find we have two intensely *black* lines upon a bright continuous spectrum, the "D" light having been absorbed by the sodium vapour. The difference between the intensities of the lights on each side of these lines and in that particular part where the lines fall is so great as to give an actual shadow, which we see as a black line. There is a well-known experiment by which we cast a shadow with a luminous object, such as a candle flame: so here, although these black lines are not wholly devoid of light, yet the light is so much less intense than in the surrounding parts, that they appear black to us.

I can illustrate this to you in another way. Here (Fig. 63) we have a large sheet of non-luminous gas flame (*bb*) burning under a tall chimney (*c*), and the flame I can colour by sodium. In front of this I am going to ignite a flame of hydrogen (*a*), and I will also place in the hydrogen flame some sodium compound; so that we shall have two sodium flames burning, one in front of the other. I want you to notice that the yellow rays passing from this large flame at the back through the hydrogen flame tinged with soda will be absorbed, and that the outer rim of this hydrogen flame will appear dark; in fact, it will look just as if the hydrogen flame was smoky,—as though we had a smoky candle burning in front of the large flame. There is no carbon in this flame to produce a smoky appearance. We shall have nothing but pure hydrogen burning. We will light our hydrogen here, but before you can see the phenomenon of absorption I must first make a large soda flame. This I do by burning a little sodium, the fumes of which I waft into the flame. Now you see the large flame is turned yellow, and you will notice that in front we get a smoky flame. It is now very distinct. If, instead of a sodium compound, I next place some lithium salt in the flame, no black rim will appear. We shall get the red colour of the lithium flame, but it will not give us any black shadow, because it has no power of absorbing the yellow light. Hence we conclude that the smoky appearance was really caused by the absorption of the yellow "D" light by the sodium vapour in a state of incandescence.

Here is another most ingenious apparatus lately sent me by my friend Professor Bunsen, for exhibiting a con-

Fig. 64.

stant black sodium flame absorbing the rays of the same degree of refrangibility as it emits. The little cap of yellow flame (*d*) which floats from the first burner in front of the larger yellow soda flame (*g*) absorbs the "D" rays, and in consequence we have the peculiar phenomenon of a constantly burning black sodium flame (Fig. 64).

I can also show you in a third way the fact that sodium vapour is opaque to the light which it gives off. I have prepared a tube containing some sodium which I can convert into vapour. By heating the tube as I am doing, it will become filled with sodium vapour, and you will see that it is perfectly colourless and transparent when we look at it with the white sunlight; but when we look at it with the yellow sodium light it will appear to be opaque. We shall then see that the tube containing the sodium vapour throws a dark shadow on the screen. [The lights were turned down, and the screen was illuminated with a yellow sodium flame.] Now the tube looks black; we cannot see through it; it throws a dark shadow. [Light was again admitted.] Now, by the daylight, it is colourless. This shows us, then, very distinctly, that the sodium vapour is opaque for the rays which itself can emit.

Thoroughly understanding, then, the nature of the phenomena with which we have to deal, let us follow Kirchhoff to the interesting conclusions which he draws from this experiment. He states that from this fact it appears likely that glowing gases have the power of especially absorbing rays of the same degree of refrangibility as those they emit; and that therefore the spectrum of such a glowing gas can be reversed, or the bright lines turned into dark ones, when light of sufficient degree of intensity, giving a continuous spectrum, is passed through it. This idea was further confirmed by substituting for the sodium flame the flame coloured by potassium, when *dark* lines appeared in the exact position of the characteristic *bright* lines of this metal. Bunsen and Kirchhoff have likewise succeeded in reversing the flames of lithium, calcium, strontium, and barium; and Dr. Miller has also reversed some of the lines in the spectrum of copper. I can here show you the reversal of the red lithium line on the screen. For this purpose I bring on to the carbon pole of the lamp some salt of lithium, together with a piece of metallic sodium. The sodium will reduce the lithium salt to the metallic state, and I can then show you that we have got not only a dark sodium band, but a dark lithium band in the red part of the spectrum. Now the reversed lines of both these metals are clearly seen.

Generalizing from these facts, Kirchhoff has arrived, by the help of theoretical considerations which I am unable now to lay before you, at a law, previously partially enunciated by Prevost of Geneva and by Prevostaye and Desains in France, and extended by Dr. Balfour Stewart in this country, which expresses the relation between the amount of heat which a body receives and that which it emits. This law has been called *the law of exchanges*. It asserts that the relation between the amount of heat emitted and that which is absorbed at any *given temperature* remains *constant for all bodies;* and that the greater the amount of heat emitted, the greater must be the amount of heat absorbed. Kirchhoff has proved that the same law holds good for light as well as for heat; that it is as true of the luminous as of the heat-giving rays; and for rays of different kinds, if we compare the same kind of rays:—for instance, if we compare red rays emitted with red rays absorbed, or yellow rays emitted with yellow rays absorbed. From this we see that an incandescent gas which is giving off only certain kinds of light—that is, whose power of emission is finite for light of certain definite degrees of refrangibility—must have the power of absorbing those kinds of light, and those kinds only. This is what we find to be the case with the luminous sodium vapour: it has a very high power of emission for the "D" rays, and it has a proportionately high power of absorption for that kind of light; but for it alone. And we see that every substance which emits at a *given temperature* certain kinds of light must possess the power, at that *same temperature*, of absorbing the same kinds of light.[1]

We must remember, however, that the emissive and absorptive powers of substances can only be compared at the same temperature. This is of very great importance, for it has been supposed that the law of exchanges does not hold good, the comparison not having been made at the same temperature. It must not be assumed that because the bright lines of the incandescent iodine spectrum, for instance, do not correspond to the dark absorption bands of the gas at a much lower temperature, therefore the law is faulty or incorrect. We must compare the lines at the same temperature.

Now we know that the same kind of law holds good with the other vibrations known to us—the vibrations of the air which we call *sound*. We are all acquainted with what is called *resonance*. When we sing a particular note in the neighbourhood of a piano, that same note is returned to us. The particular vibrating string which can emit that note has the power also of absorbing

[1] Report on the Theory of Exchanges, by B. Stewart (Brit. Assoc. 1861); Kirchhoff on the History of the Analysis of the Solar Atmosphere (Phil. Mag. Fourth Series, vol. xxv. p. 256).

vibrations of that particular kind, when proceeding in a straight line, and emitting them again in all directions. We are not, therefore, without analogy, in the case of sound, for the absorption and emission of the same kind of undulation by the same substance.

We will now pass to the application of this principle of the reversibility of the spectra of luminous gases to the foundation of a solar and stellar chemistry. How does this principle assist us in our knowledge of the constitution of the solar atmosphere?

In order to map and determine the positions of the bright lines found in the electric spectra of the various metals, Kirchhoff, as I have already stated, employed the dark lines in the solar spectrum as his guides. Judge of his astonishment, when he observed that dark solar lines occur in positions coincident with those of all the bright iron lines! Exactly as the sodium lines were identical with Fraunhofer's lines D, so for each of the iron lines, of which Kirchhoff and Ångström have mapped no less than 460, a dark solar line was seen to correspond. Not only had each iron line its dark representative in the solar spectrum, but the breadth and degree of shade of the two sets of lines were seen to agree in the most perfect manner, the brightest iron lines corresponding to the darkest solar lines.

To those who have not themselves witnessed this coincidence it is impossible to give an adequate idea, by words, of the effect produced on the beholder, when looking into the spectroscope he sees the coincidence of every one of perhaps a hundred of the iron lines with a dark representative in the sunlight, and the idea that iron is contained in the solar atmosphere flashes at once on his mind.

These hundreds of coincidences cannot be the mere effect of chance; in other words, there must be some causal connexion between these dark solar lines and the bright iron lines. That this agreement between them cannot be simply fortuitous is proved by Kirchhoff, who calculates—from the number of the observed coincidences, the distance between the several lines, and the degree of exactitude with which each coincidence can be determined—the fraction representing the chance or probability that such a series of coincidences should occur without the two sets of lines having any common cause: this fraction he finds to be less than $\frac{1}{1000000000000000000}$: or, in other words, it is practically certain that these lines have a common cause. "Hence this coincidence," says Kirchhoff, "must be produced by some cause; and a cause can be assigned which affords a perfect explanation of the phenomenon. The observed phenomenon may be explained by the supposition that the rays of light which form the solar spectrum have passed through the vapour of iron, and have thus suffered the absorption which the vapour of iron must exert."

"As this is the only assignable cause of this coincidence, the supposition appears to be a necessary one. These iron vapours might be contained either in the atmosphere of the sun or in that of the earth. But it is not easy to understand how our atmosphere can contain such a quantity of iron vapour as would produce the very distinct absorption lines which we see in the solar spectrum; and this supposition is rendered still less probable by the fact that these lines do not appreciably alter when the sun approaches the horizon. It does not, on the other hand, seem at all unlikely, owing to the high temperature which we must suppose the sun's atmosphere to possess, that such vapours should be present in it. Hence the observations of the solar spectrum appear to me to prove the presence of iron vapour in the solar atmosphere with as great a degree of certainty as we can attain in any question of natural science." This statement is, I believe, not one jot more positive than the facts warrant. For what does any evidence in natural science amount to, beyond the expression of a probability? A mineral sent to me from New Zealand is examined by our chemical tests, of which I apply a certain number; and these show me that the mineral contains iron: and no one doubts that my conclusion is correct. Have we, however, in this case, proof positive that the body really is iron? May it not turn out to be a substance which in these respects resembles, but in other respects differs from, the body which we designate as iron? Surely. All we can say is, that in each of the many comparisons which we have made the properties of the two bodies prove identical, and it is solely this identity of the properties which we express when we call both of them iron.

Exactly the same reasoning applies to the case of the existence of these metals in the sun. Of course the metals present there, causing these dark lines, *may* not be identical with those we have on earth; but the evidence of their being the same is as strong and cogent as that which is brought to bear upon any other question of natural science the truth of which is generally admitted.

I do not think I can give you a more clear or succinct account of the development of this great discovery than by quoting from Kirchhoff's admirable memoir the following passage:—"As soon as the presence of *one* terrestrial element in the solar atmosphere was thus determined, and thereby the existence of a large number of Fraunhofer's lines explained, it seemed reasonable to suppose that other terrestrial bodies occur there, and that, by exerting their absorptive power, they may cause the production of other Fraunhofer's lines. For it is very probable that elementary bodies which occur in large quantities on the earth, and are likewise distinguished by special bright lines in their spectra, will, like iron, be visible in the solar atmosphere. This is found to be the case with calcium, magnesium, and sodium. The number of bright lines in the spectrum of each of these metals is indeed small, but those lines, as well as the dark lines in the solar spectrum with which they coincide, are so uncommonly distinct that the coincidence can be observed with great accuracy. In addition to this, the circumstance that these lines occur in groups renders the observation of the coincidence of these spectra more exact than is the case with those composed of single lines. The lines produced by chromium, also, form a very characteristic group, which likewise coincides with a remarkable group of Fraunhofer's lines: hence I believe that I am justified in affirming the presence of chromium in the solar atmosphere. It appeared of great interest to determine whether the solar atmosphere contains nickel and cobalt, elements which invariably accompany iron in meteoric masses. The spectra of these metals, like that of iron, are distinguished by the large number of their lines. But the lines of nickel, and still more those of cobalt, are much less bright than the iron lines; and I was therefore unable to observe their position with the same degree of accuracy with which I determined the position of the iron lines. All the brighter lines of nickel appear to coincide with dark solar lines; the same was observed with respect to some of the cobalt lines, but was not seen to be the case with other equally bright lines of this metal. From my own observations I consider that I am entitled to conclude that nickel is visible in the solar atmosphere. I do not, however, yet express an opinion as to the presence of cobalt. Barium, copper, and zinc appear to be present in the solar atmosphere, but only in small quantities; the brightest of the lines of these metals correspond to distinct lines in the solar spectrum, but the weaker lines are not noticeable. The remaining metals which I have examined—viz. gold, silver, mercury, aluminium, cadmium, tin, lead, antimony, arsenic, strontium, and lithium —are, according to my observation, not visible in the solar atmosphere."

The lines of the following metals have their dark representatives in the sunlight:—

1. Sodium.	5. Iron.	9. Zinc.	13. Hydrogen.[1]
2. Calcium.	6. Chromium.	10. Strontium.	14. Manganese.
3. Barium.	7. Nickel.	11. Cadmium.	15. Aluminium.[2]
4. Magnesium.	8. Copper.	12. Cobalt.	16. Titanium.[3]

The coincidences in the case of some of these metals are not so numerous as with iron; still they are so

[1] The conclusion that the lines C, F, a line marked 2796 on Kirchhoff's maps lying near G, and *h* are due to the absorption of hydrogen in the sun's atmosphere, and are not caused by the presence of aqueous vapour in our own, is proved by the fact that in the spectra of certain stars these lines are altogether wanting.

[2] Aluminium has been found by Ångström to be contained in the solar atmosphere, and two of its lines form a portion of the solar bands H.

[3] According to Thalén, the metal titanium is also present in the solar atmosphere, giving lines which were formerly supposed to be due to calcium. Of these no less than 170 have been seen to be coincident with dark lines in the solar spectrum.

characteristic and distinct as to leave no doubt of the presence of these metals in the solar atmosphere. In the cases of cadmium, strontium, and cobalt, there may be some doubt, either because only a few coincidences have been observed, or because one or more prominent metal lines are not seen in the solar spectrum. The following metals appear to be either altogether absent, or present in a very small quantity in the solar atmosphere :—

1. Gold.	6. Potassium.	11. Silicium.	16. Ruthenium.
2. Silver.	7. Lead.	12. Glucinium.	17. Iridium.
3. Mercury.	8. Antimony.	13. Cerium.	18. Palladium.
4. Rubidium.	9. Arsenic.	14. Lanthanum.	19. Platinum.
5. Cæsium.	10. Lithium.	15. Didymium.	20. Thallium.

I will now show you these bright lines of some of the metals contained in the solar atmosphere. Here we have the green magnesium lines, and I can point out to you the dark lines in the solar atmosphere which are coincident with these green lines. You see these two dark lines on the upper part in the right-hand corner (Plate IV. in Kirchhoff's map): these are the bands which Fraunhofer called *b*, and some of them at least are caused by magnesium. Hence you see that the *b* lines are caused by the presence of iron and magnesium in the solar atmosphere. I have written down here a short *résumé* of Kirchhoff's experiments and reasoning on this subject.

Sodium and Iron in the Sun's Atmosphere.

1. The light emitted by luminous sodium vapour is homogeneous. The sodium spectrum consists of one double bright yellow line.

2. This bright double sodium line is exactly coincident with Fraunhofer's dark double line D.

3. The spectrum of a Drummond's light is continuous; it contains no dark lines or spaces.

4. If between the prism and the Drummond's light a soda flame be placed, a dark double line identical with Fraunhofer's double line D is produced.

5. If, instead of using Drummond's light, we pass sunlight through the sodium flame, we see that the line D becomes much more distinct than when sunlight alone is employed.

6. The sodium flame has, therefore, the power of absorbing the same kind of rays as it emits. It is opaque for the yellow D rays.

7. Hence we conclude that luminous sodium vapour in the sun's atmosphere causes Fraunhofer's dark double line D; the light given off from the sun's body giving a continuous spectrum.

8. Kirchhoff found that each and all of the bright lines produced in the spectra of certain metals—for instance, of iron, magnesium, and chromium—coincide exactly with dark lines in the solar spectrum.

9. Hence it is certain that these bright metallic lines must be connected in some way with the dark solar lines.

10. The connexion is as follows: each of the coincident dark lines in the solar spectrum is caused by the absorption effected in the solar atmosphere by the glowing vapour of that metal which gives the corresponding bright line.

There are a great many very interesting points which I should like to show you with regard to Kirchhoff's map. Here, for instance, is an extract from the tables which accompany these diagrams, complete copies of which are found at the end of this volume. You see in column I. the numbers representing the lines which refer to his arbitrary scale of millimetres on the top line of his drawing. In column II. we have the thickness and darkness of the lines, represented respectively by letters, *a* to *g*, and by numbers, 1 to 6, *a* being the smallest and 1 the lightest; whilst column III. gives the metallic lines which are coincident with certain solar lines. For instance, the dark line numbered 1648·8 is coincident with a magnesium line, 1627·2 with a calcium line, 1622·3 with an iron line. Here you see this one line 1653·7 belongs both to iron and to nickel, and 1655·6 is both an iron and a magnesium line.

It is a singular fact that quite recently it has been noticed by Ångström and Thalén that many of the lines which have been classed as calcium lines are really due to titanium, a metal which is of but comparatively rare occurrence on the earth.

Extract from the Index Table of Kirchhoff's Maps, showing the coincidences of the dark Solar and bright Metallic Lines.

I.	II.	III.	I.	II.	III.
1621·5	1*b*		1648·4	4*e*	
1622·3	5*c*	*Fe*	1648·8	6*f*	*Mg*
1623·4	5*b*	*Fe*	1649·2	4*e*	
1627·2	5*b*	*Ca*	1650·3	6*b*	
1628·2	1*b*		1653·7	6*b*	*Fe, Ni*
1631·5	1*b*		1654·0	4*c*	
1633·5	4*g*		1655·6	6*c*	*Fe, Mg*
1634·1	6*g*	*Mg*	1655·9	4*d*	
1634·7	4*g*		1657·1	5*b*	
1638·7	1*b*		1658·3	2*b*	
1642·1	1*b*		1659·4	1	
1643·0	1*b*	*Ni*	1662·8	5*b*	*Fe*
1647·3	3*a*				

Whether these apparently coincident lines will prove to be absolutely identical is a matter which we cannot as yet decide. Kirchhoff thinks it is necessary, for the purpose of settling this question, to use a much more delicate apparatus than even that which he employed.

Fraunhofer's line D corresponds on Kirchhoff's map to the lines 1002·8 and 1006·8; Fraunhofer's E to the lines 1523·7 and 1522·7; and Fraunhofer's *b* to the lines 1633·4, 1648·3, and 1655·0. Kirchhoff observed the

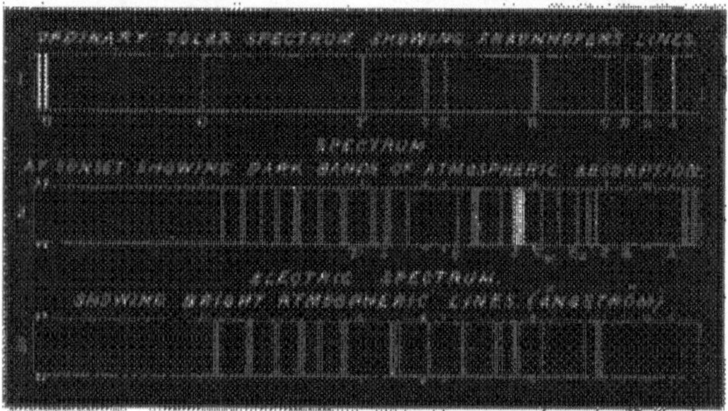

FIG. 65.

traces of many lines and nebulous bands, which the power of even his instrument did not prove adequate to resolve. He adds: "The resolution of these nebulous bands appears to me to possess an interest similar to that of the resolution of the celestial nebulæ, and the investigation of the solar spectrum to be of no less importance than the examination of the heavens themselves." It is important to remark that it is by no means the case that all the lines have been identified; the cause of many of them is known, but a still greater number yet remain for identification. I may again remind you, that the well-known double line D is caused by sodium, and we learn from the exact observations of Mr. Huggins, that not only the line D, but several other less distinct lines (seen on the Maps following Lecture IV.), one lying nearly neutral between the D lines, are produced by sodium in the sun. The line E is an iron line, and the lines C, F, a line near G, and *h* are hydrogen lines; the line *b* is a line of magnesium, and the line H appears from the researches of Ångström to be at any rate partly produced by calcium. Many, however, of the lines frequently seen in the solar spectrum are not due to the presence of metals in the sun, but are caused by the absorption occurring in our own atmosphere. The existence of dark bands caused by atmospheric absorption was first pointed out by Brewster in 1833, and a map of these bands was subsequently published by Sir David Brewster and Dr. Gladstone. Fig. 65 shows the chief of these lines compared with the solar lines and the bright lines of nitrogen and oxygen. These telluric or atmospheric lines are most plainly seen when the sun is low on the horizon, because the column of air which the rays have to traverse is then the longest.

Some very interesting experiments were made in 1866 by the French physicist, M. Janssen: he observed that if light from 16 jets of coal-gas be passed through a long column of steam 37 metres in length, under a pressure of 7 atmospheres, the steam exerts a strong absorptive power, and groups of dark lines appear in the spectrum between the extreme red and the line D. These lines are found to coincide with lines in the solar spectrum which become intense when the sun is near the horizon, and are therefore due to absorption in the aqueous vapour of our own atmosphere. An accurate map of the *telluric* lines between D and C, quite recently published by Janssen, is given in Fig. 66. The important results of his researches on this subject are (1) that Brewster's dark bands are resolved into fine lines comparable with Fraunhofer's lines, and (2) that the terrestrial atmosphere produces in the spectrum a system of fine lines, so that the absorptive action exerted by our atmosphere is analogous to that of the sun in spite of the enormous difference of temperature. All the dark lines seen in the lower but not found in the upper spectrum (Fig. 66) have a telluric origin, and they have been designated by the Greek

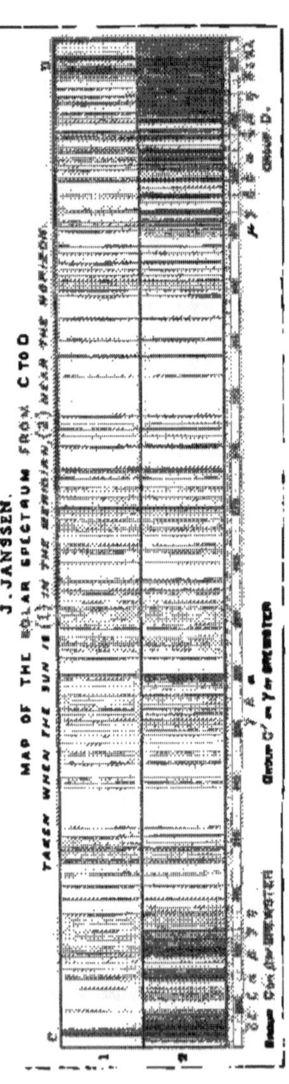

Fig. 66.

letters, and are classed in groups according to their position with regard to well-known solar lines.

I do not know that I can do better in concluding this portion of my subject than give you Professor Kirchhoff's exact opinions, by reading a short extract from his chapter on the "Physical Constitution of the Sun." "In order to explain," he says, "the occurrence of the dark lines in the solar spectrum, we must assume that the solar atmosphere encloses a luminous nucleus, producing a continuous spectrum,[1] the brightness of which exceeds a certain limit. The most probable supposition which can be made respecting the sun's constitution is, that it consists of a solid or liquid nucleus heated to a temperature of the brightest whiteness, surrounded by an atmosphere of somewhat lower temperature. This supposition is in accordance with Laplace's celebrated nebular theory respecting the formation of our planetary system. If the matter now concentrated in the several heavenly bodies existed in former times as an extended and continuous mass of vapour, by the contraction of which sun, planets, and moons have been formed, all these bodies must necessarily possess mainly the same constitution. Geology teaches us that the earth once existed in a state of fusion: and we are compelled to admit that the same state of things has occurred in the other members of our solar system. The amount of cooling which the various heavenly bodies have undergone, in accordance with the laws of radiation of heat,

[1] This continuous spectrum is most probably derived from incandescent solids or liquids, but may, under certain conditions, be given off by luminous gases. Kirchhoff, as will be seen, has carefully guarded himself from expressing a definite opinion as to the exact condition of the luminous portion of the sun's body.

differs greatly, owing mainly to the difference in their masses. Thus, whilst the moon has become cooler than

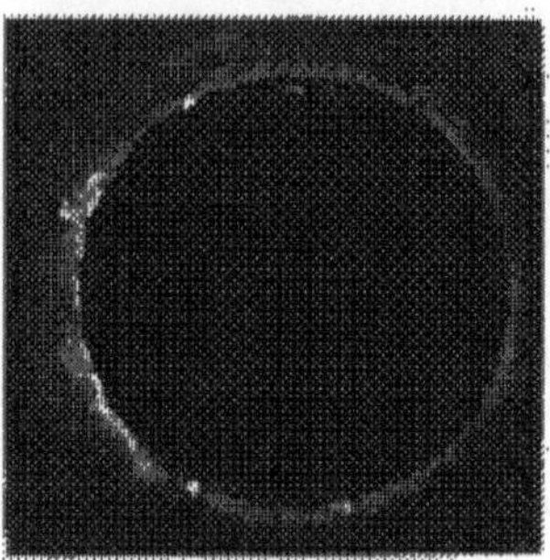

FIG. 67.

the earth, the temperature of the surface of the sun has not yet sunk below a white heat. Our terrestrial atmosphere, in which now so few elements are found, must

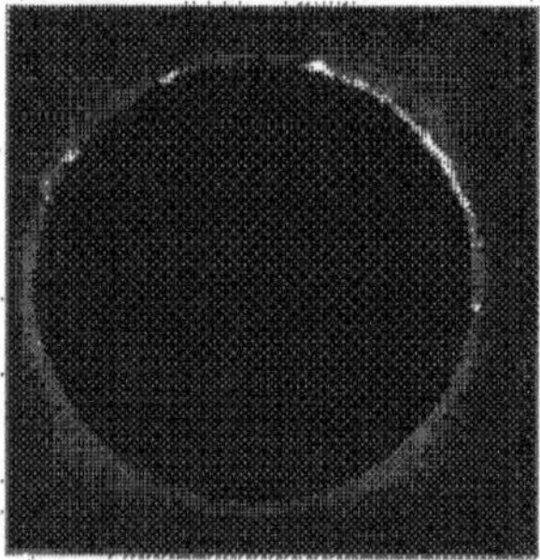

FIG. 68.

have possessed, when the earth was in a state of fusion, a much more complicated composition, as it then contained all those substances which are volatile at a white heat. The solar atmosphere at this time possesses a similar constitution."

Within the last eighteen months our knowledge concerning the physical constitution of the sun has received additions second only in importance to the original discovery of Kirchhoff. The spectroscope in this case is again the instrument by which the extraordinary phenomena of solar physics have been revealed, and the first step towards the extension of our knowledge has been the examination of the light emitted by those remarkable protuberances or red flames which, during a total eclipse, are seen to dart out from the surface of the sun to the enormous height of some 60,000 to 90,000 miles. The appearance of the sun during the total eclipse of 1860 is represented by Figs. 67 and 68, which are copies of photographs taken by Mr. De la Rue in Spain during the eclipse. The first one of these was taken immediately after the total obscuration, and the second just previous to the reappearance of the sun. Fig. 69 gives a representation of the total eclipse of 1869, copied from the photographic registrations taken at Burlington in the United States by Dr. Mayer. In this drawing both the prominences and the coronal rays are seen, and the peculiarly indented appearance of the moon's dark limb in the neighbourhood of the prominences is well shown. The existence of these flames proves that the sun's incandescent atmosphere extends to a very great height above the ordinary and visible portion, and it is very remarkable that certain protuberances which were not visible to the naked eye are found in the photograph; the flames emitting rays of a high degree of refrangibility so weak as not to act upon the retina, although strong enough to produce an image on the sensitive plate.

A most striking feature of the discovery of the nature of the material composing these red protuberances is that it was made independently and nearly simultaneously by two observers many thousands of miles apart, namely, by Mr. J. Norman Lockyer in England, and by the French

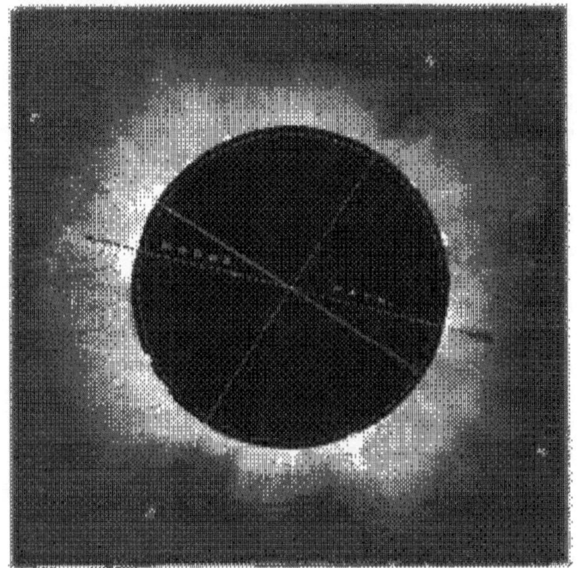

Fig. 89.

physicist M. Janssen in India. More than three years ago[1] Mr. Lockyer suggested that it might be possible, by the use of the spectroscope, to obtain evidence, under the ordinary conditions of the solar disc, of the red pro-

[1] Proc. Roy. Soc. Oct. 11, 1866.

minences which had hitherto only been seen during total eclipses. After many fruitless attempts Mr. Lockyer at last succeeded in seeing the prominences with an unobscured sun on October 20, 1868, and he ascertained that the spectrum of the prominences is discontinuous, consisting of three bright bands, viz.—1, absolutely coincident with C; 2, nearly coincident with F; 3, near D. The principle adopted by Lockyer, by which these bright lines, proving the *gaseous* nature of the prominence, were rendered visible, was that of employing a spectroscope with a heavy battery of prisms, and possessing a strong dispersive power. The light from the body of the sun, producing an almost continuous spectrum, was in this way much spread out and thereby weakened; whilst the luminous intensities of the monochromatic rays emitted by the glowing gas were but slightly diminished, and thus the light from the prominences became visible without being interfered with by that emanating from the body of the sun. A simple experiment will render this important point clear. If I throw the light from the incandescent carbon points on to the screen by passing the rays through a flint glass prism, you will observe a short but very bright and perfectly continuous spectrum; if I next substitute for the glass prism two prisms filled with carbon disulphide, you see that the spectrum becomes very much elongated and its luminous intensity correspondingly diminished, so that now you can distinctly observe the narrow yellow sodium band and many other bright lines due to impurities in the carbon. These lines were present when the glass prism was used, but they were rendered invisible by the greater brightness of the continuous spectrum.

Whilst Mr. Lockyer was experimenting in England, M. Janssen had been sent out by the French Government to Guntoor, in India, to observe the spectroscopic appearances presented by the sun on the total eclipse of August 18, 1868. On that occasion he saw and measured the position of the bright lines above referred to; but, struck by their intensity, he likewise conceived the idea that they might be seen when the sun was un-eclipsed, and cried out, as he was looking through his telescope, "Je reverrai ces lignes là!" On the next morning, as soon as the sun rose out of the bank of clouds which lay on the horizon, he succeeded in his endeavour,—he saw the protuberances plainly, and was able to do what he failed to accomplish in the hurry and excitement of the eclipse; namely, to measure the exact position of the bright lines. "So that," he writes, "the last seventeen days have been to me like a perpetual eclipse.

The announcement of M. Janssen's independent observation was received by the French Academy on October 26, 1868, a few days after Mr. Lockyer's discovery had been made known to the Royal Society. As regards the claims of priority of this discovery, you will all, I am sure, feel inclined to agree with the following eloquent words of M. Faye when speaking on this subject in the French Academy on October 26, 1868:—

"Mais au lieu de chercher à partager, et par conséquent à affaiblir le mérite de la découverte, ne vaut-il pas mieux en attribuer indistinctement l'honneur entier à ces deux hommes de science qui ont eu séparément, à plusieurs milliers de lieues de distance, le bonheur d'aborder l'intangible et l'invisible par la voie la plus étonnante peut-être que le génie de l'observation ait jamais conçue?"

Lockyer's investigations have not only proved that these singular red prominences consist of glowing gaseous hydrogen, but have revealed the existence of an atmosphere, chiefly consisting of incandescent hydrogen, extending all round the sun's surface. The prominences are only local aggregations of this envelope of glowing hydrogen, which extends for 5,000 miles in height, and has been termed the *Chromosphere*, to distinguish it from the cooler absorbing atmosphere on the one hand, and the light-giving photosphere on the other. Under proper instrumental and atmospheric conditions the spectrum of the chromosphere is always visible in every part of the sun's periphery. Fig. 70 gives a representation, taken

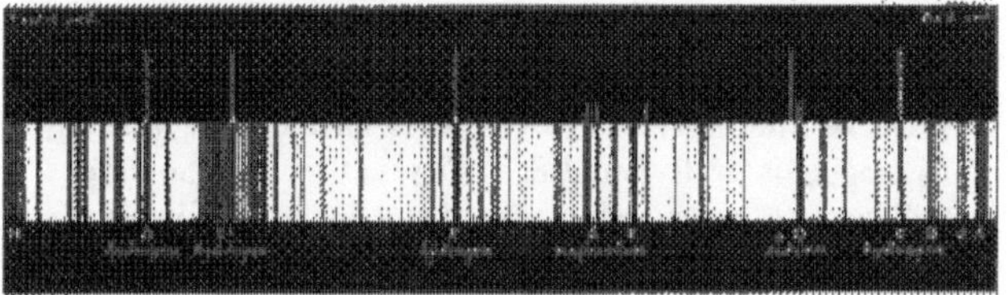

FIG. 70.

from one of Mr. Lockyer's drawings, of the spectrum of the edge of the sun's limb, and of that of the outlying chromosphere. The bright lines of hydrogen, sodium, magnesium, and iron are here seen to be coincident with the corresponding dark lines in the solar spectrum placed below. Another observation of the greatest importance we likewise owe to Lockyer, viz. that in examining the bright line coincident with Fraunhofer's F, the breadth or strength of the line is seen to expand or increase as the sun's limb is approached, whilst the line coincident with C and that near to D do not suffer any change of breadth. This remarkable fact, placed in connexion with an observation of Plücker and Hittorf (since confirmed by Huggins, Lockyer, and Frankland), viz. that a similar expansion of the F line in the hydrogen spectrum occurs when the pressure of the incandescent hydrogen is considerably increased, suggested the possibility of ascertaining the absolute pressure under which the hydrogen of the prominences exists; and it thus appears probable that the tension at the lowest portion of the chromosphere is very much less than that of our atmosphere, whereas at the higher parts of a prominence the pressure amounts only to a fraction of a millimetre of mercury.

Lockyer has succeeded in detecting the third (blue) line of hydrogen (viz. 2796 on Kirchhoff's scale), as well as the violet line known as *h* in the light of the chromosphere; and we must therefore admit that the proof of the existence in the solar envelope of glowing hydrogen is well founded. The nature of the bright yellow line, more refrangible by 8 or 9 of Kirchhoff's degrees than the D lines, remains as yet a mystery. This line, however, appears always most strongly at the lowest, and therefore at the hottest, portion of the prominence or chromosphere. Lockyer has, however, proved that the ordinary solar spectrum contains a dark absorption line coincident with this bright orange line in the chromosphere. Fig. 70 shows the position of this yellow line; Fig. 76, p. 240, exhibits the appearance of the bright F line, which is seen to thicken out or become wedge-shaped at the point where it touches the ordinary solar spectrum.

By employing a wide slit, with a screen of ruby-red glass, Huggins, Lockyer, and Zöllner succeeded in seeing the *form* of a prominence, and thus clearly ob-

Forms of Eruptions & cloud like prominences
as observed and drawn by Prof Zöllner

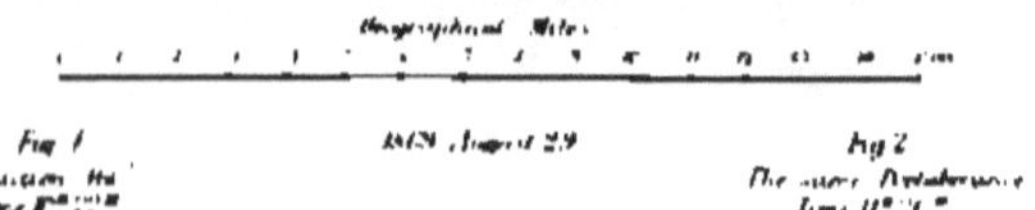

Fig 1
[illegible]
Time [illegible]

1869 August 29

Fig 2
The same Protuberance
Time 11h 21m

serving the singularly rapid changes in shape as well as in intensity which they undergo. Lockyer describes some of these enormous flames of incandescent hydrogen 27,000 miles in height, which totally disappeared in less than 10 minutes! "During the last few days," he writes, "I have been perfectly enchanted with the sight which my spectroscope has revealed to me. The solar and atmospheric spectra being hidden, and the image of the wide slit alone being visible, the telescope or slit is moved slowly, and the strange shadow-forms flit past. Here one is reminded by the fleecy, infinitely delicate cloud-forms, of an English hedge-row with luxuriant elms; here of a densely intertwined tropical forest, the intimately interwoven branches threading in all directions, the prominences generally expanding as they mount upwards, and changing slowly, indeed almost imperceptibly. By this method the smallest details of the prominences, and of the chromosphere itself, are rendered perfectly visible and easy of observation."

Zöllner has also made similar observations, and has published striking drawings of some of these protuberances, which I here have the pleasure of showing you (Fig. 71). These drawings represent one and the same protuberance observed on July 1, 1869, which underwent the singular changes here seen between the hours of 6h. 45′ and 7h. 8′ A.M., some of which remind one of the outbursts of a volcano, or the eruptive discharges of a geysir.[1] In one case a flickering flame-like motion was observed to pass in a few seconds up and down a horn shooting up to the height of 50,0000 miles (Fig. 72); whilst other protuberances formed clouds

[1] See Appendix C.

which seem to have been shot upwards by a kind of explosion (Figs. 74 & 75).

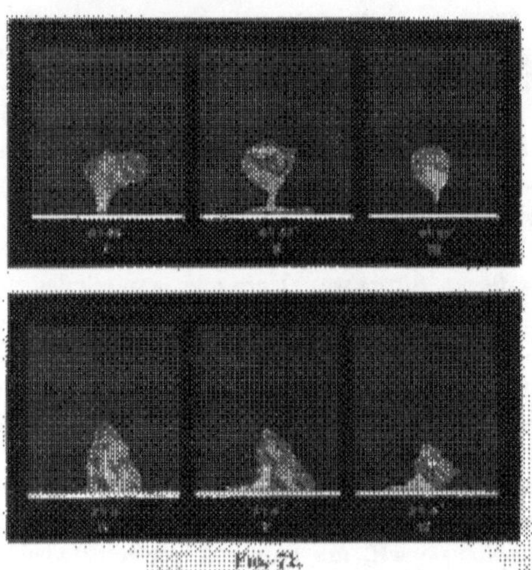

FIG. 71.

The startling rapidity with which these gigantic masses of hydrogen appear and disappear proves that currents,

FIG. 72.

FIG. 73.

storms, and hurricanes are of constant occurrence in the chromosphere. These changes are doubtless brought about there, as in our own atmosphere, by local variations in temperature. "But," as Kirchhoff remarks, "the differences in temperature which produce these winds may amount to thousands of degrees, and therefore the force of the currents must far exceed that of the most violent terrestrial tornadoes." Both telescopic and spectroscopic observations of phenomena which have long been known as peculiarities of the solar disc fully bear out this conclusion. You will have anticipated me when I say, that I refer to the dark sun-spots and to the bright stripes or faculæ which are always more or less visible on the solar surface.

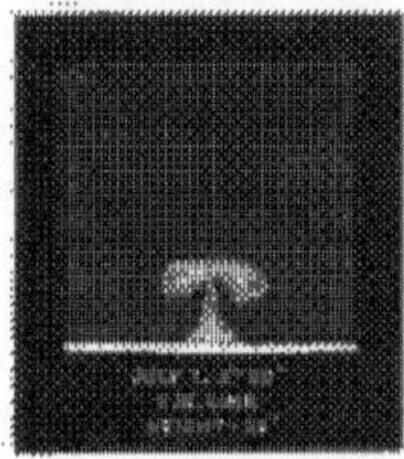

FIG. 74.

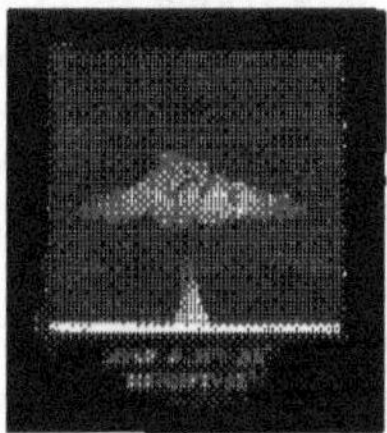

FIG. 75.

Astronomers formerly supposed that the sun-spots were holes in the photosphere through which the dark body of the sun was visible. Kirchhoff proved that this theory could not be correct, inasmuch as the interior portion of the sun's body must be white-hot, and he threw out the idea that these spots were clouds floating in the solar atmosphere. Subsequent research has somewhat modified these views of Kirchhoff's, for we now know that the sun-spots are really hollows or cavities in the solar atmosphere where the temperature of the glowing gases has been reduced. These spots look to us black, because they give off less light than the surrounding portions of the sun's surface, but they are not therefore non-luminous; on the contrary, they emit much light though they appear dark, just as a candle-flame looks black when held in front of the much more intensely-heated surface of the lime-light. The telescope has long ago told us of the gigantic changes, sometimes extending over even thousands of millions of square miles, which these spots undergo, but it is the spectroscope which admits us into the secrets of the constitution of these most singular phenomena. This instrument has shown us that the gases of the whole chromosphere are constantly in a state of the most violent motion, and has proved that a spot is a region of greater absorption. The faculæ, on the contrary, are caused by the incandescent vapours of sodium, iron, magnesium, and barium, thrown up by a sort of volcanic action from the lower to the higher regions of the solar atmosphere; for the bright lines, indicative of the presence of these bodies, occasionally make their appearance, not only at the edge of the sun's disc, but also in the centre of the sun's surface near the bright faculæ, or in the hotter ascending currents of the ignited gases.[1] Lockyer even describes a cloud of incandescent magnesium that he saw floating high up above a prominence,

[1] The following is a list of the lines seen by Mr. Lockyer in the chromosphere, with the dates of discovery. The lines C and F were first seen by M. Janssen:—

Hydrogen—C, Oct. 20, 1868; F, Oct. 20, 1688; near D, Oct. 20, 1868; near G, Dec. 22, 1868; h, March 14, 1869.

Sodium—D, Feb. 28, 1869.

Barium—1985·5 (Kirchhoff's scale), March 15, 1869; 2031·2 (ditto), July 5, 1869.

Magnesium and included line—b_1, b_2, b_3, b_4, Feb. 21, 1869.

Other lines—Iron, 1474, June 6, 1869; 1515·5, June 6, 1869.

Bright lines unknown—1529·5, 1567·5, 1613·8, 1871·5, 2054·0; Iron, 2001·5, 2003·4, June 26, 1869.

as in Fig. 75. What an insight this single observation gives us into the condition of solar physics!

These bright metallic lines, some of which may generally be seen when carefully looked for in the ordinary solar spectrum, are always *thinner* than the corresponding Fraunhofer's lines: this is accounted for by the fact that the glowing gases form part of an *upward* current, and that they therefore are in a condition of extreme tenuity when we know the lines are always thinnest. In the sun-spots, on the other hand, the opposite state of things occurs; there a *downward* current draws the cooler gases and vapours into a lower position, where they become condensed, and exert the powerful absorption on the rays of light which we observe in the darkened area of the spot. The lines of sodium, magnesium, and barium accordingly always appear in a spot-spectrum *darker* and *broader* than the corresponding Fraunhofer's lines as seen in the ordinary solar spectrum. A simple but very beautiful experiment, first proposed by Dr. Frankland, will serve to prove to you that the power of absorption, exerted by the vapours of the metals, increases with the density of the vapour. For this purpose I have here a tube filled with hydrogen hermetically sealed, at the bottom of which I have placed a few small pieces of sodium. If I now bring this tube before the slit of our electric lamp, and heat the metallic sodium, you will see a fine black line make its appearance on the screen: this line is caused by absorption of the yellow rays by the vapour which is now beginning to come off from the heated metal. As the density of the sodium vapour increases, you observe that the black line assumes a wedge or V shape, the absorption being greatest where the gas is most dense. Here, then, we have a precisely similar phenomenon to that observed by Lockyer in the widening out of the F line seen in the chromosphere and in the sun-spots.

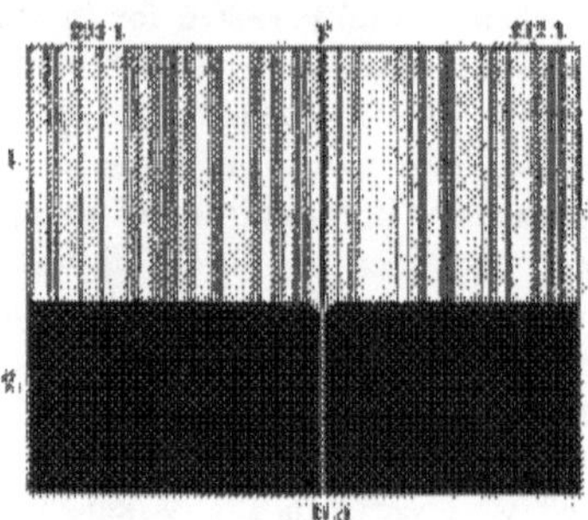

FIG. 76.

This arrow-headed form of the bright F line close to the sun's edge is shown in Fig. 76; the upper part of the diagram gives part of the ordinary solar spectrum, whilst on the lower is seen the bright hydrogen line $H\beta$.

Another most interesting subject has reference to the spectrum of the corona or the halo of silver-white light

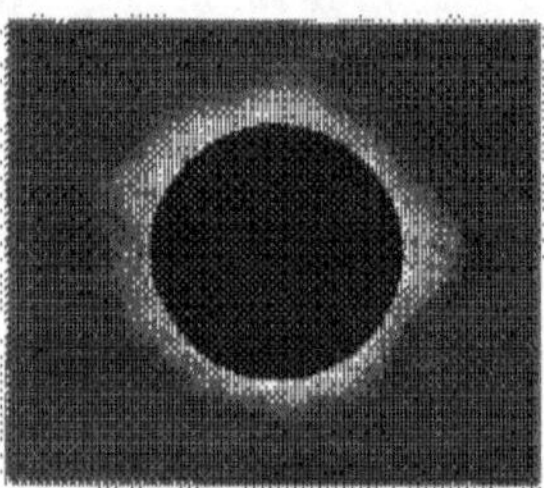

FIG. 77.

which surrounds the moon on all sides during a total eclipse of the sun (Fig. 77). The spectrum of this corona has recently been examined, in 1868 by Major Tennant[1] in India, and in 1869 by several American astronomers; but we do not find ourselves as yet in possession of data sufficient to enable us to decide positively concerning the nature of this coronal light, whether it arises in the solar atmosphere or is a terrestrial phenomenon. Major Tennant states that the spectrum of the corona is the ordinary solar spectrum, whilst Professor Young in 1869 appears to have seen three bright lines in the portion of the sky beyond the prominences, but he has some doubts as to whether these bright bands may not have been produced by an outlying and nebulous portion of the chromosphere; Professor Pickering, on the other hand, saw only a continuous spectrum. We must, therefore, patiently wait for the eclipse of December 1870 to settle this matter. It is very singular that the positions of these three lines (1250, 1350, and 1474, on Kirchhoff's scale) coincide, within the probable errors of observation, with three lines observed by Professor Winlock in the spectrum of the aurora borealis. The line 1474 is due to iron. Can we suppose that Dalton's old speculations as to the nature of the aurora have after all some foundation, and that this beautiful phenomenon is due "to the presence of some elastic fluid substance, probably of a ferruginous nature, existing in the higher regions of our atmosphere?"

Ångström[2] also observed the spectrum of the aurora in the winter of 1867–8; he saw only one very bright band, with traces of two others: but the most interesting discovery which he made on this branch of the subject was, that he succeeded in observing the same bright band

[1] See Appendix B.   [2] See end of Appendix A.

in the spectrum of the zodiacal light; and even on a star-light night, when the whole sky seemed almost phosphorescent, Ångström saw traces of these bands in light from all parts of the heavens! Strangely enough, this band, which links together the apparently unconnected phenomena of the solar corona, the zodiacal light, and the aurora, does not, according to Ångström, appear to correspond to the lines of any known substance.

From what I have already said, you will see that the spectroscope has already become an instrument as essential to the astronomer as to the chemist. You will, however, be more forcibly convinced of the power of this new aid to astronomy when I explain, as I hope to do in the next lecture, that it has not only given us information as to the chemical and physical condition of the stars, comets, and nebulæ, but that it has actually enabled us to ascertain with accuracy the rate with which the ignited gases of the solar atmosphere rush forward, and to measure the speed of motion of the fixed stars.

LECTURE V.—APPENDIX A.

ON THE NORMAL SOLAR SPECTRUM.

BY A. J. ÅNGSTRÖM.

A MOST valuable memoir on the Normal Spectrum of the Sun has recently been published by Professor Ångström, of Upsala, accompanied by an atlas of six magnificent plates, exhibiting the lines in the whole length of the solar spectrum from A to H. The positions of these lines are mapped according to their *wave-lengths*, which have been calculated from observations most carefully made with diffraction spectra. The bright metallic lines coincident with those of Fraunhofer are also given. The following table gives a *résumé* of the solar lines shown on his maps as produced by known elements:—

Substances.	Number of Lines.	Substances.	Number of Lines.
Hydrogen	4	Manganese	57
Sodium	9	Chromium	18
Barium	11	Cobalt	19
Calcium	75	Nickel	33
Magnesium	4 + 3 (?)	Zinc	2 (?)
Aluminium	2 (?)	Copper	7
Iron	450	Titanium	118[2]

The total number of these coincident metallic lines amounts to close upon 800, and this number might be easily increased by using more powerful means of raising the temperature of the substances under examination. "Nevertheless the number already mapped suffices to show that to account for the origin of almost all the more prominent rays in the solar spectrum, and in con-

[1] "Recherches sur le Spectre Solaire," par A. J. Ångström; "Spectre normal du Soleil, Atlas de six planches," Upsala, 1868.

[2] The presence of titanium in the solar atmosphere was discovered by Thalén.

firmation of an opinion expressed by me in a former research, we must assume that the substances constituting the chief mass of the sun are without doubt the same substances as exist on our planet. It must, however, not be forgotten that nearly equidistant between F and G there exist certain prominent rays whose nature is as yet unknown: still, any conclusions concerning the presence of substances in the sun which are unknown on the earth are certainly premature. We may nevertheless notice as a singular fact, that one of the darkest of these unknown lines coincides with a prominent line in the spectrum of bromine; but, as chlorine exhibits no coincidences with Fraunhofer's lines, it is not likely that this correspondence is really due to bromine.

"Aluminium undoubtedly exhibits several bright lines in various parts of the spectrum, but the two lines situated between the two H bands are the only ones which have been observed to be coincident with Fraunhofer's lines. To explain this singular fact it must be remembered that the violet rays of this metal are much the most intense. By observing the ultra-violet rays of this metal it will be possible to ascertain whether these two lines are caused by aluminium, as the ultra-violet bands ought to coincide with the invisible dark lines in the chemically active portion of the solar spectrum."

"To the two zinc lines which I have indicated on my map as coincident with dark solar lines I have to add a third, situated at 4809·7, but two other bright and broad lines of this metal which appear somewhat nebulous do not coincide with Fraunhofer's lines; and hence I consider that the presence of zinc in the solar atmosphere is very doubtful. At the same time I may mention that there are also three nebulous bands due to magnesium, none of which are seen in the sun, although the presence of this element in the solar atmosphere does not admit of a doubt.

"Of all the elements, iron certainly contains the largest number of lines visible in the solar spectrum. The iron lines which are not symmetrically distributed throughout the spectrum exhibit two maxima; the one situated near E, and the other near

c. Some of the iron lines appear to be coincident with those of calcium, but such coincidences are often only apparent. Thus, for example, there is a strong iron line between E and *b* (wave-length = 5226[1]), which is drawn as a single line on both Kirchhoff's maps and my own. Nevertheless M. Thalén has proved, by increasing the dispersive power of his instrument by using six flint-glass prisms, that this ray is in reality a triple one, and that its constituent lines are produced, one by iron and one by titanium.

"Among the metalloids, hydrogen is the only one indicated by spectrum analysis as existing in the sun: the other substances, such as oxygen, nitrogen, and carbon, which exist in such large quantities on the earth, can never be discovered in the sun by this process. Still, in spite of the almost complete want of coincidences between the solar lines and those of oxygen and nitrogen, we have no right to pronounce definitively upon the absence of these two bodies in the sun. And for this reason: the air spectrum cannot be observed even between the carbon poles of a battery of fifty cells, and in general is not seen when the electricity passes by what may be termed the electrolytic discharge. These spectra need for their production the disruptive discharge, as is seen clearly in the experiments with Geissler's tubes containing these two gases. In fact, when the discharge is accompanied by electrolysis, the spectra obtained in rarefied gases are those of *compound bodies*; and thus Plücker is incorrect in naming these the spectra of the first order: on the contrary, when the discharge becomes disruptive, as by using the condenser, the spectra of the elementary bodies at once become visible. This fact possesses a great degree of importance for the true interpretation of the spectra of the sun and stars, as it points out to us as very probable that the high temperature of the sun is insufficient to produce the brilliant rays of oxygen and nitrogen. . . . .

"In a memoir on the double spectra of the elementary bodies, which M. Thalén and I are about to publish, we treat of the important points of this interesting subject. Let it suffice for

[1] Ten-millionths of a millimetre.

me here to remark, that the results to which we have arrived in no way bear out the opinion of Plücker, that one element can give totally different spectra. The exact reverse of this is the truth. By successively augmenting the temperature we find that the intensity of the rays varies in a most complicated manner, and that, accordingly, even new rays can make their appearance if the temperature is sufficiently raised. But, independently of all these mutations, the spectrum of each substance always preserves its individual character."

SPECTRA OF THE AURORA BOREALIS AND OF THE ZODIACAL LIGHT.

BY A. J. ÅNGSTRÖM.

"During the winter 1867-8 I have several times observed the spectrum of the luminous arc which bounds the dark circle, and is always seen in feeble auroras. The light of this arc is almost monochromatic, and exhibits a single brilliant band, situated to the left of the well-known group of calcium lines. By measuring its distance from this group I have determined its wave-length to be = 5567. In addition to this ray, of which the intensity is relatively high, I have also observed, by widening the slit, traces of three very feeble bands situated near to F. Another circumstance gives a greater, and indeed an almost cosmical, importance to this observation of the auroral spectrum. During the month of March 1867 I succeeded in observing the same bright band in the spectrum of the zodiacal light, which was at that time seen of great intensity. Indeed during a starlight night, when the sky was almost phosphorescent, I found traces of this band visible from all parts of the heavens. It is a remarkable fact that this bright band does not coincide with any of the known rays of simple or compound gases which I have as yet examined."

APPENDIX B.

THE INDIAN TOTAL SOLAR ECLIPSE.

EXTRACTS FROM THE REPORT OF THE COUNCIL OF THE ROYAL ASTRONOMICAL SOCIETY TO THE 49TH ANNUAL GENERAL MEETING.

Solar Eclipse of 1868, *August* 18.

"The results obtained by the different observers are of such interest and importance that the principal observations which would not otherwise appear in our 'Transactions' are given in considerable detail in the observers' own words.

"It is with great satisfaction that the Council call the attention of the Fellows of the Society to the complete success of their own expedition;—a success for which the Fellows are much indebted to the skill and energy of the Superintendent, Major Tennant.

The Astronomical Society's Expedition.

"It will be in the recollection of our Fellows that at the last anniversary meeting it was stated that preparations had been made at the recommendation of the Council of our Society for the observation of the total eclipse of the sun in India. The Astronomer Royal took a warm interest in the proposed observations, and addressed the Secretary of State for India on the subject. It was ultimately arranged that the expense of the expedition should be borne jointly by the Government of India and the Imperial Government. The superintendence of the expedition was entrusted to Major Tennant. It is with great satisfaction that the Council is able to announce that Major Tennant has been most deservedly and eminently successful.

"The Report of Major Tennant's observations is now in the hands of the Society, and it is intended that it shall appear in the forthcoming volume of the 'Transactions,' fully illustrated with fac-similes of the photographs taken at Guntoor, which it is proposed to enlarge photographically, in order that the details of the prominences may be seen more clearly than is possible in the small copies which accompany the paper. Mr. De la Rue, who evinced considerable interest in the expedition, and afforded facilities to Major Tennant for familiarizing himself with astronomical photography before he started, has undertaken to see that the photographs are properly enlarged and copied.

"It is here proper to state that to Major Tennant is due the credit of having first called attention to the peculiarly favourable conditions which would be presented by the solar eclipse of August 1868.[1]

"It is only justice also to mention that, as far as regards the part which England took in the observations, it was mainly attributable to the energetic, active, and untiring zeal of Major Tennant, who happened to be in England on leave during the greater part of 1867, and who devoted much time in promoting the observations which, in spite of many difficulties, have been so successfully undertaken and carried out.

"It will be recollected that Major Tennant, after consulting with the Astronomer Royal and other Fellows of the Society, undertook the following work. It was most comprehensive, and entailed possibly almost too much responsibility for the director of a single expedition.

"1. *The Determination of the Geographical Position of the Station.*—This was successfully accomplished by means of a repeating circle, although, in consequence of bad weather, there were not many available days between the arrival of the observers and instruments at Guntoor and the day of the eclipse.

[1] Monthly Notices, vol. xxvii. pp. 79, 174.

"The position was found to be, Latitude N. 16° 17′ 29·23″, and Longitude E. 5h. 21m. 48·6s.

"Captain Branfill, R.E., subsequently connected the station with the marks of the Great Trigonometrical Survey, and deduced the following result: Lat. N. 16° 17′ 34·3″, and Long. E. 5h. 21m. 46·5s.

"2. *Spectroscopic Observations.*—These were undertaken by Major Tennant himself, by means of the Sheepshanks equatorial, of 4·6 inch aperture and 5 feet focal length. This had been mounted equatorially by the late Mr. Cooke, and was suitable for all latitudes in the British Isles, but it had to be altered to suit the more southern stations of India. The spectroscope employed with the telescope was made by Messrs. Troughton and Simms, and was provided with a scale of equal parts, which was illuminated by means of a lamp. The addition of this spectroscope threw additional work on the driving clock beyond that for which it was originally calculated, and, in consequence, some difficulties were experienced just at the critical time of observation from the irregularity of its going.

"In spite, however, of this and other mishaps, Major Tennant was able to carry out his observations, and ascertained, 1st, that the corona only gave the continuous solar spectrum; 2d, that the light of the prominences was resolvable into certain bright lines of definite refrangibility, showing that these appendages consist of gaseous matter at a very high temperature. Major Tennant states that the Great Horn gave a beautiful line in the red, a line in the orange, and one in the green, which appeared multiple, also a line seen with difficulty near F; he says the red and yellow lines were evidently C and D: the reading of the bright line coincides with that of the brightest line in *b*. The line near to F was, in all probability, F itself; E, he says, was certainly not seen by him, and that, as regards the line in the blue, it was useless from his data to speculate upon it.

"We now have more precise information from the researches of M. Janssen and Mr. Lockyer respecting the position of the bright lines, and the probable nature of the sun's appendages; but it must be admitted that Major Tennant did this part of his work well, especially when the scope of the instruments at disposal is taken into account.

"3. Major Tennant noted the time of first contact of the sun and moon's limbs by means of the repeating circle at 6h. 2m. 12·60s. sidereal time, and that of the last contact at 8h. 47m. 24·62s. sidereal time.

"4. *Polariscope.*—This part of the work was most ably performed by Captain Branfill, who joined Major Tennant early in August, and immediately set to work to familiarize himself with the phenomena produced by polarized light in the telescope. This instrument was one of the old collimators of the Great Transit circle of Greenwich, and was lent by the Astronomer Royal. It was mounted on a polar axis, so that with one movement it could be made to follow the apparent motion of the sun; but it was not provided with a driving clock. To this telescope a polariscope eyepiece had been fitted by Mr. Ladd. The polarizing apparatus comprised several combinations which could readily and rapidly be substituted one for the other. All these concurred in showing that the prominences (the Great Horn was chiefly observed) gave no indication of polarized light; on the other hand, every arrangement brought out the fact that the light of the corona was polarized in a plane passing through the sun's centre. These observations were therefore fully and successfully carried out.

"5. We now come to the photographic observations: these were under the immediate direction of Sergeant Phillips, who is not only a skilled photographer, but also had the advantage (as well as the Sappers who aided him) of working in Mr. Warren De la Rue's observatory at Cranford. The telescope employed is a Newtonian with a silvered-glass mirror, 9 inches in diameter, by With, and specially mounted by Mr. Browning. Preparations had been made for having a very large field, in order that the corona might be depicted as well as the prominences. Unfortunately the sun was covered with cumulo-stratus clouds, which diminished the actinic power of the light of the corona so much that it was not recorded. In other respects the photographs (six in number) were eminently successful.

Interval. s	No.		Exposed at h. m. s.		S. T. for s.
	No. 1	Exposed at	7 17 2·9	S. T. for	1
58·5	,, 2	,,	7 18 1·4	,,	5
61·4	,, 3	,,	7 19 2·8	,,	10
58·1	,, 4	,,	7 20 0·9	,,	5
63·2	,, 5	,,	7 21 4·1	,,	1
90·7	,, 6	,,	7 22 34·8	,,	1

"Paper copies of these, about two inches in diameter, accompany the Report; and Mr. James, one of Major Tennant's assistants, made excellent drawings of the Great Horn and other prominences seen in the photograph, by means of the microscope. Since the arrival of the memoir Sergeant Phillips has brought safely to England eight sets of transparent copies on glass, which have been distributed to individuals and learned bodies; amongst others, to the Royal Society and the French Academy of Sciences. On the occasion of a lecture given by Prof. Herschel at the Royal Institution, on January 22, these were shown by means of the electric lamp, and projected on a screen, on a scale of about 5 feet for the moon's diameter. The amount of detail visible under these circumstances was very remarkable. The spiral structure of the Great Horn, to which Major Tennant has called attention, was very evident. This spiral formation Major Tennant ascribes to the confliction of an ascending current and one at right angles to it. Since then Mr. Warren De la Rue has procured some very beautiful copies, about $6\frac{1}{2}$ inches diameter. He has also discussed, graphically, the small paper photographs, and communicated the results to the Society.[1] The diagram accompanying his paper shows fairly the form and relative position of these appendages with respect to the sun. Mr. Warren De la Rue thought that he had detected a rotation of the Great Horn on its axis during the intervals between occurrence of totality at the various stations along the line of the eclipse. He has since been favoured by Prof. Foerster, Director of the Berlin Observatory, with a copy of the first Aden photograph, and informs the Council that there does not appear to be any very

[1] Monthly Notices, vol. xxix. p. 73.

great change in the appearance of the Great Horn at Aden and at Guntoor. This comparison of results brings out forcibly the great value of photography for this class of observations, for the most careful and collected observer is liable to make an error in recording eye-observations.

"In justice, however, to Col. Addison, whom Major Tennant had induced to make observations at Aden, it must be stated that a drawing of the prominences which he sent to Major Tennant led that gentleman to conclude that no change had occurred in them between the epoch of Aden and that of Guntoor.

"Major Tennant reached Aden on the 25th of January: as this was nearly the first place where observations of the totality could be made, he enlisted the services of Captain Davis, the Peninsular and Oriental Company's agent, and of an old companion, Major Napier, R.A. Both these gentlemen promised their aid, and he learnt from them that Col. Addison and Major Weir, H.M. 2d Royal Regiment, would be likely to be valuable coadjutors. Unfortunately, the contemplated observations with the polariscope, spectroscope, and intended drawings of the corona were rendered impossible, in consequence of clouds. But the prominences were seen and recorded with great accuracy, as it has been before stated.

"The Council have every reason to feel satisfied with the steps they took in conjunction with the Astronomer Royal in furthering Major Tennant's views, and in thus securing a most valuable series of observations."

Lieut. J. Herschel's Account. Position, Jamkandi.

"The totality commenced unseen. 'A few seconds more, and the spectrum of diffuse light vanished also, and told me the eclipse was total, but behind a cloud. I went to the finder, removed the dark glass, and waited, how long I cannot say, perhaps half a minute. Soon the cloud hurried over, following the moon's direction, and therefore revealing, first, the upper limb, with its scintillating corona, and then the lower. Instantly I marked a prominence near the needle point, an object so conspicuous that I felt there was no need to take any precautions to secure identification. It was a long finger-like projection from the (real) lower left-hand portion of the circumference. A rapid turn of the declination screw covered it with the needle point, and in another instant I was at the spectroscope. A single glance, and the problem was solved.

"'*Its Spectrum.*—Three vivid lines, red, orange, blue; no others, and no trace of a continuous spectrum.

"'When I say the problem was solved, I am, of course, using language suited only to the excitement of the moment!. It was still very far from solved, and I lost no time in applying myself to measurement. And here I hesitate, for the measurement was not effected with anything like the ease and certainty which ought to have been exhibited. Much may be attributed to haste and unsteadiness of hand, still more to the natural difficulty of measuring intermittent glimpses; but I am bound to confess that these causes were supplemented by a failure less excusable. I have no idea how those five minutes passed so quickly! Clouds were evidently passing continually, for the lines were only visible at intervals—not for one half the time certainly—and not always bright; but still I ought to have measured them all. My failure was insufficient illuminating power; but why, I cannot tell. I never experienced any difficulty of the kind with the nebulæ, which required that I should flash in light suddenly over and over again. I had found the hand-lamp the surest way, but it failed me here in great measure. The red line must have been less vivid than the orange, for after a short attempt to measure it I passed on to secure the latter. In this I succeeded to my satisfaction, and accordingly tried for the blue line. Here I was not so successful. The glimpses of light were rarer and feebler, the line itself growing shorter, and what remained of it further from the cross. I did, however, place the cross wires in a position certainly very near the true one, and got a reading before the re-illumination of the field told me that the sun had reappeared on the other limb. These readings were called out, as those on the solar lines had been, to my recorder, and it was only afterwards that I compared them

"'I need not dwell on the feelings of distress and disappointment which I experienced on realizing the fact that the long-anticipated opportunity was gone, and, as it seemed to me then, *wasted*. I seemed to have failed entirely. Almost mechanically I directed the telescope to the brightened limb, to verify the readings of the solar lines, and in doing so my interest was again awakened by the near coincidence, as it seemed, of the line F with the position of the wires; but a little reflection convinced me that the distance of the former was greater than the error which I might have made in intersecting the blue line. I read F, and then D and C. The following were my readings up and down :—

	C	D.	*b*	F.
Before	1·91	2·96	4·58	5·64
	1·90	2·94	4·58	5·61
	1·93	2·98	4·60	5·65
	1·92	2·97	4·58	5·62
Bright lines		[3·00]		[5·56]
After	1·93	3·00		5·65

"'I consider that there can be no question that the orange line was identical with D, so far as the capacity of the instrument to establish any such identity is concerned. I also consider that the identity of the blue line with F is not established; on the contrary, I believe that the former is less refracted than F, but not much. With regard to the red line, I hesitate very much in assigning an approximate place; B and C represent the limits: it might have been near C; I doubt it being so far as B. I am not prepared to hazard any more definite opinion about it. Its colour was a bright red. This estimate of its place is absolutely free from any reference to the origin of the lines C and F.'

"The spectrum of the corona does not appear to have been specially examined."

Lieut. Campbell's Report.

"The instruments in question were as follow: a telescope of 3-inch aperture, mounted on a rough double axis, admitting of motions in azimuth and altitude by hand only, unaided by any appliance for clamping and slow motion. The telescope was provided with three eyepieces of magnifying powers 27, 41, and 98; and with it were furnished two analysers for polarized light, viz. a double-image prism and a Savart's polariscope.

"On the first opportunity after the commencement of the total phase of the eclipse I turned on the double-image prism with the eyepiece of 27 magnifying power, as recommended in the Instructions, which gave a field of about 45' diameter. A most decided difference of colour was at once apparent between the two images of corona; but I could not make certain of any such difference in the case of a remarkable horn-like protuberance, of a bright red colour, situated about 210° from the vertex, reckoned (as I have done in all cases) with reference to the actual, not the inverted image, and with direct motion. I then removed the double-image prism and applied Savart's polariscope, which gave bands at right angles to a tangent to the limb, distinct, but not bright, and with little, if any, appearance of colour. On turning the polariscope in its cell, the bands, instead of appearing to revolve on their own centre, passing through various phases of brightness, arrangement, &c., travelled bodily along the limb, always at right angles thereto, and without much change in intensity, or any at all in arrangement. The point at which they seemed strongest was about 140° from the vertex, and I recorded them as black central. Believing that with a higher power and a smaller field I should find it easier to fix my attention on one point of the corona, and observe the phases of the bands at that point, I changed eyepieces, applying that of 41 power. With this eyepiece the first clear instant showed the bands much brighter than before, coloured, and as tangents to the limb at a point about 200° from the vertex: but before I could determine anything further a cloud shut out the view, and a few seconds later a sudden rush of light told that the totality was over, though it was difficult to believe that five minutes had flown by since its commencement. I experienced a strong feeling of disappointment and want of success; the only points on which I can speak with any confidence being as follows:—

"(1) When using the double-image prism, the strong difference of colour of the corona, and the absence of such difference in the case of the most prominent red flame. (2) With Savart's polariscope the bands from the corona were decided: with a low power they were wanting in intensity and colour: excepting alternate black and white, making it difficult to specify the nature of the centre: and their position was at right angles to the limb, extending over about 30° of the circumference. When the polariscope was turned, the bands travelled bodily round the limb without other changes in position or arrangement, as if, indeed, they were revolving round the centre of the sun as an axis. With a higher power, when a smaller portion of the corona was embraced, the bands were brighter-coloured, and seen in a different position, viz. tangents to the limb.

"The appearance observed with a low power seems exactly what might be expected supposing the bands to be brightest at every point when at right angles to the limb, in which case the bands growing into brightness at each succeeding point of the limb would distract attention from those fading away at the points passed over as the analyser revolved."

APPENDIX C.

SPECTROSCOPIC OBSERVATIONS OF THE SUN.

1. LOCKYER AND JANSSEN'S DISCOVERY.

Since this lecture was delivered an observation has been made with respect to the sun only second in interest and importance to the results of Kirchhoff's celebrated discovery of the coincidence of the bright iron and dark solar lines and the reversal of the sodium spectrum. The striking nature of this discovery is rendered more evident by its having been made independently by two observers situated thousands of miles apart—by M. Janssen in India and Mr. Norman Lockyer in London. No less than two years ago[1] Mr. Lockyer suggested that it might be possible by the use of the spectroscope to obtain evidence of the presence of the red prominences which total eclipses have revealed to us in the solar atmosphere, although they escape all other means of observation at other times. After many fruitless attempts to realize his hopes, Mr. Lockyer at last succeeded, on October 20, 1868, in obtaining the spectrum of a solar prominence; and he thus announces his important observation to the Royal Society, through Dr. Sharpey :—

"SIR,—I beg to anticipate a more detailed communication by informing you that, after a number of failures, which made the attempt seem hopeless, I have this morning perfectly succeeded in obtaining and observing part of the spectrum of a solar prominence.

"As a result I have established the existence of three bright lines in the following positions:—

"I. Absolutely coincident with C.

"II. Nearly coincident with F.

"III. Near D.

[1] Proc. Roy. Soc. Oct. 11, 1866.

"The third line (the one near D) is more refrangible than the two darkest lines by eight or nine degrees of Kirchhoff's scale. I cannot speak with exactness, as this part of the spectrum requires remapping.

"I have evidence that the prominence was a very fine one.

"The instrument employed is the solar spectroscope, the funds for the construction of which were supplied by the Government Grant Committee. It is to be regretted that its construction has been so long delayed.

"I have, &c.

"J. NORMAN LOCKYER.

"*The Secretary of the Royal Society.*"

M. Janssen was sent by the French Government to observe the total eclipse at Guntoor in India, and on August 18, when examining the bright lines exhibited by the spectra of the prominences visible during the totality, the thought struck him that it might be possible to see these lines when the sun was unobscured, and on trying the experiment on the next day he succeeded in his endeavour, "so that," he writes, "for the last seventeen days I have been working as in a perpetual eclipse." The results of his observations were communicated (Oct. 26, 1868) to the French Academy in the following words:—

"La station de Guntoor a été sans doute la plus favorisée: le ciel a été beau, surtout pendant la totalité, et mes puissantes lunettes de près de trois mètres de foyer m'ont permis de suivre l'étude analytique de tous les phénomènes de l'éclipse.

"Immédiatement après la totalité, deux magnifiques protubérances ont apparu: l'une d'elles, de plus de trois minutes de hauteur, brillait d'une splendeur qu'il est difficile d'imaginer. L'analyse de sa lumière m'a immédiatement montré qu'elle était formée par une immense colonne gazeuse incandescente, principalement composée de gaz hydrogène.

"L'analyse des régions circumsolaires, où M. Kirchhoff place l'atmosphère solaire, n'a pas donné des résultats conformes à la théorie formulée par ce physicien illustre; ces résultats me paraissent devoir conduire à la connaissance de la véritable constitution du spectre solaire.

"Mais le résultat le plus important de ces observations est la découverte d'une méthode, dont le principe fut conçu pendant l'éclipse même, et qui permet l'étude des protubérances et des régions circumsolaires en tout temps, sans qu'il soit nécessaire de recourir à l'interposition d'un corps opaque devant le disque du soleil. Cette méthode est fondée sur les propriétés spectrales de la lumière des protubérances, lumière qui se résout en un petit nombre de faisceaux très-lumineux, correspondant à des raies obscures du spectre solaire.

"Dès le lendemain de l'éclipse la méthode fut appliquée avec succès, et j'ai pu assister aux phénomènes présentés par une nouvelle éclipse qui a duré toute la journée. Les protubérances de la veille étaient profondément modifiées. Il restait à peine quelques traces de la grande protubérance et la distribution de la matière gazeuse était tout autre.

"Depuis ce jour, jusqu'au 4 septembre, j'ai constamment étudié le soleil à ce point de vue. J'ai dressé des cartes des protubérances, qui montrent avec quelle rapidité (souvent en quelques minutes) ces immenses masses gazeuses se déforment et se déplacent. Enfin, pendant cette période, qui a été comme une éclipse de dix-sept jours, j'ai recueilli un grand nombre de faits, qui s'offraient comme d'eux-mêmes, sur la constitution physique du soleil.

"Je suis heureux d'offrir ces résultats à l'Académie et au Bureau des Longitudes, pour répondre à la confiance qui m'a été témoignée et à l'honneur qu'on m'a fait en me confiant cette importante mission."

The following abstracts of Mr. Lockyer's various communications to the Royal Society give the latest results of his observations, and clearly indicate the important additions to our knowledge of solar physics to which these researches have already led.

2. LOCKYER: SPECTROSCOPIC OBSERVATIONS OF THE SUN, NO. II.[1]

"The author, after referring to his ineffectual attempts since 1866 to observe the spectrum of the prominences with an instrument of small dispersive power, gave an account of the delays which had impeded the construction of a larger one (the funds for which were supplied by the Government Grant Committee early in 1867), in order that the coincidence in time between his results and those obtained by the Indian observers might not be misinterpreted.

"Details are given of the observations made by the new instrument, which was received incomplete on the 16th of October. These observations include the discovery, and exact determination of the lines, of the prominence spectrum on the 20th of October, and of the fact that the prominences are merely local aggregations of a gaseous medium which entirely envelopes the sun. The term *chromosphere* is suggested for this envelope, in order to distinguish it from the cool-absorbing atmosphere on the one hand, and from the white light-giving photosphere on the other. The possibility of variations in the thickness of this envelope is suggested, and the phenomena presented by the star in Corona are referred to.

"It is stated that, under proper instrumental and atmospheric conditions, the spectrum of the chromosphere is always visible in every part of the sun's periphery: its height, and the dimensions and shapes of several prominences, observed at different times, are given in the paper. One prominence, three minutes high, was observed on the 20th October.

"Two of the lines correspond with Fraunhofer's C and F; another lies 8° or 9° (of Kirchhoff's scale) from D towards E. There is another bright line, which occasionally makes its appearance near C, but slightly less refrangible than that line. It is remarked that the line near D has no corresponding line ordinarily visible in the solar spectrum. The author has been led by his observations to ascribe great variation of brilliancy to the lines. On the 5th of November a prominence was observed

[1] Proc. Roy. Soc. vol. xvii. p. 131.

in which the action was evidently very intense; and on this occasion the light and colour of the line at F were most vivid. This was not observed all along the line visible in the field of view of the instrument, but only at certain parts of the line, which appeared to widen out.

"The author points out that the line F invariably expands (that the band of light gets wider and wider) as the sun is approached, and that the C line and the D line do not; and he enlarges upon the importance of this fact, taken in connexion with the researches of Plücker, Hittorf, and Frankland on the spectrum of hydrogen—stating at the same time that he is engaged in researches on gaseous spectra which, it is possible, will enable us to determine the temperature and pressure at the surfaces of the chromosphere, and to give a full explanation of the various colours of the prominences which have been observed at different times.

"The paper also refers to certain bright regions in the solar spectrum itself.

"Evidence is adduced to show that possibly a chromosphere is, under certain conditions, a regular part of star economy; and the outburst of the star in Corona is especially dwelt upon."

3. LOCKYER: SPECTROSCOPIC OBSERVATIONS OF THE SUN, NO. III.[1]

"In my former paper it was stated that a diligent search after the known third line of hydrogen in the spectrum of the chromosphere had not met with success. When, however, Dr. Frankland and myself had determined that the pressure in the chromosphere even was small, and that the widening out of the hydrogen lines was due in the main, if not altogether, to pressure, I determined to seek for it again under better atmospheric conditions; and I succeeded after some failures. The position of this third line is at 2796 of Kirchhoff's scale. It is generally excessively faint, and much more care is required to see it than is necessary in the case of the other lines; the least haze in the sky puts it out altogether. Hence, then, with the

[1] Proc. Roy. Soc. March 4, 1869, vol. xvii. p. 350.

exception of the bright yellow line, the observed spectra of the prominences and of the chromosphere correspond exactly with the spectrum of hydrogen under different conditions of pressure—a fact not only important in itself, but as pointing to what may be hoped for in the future. As this yellow line may be possibly caused, as Frankland and I have suggested, by the radiation of a great thickness of hydrogen, it became a matter of importance to determine whether, like the red and green lines (C and F) it could be seen extending on to the limb. I have not observed this; it has always in my instrument appeared as a very fine sharp line resting absolutely on the solar spectrum, and never encroaching on it.

"Dr. Frankland and myself have pointed out, that although the chromosphere and the prominences give out the spectrum of hydrogen, it does not follow that they are composed merely of that substance; supposing others to be mixed up with hydrogen, we might presume that they would be indicated by their selective absorption near the sun's limb. In this case the spectrum of the limb would contain additional Fraunhofer lines. I have pursued this investigation to some extent with, at present, negative results; but I find that special instrumental appliances are necessary to settle the question, and these are now being constructed. If we assume, as already suggested by Dr. Frankland and myself, that no other extensive atmosphere besides the chromosphere overlies the photosphere, the darkening of the limb being due to the general absorption of the chromosphere, it will follow:—

"1. That an additional selective absorption near the limb is extremely probable.

"2. That the hydrogen Fraunhofer lines indicating the absorption of the outer shell of the chromosphere will vary somewhat in thickness: this I find to be the case to a certain extent.

"3. That it is not probable that the prominences will be visible on the sun's disc.

"In connexion with the probable chromospheric darkening of the limb, an observation of a spot on February 20th is of importance. The spot observed was near the limb, and the absorption was much greater than anything I had seen before; so great, in fact, was the *general* absorption, that the several lines could only be distinguished with difficulty, except in the very brightest region. I ascribe this to the greater length of the absorbing medium in the spot itself in the line of sight, when the spot is observed near the limb, than when it is observed in the centre of the disc—another indication of the great general absorbing power of a comparatively thin layer, on rays passing through it obliquely. I now come to the selective absorption in a spot. I have commenced a map of the spot spectrum, which, however, will require some time to complete. In the interim, I may state that the result of my work up to the present time in this direction has been to add magnesium and barium to the material (sodium) to which I referred in my paper in 1866, No. I. of the present series; and I no longer regard a spot simply as a cavity, but as a place in which principally the vapours of sodium, barium, and magnesium (owing to a downrush) occupy a lower position than they do ordinarily in the photosphere. I do not make this assertion merely on the strength of the lines observed to be thickest in the spot-spectrum, but also upon the following observations on the chromosphere made on the 21st and 28th ultimo.

"On both these days the brilliancy of the F line taught me that something unusual was going on; so I swept along the spectrum to see if any materials were being injected into the chromosphere. On the 21st I caught a trace of magnesium; but it was late in the day, and I was compelled to cease observing by houses hiding the sun.

"On the 28th I was more fortunate. If anything, the evidences of intense action were stronger than on the 21st, and after one glanceat the F line, I turned at once to the magnesium lines. I saw them appearing short and faint at the base of the chromosphere. My work on the spots led me to imagine I should find sodium-vapour associated with the magnesium; and, on turning from *b* to D, I found this to be the case. I afterwards reversed barium in the same way. The spectrum of the chromosphere seemed to be full of lines, and I do not think the three substances I have named accounted for all of them. The observation was one of excessive delicacy, as the lines were short, and *very thin*. The prominence was a small one, about twice the usual height of the chromosphere; but the hydrogen lines towered high above those due to the newly injected materials. The lines of magnesium extended perhaps one-sixth of the height of the F line, barium a little less, and sodium least of all.

"We have, then, the following facts :—

"1. The lines of sodium, magnesium, and barium, when observed in a spot, are thicker than their usual Fraunhofer lines.

"2. The lines of sodium, magnesium, and barium, when observed in the chromosphere, are thinner than their usual Fraunhofer lines.

"For some time past I have been engaged in endeavouring to obtain a sight of the prominences, by using a very rapidly oscillating slit; but although I believe this method will eventually succeed, the spectroscope I employ does not allow me to apply it under sufficiently good conditions, and I am not at present satisfied with the results I have obtained.

"Hearing, however, from Mr. De la Rue, on February 27th, that Mr. Huggins had succeeded in anticipating me by using absorbing media and a wide slit (the description forwarded to me is short and vague), it immediately struck me, as possibly it had struck Mr. Huggins, that the wide slit is quite sufficient without any absorptive media; and during the last few days I have been perfectly enchanted with the sight which my spectroscope has revealed to me. The solar and atmospheric spectra being hidden, and the image of the wide slit alone visible, the telescope or slit is moved slowly, and the strange shadow-forms flit past. Here one is reminded, by the fleecy, infinitely delicate cloud films, of an English hedgerow with luxuriant elms; here, of a densely intertwined tropical forest, the intimately interwoven branches threading in all directions, the prominences generally expanding as they mount upwards, and changing slowly, indeed almost imperceptibly. By this method, the smallest details of the prominences and of the chromosphere itself are rendered perfectly visible and easy of observation."

4. LOCKYER: SPECTROSCOPIC OBSERVATIONS OF THE SUN, NO. IV.[1]

"The following observations were made on April 11th, 1869, near a fine spot situated not very far from the sun's limb:—

"1. Under certain conditions, the C and F lines may be observed *bright on the sun;* and in the spot-spectrum also, as in prominences or in the chromosphere.

"2. Under certain conditions, although they are not observed as bright lines, the corresponding Fraunhofer's lines are blotted out.

"3. The accompanying changes of refrangibility of the lines in question show that the absorbing material moves upwards and downwards as regards the radiating material, and that these motions may be determined with considerable accuracy.

"4. The bright lines observable in the ordinary spectrum are sometimes interrupted by the spot-spectrum, *i.e.* they are only visible in those parts of the solar spectrum near, and away from, spots.

"5. The C and F lines vary excessively in thickness over and near a spot; and on the 11th, in the deeper portion of the spot, they were much thicker than usual."

5. ON A POSSIBLE METHOD OF VIEWING THE RED FLAMES WITHOUT AN ECLIPSE. BY WILLIAM HUGGINS, F.R.S.[2]

"In the report of my Observatory at the last anniversary (p. 88 of the last volume), it is stated that 'during the last two years numerous observations have been made for the purpose of obtaining a view of the red prominences seen during a solar eclipse. If these bodies are gaseous, their spectra would consist of bright lines. With a powerful spectroscope, the light reflected from our atmosphere near the sun's edge would be greatly reduced in intensity by the dispersion of the prisms, while the bright lines of the prominences, if such be present, would remain but little diminished in brilliancy. This principle has been

[1] Proc. Roy. Soc. April 14, 1869, vol. xvii. p. 415.

[2] Monthly Notices of the Royal Astronomical Society, Nov. 13, 1868.

carried out by various forms of prismatic apparatus, and also by other contrivances, but hitherto without success.' The observations of the eclipse of August last having shown the position in the spectrum of the bright lines of the red flames, Mr. Lockyer and M. Janssen succeeded independently, by a similar method, in viewing the spectra of these objects.

"My object in this note is to describe one of the 'other contrivances' mentioned in the report.

"The apparatus consisted of screens of coloured glasses and other absorptive media, by which I was able to isolate portions of the spectrum. It appeared highly probable, that if the parts of the spectrum which then alone remained were identical with those in which the bright lines of the flames occur, these objects would become visible.

"For this inquiry I obtained a great variety of coloured glasses and other absorptive media. I first examined them with a prism to learn the absorptive power which they exercised on different parts of the spectrum. I then combined them in various ways. These glasses were sometimes employed before the eye, but more frequently by projecting the image of the sun's edge upon a screen, after the light had been sifted by the coloured media. In making these experiments, means were taken that the whole of the sun's image should be got rid of, in order that the eye, kept in comparative darkness, might be more sensitive to the greatly feebler illumination of the objects sought for. As I had no knowledge of the position in the spectrum of the bright lines, it would have been by accident only if I had succeeded in obtaining a view of the flames.

"Now that the positions of these lines are known, this method appears to be very promising. Perhaps the light about the red line at C will be most easily isolated. I have a deep ruby glass which cuts off all the spectrum except the extreme red. I have since the observations only been able to make one attempt, when the state of the atmosphere was unfavourable.

"It is obvious that by this method the form and appearance of these flames could be observed, and the objects measured with accuracy."

6. NOTE ON A METHOD OF VIEWING THE SOLAR PROMINENCES WITHOUT AN ECLIPSE. BY WILLIAM HUGGINS, F.R.S.[1]

"Last Saturday, February 13, 1869, I succeeded in seeing a solar prominence so as to distinguish its form. A spectroscope was used; a narrow slit was inserted after the train of prisms before the object-glass of the little telescope. This slit limited the light entering the telescope to that of the refrangibility of the part of the spectrum immediately about the bright line coincident with C. The slit of the spectroscope was then widened sufficiently to admit the form of the prominence to be seen. The spectrum then became so impure that the prominence could not be distinguished. A great part of the light of the refrangibilities removed far from that of C was then absorbed by a piece of deep ruby glass. The prominence was then distinctly seen."

7. EXTRACT FROM ZÖLLNER—DESCRIPTION OF PROTUBERANCES.[2]

(*See Figs.* 71 *to* 75.)

"One of the most remarkable forms is that shown on Fig. 72. I scarcely believed my eyes when I noticed in this one the flickering motion of a flame. This motion was, however, slower in proportion to the dimensions of the protuberance than that of a large mass of ordinary flame. The time needed for the propagation of this flame-wave from the base to the point of the prominence amounted to about from two to three seconds. Fig. 71 exhibits good examples of the rate of change which the form and intensity of these prominences undergo. The time at which each form was observed is given underneath each figure. Most of the prominences exhibit forms analogous to those of the various clouds and mists occurring in our atmosphere; of these, the cumulus type is the commonest. The flame-like protuberance Fig. 72 is an exception to the ordinary form; and in the forms of Figs. 73, 74, and 75, it is almost impossible to help believing that the masses which are seen to rise from the sun's surface are immediately connected with the cloudy portions which float above, and one is forcibly reminded of the phenomena of the eruptions of volcanoes or geysirs."

[1] Proc. Roy. Soc. vol. xvii. p. 302. [2] Pogg. Ann. cxxxvii. p. 624.

LECTURE VI.

Planet and Moonlight.—Stellar Chemistry.—Huggins and Miller.—Spectra of the Fixed Stars.—Difficulties of Observation.—Methods employed.—Variable Stars.—Double Stars.—Temporary Bright Stars.—Nebulæ.—Comets.—Motion of the Stars (Huggins).—Determinations of Velocity of Solar Storms (Lockyer).

Appendix A.—Extract from a Memoir "On the Spectra of some of the Fixed Stars."

Appendix B.—"On the Spectrum of Mars, with some Remarks on the Colour of that Planet."

Appendix C.—"On the Occurrence of Bright Lines in Stellar Spectra," and "On the Spectra of Variable Stars."

Appendix D.—"Further Observations on the Spectra of some of the Stars and Nebulæ, with an attempt to determine therefrom whether these Bodies are moving towards or from the Earth; also Observations on the Spectra of the Sun and of Comet II. 1868."

Appendix E.—"Researches on Gaseous Spectra in relation to the Physical Constitution of Sun, Stars, and Nebulæ."

Appendix F.—"On a new Spectroscope, and Contributions to a knowledge of the Spectrum Analysis of the Stars." Notice of Browning's new Automatic Spectroscope.

In the last lecture I endeavoured to point out to you the principles upon which Professor Kirchhoff arrived at the remarkable conclusion that certain metals well known on earth are contained in the solar atmosphere. I have to-day to bring before you facts which are still more interesting, with regard to the chemical composition of the stars and the nebulæ; and if in the former lectures I had to couple the names of two great German philosophers, I have to-day to bring before your notice the

SPECTRA OF THE STARS AND NEBULÆ.

Compared with the Solar Spectrum and the Spectra of some of the Non-Metallic Elements.

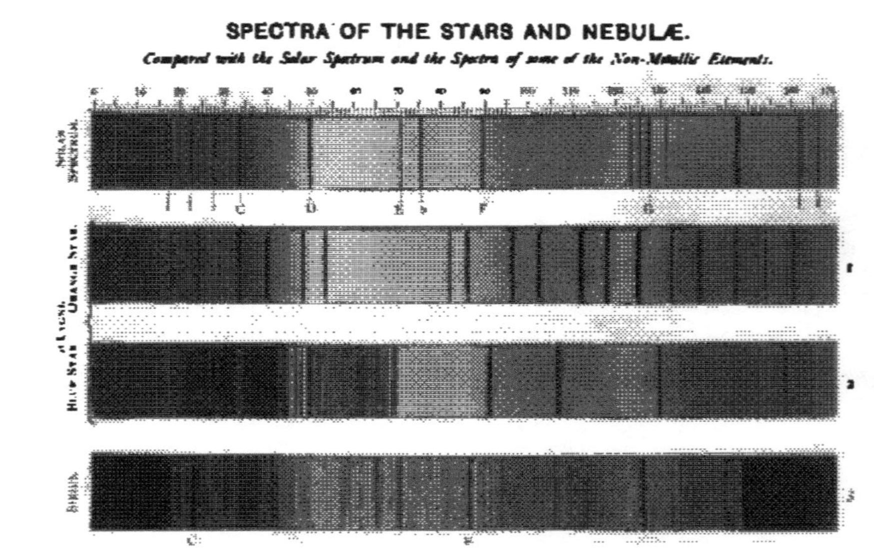

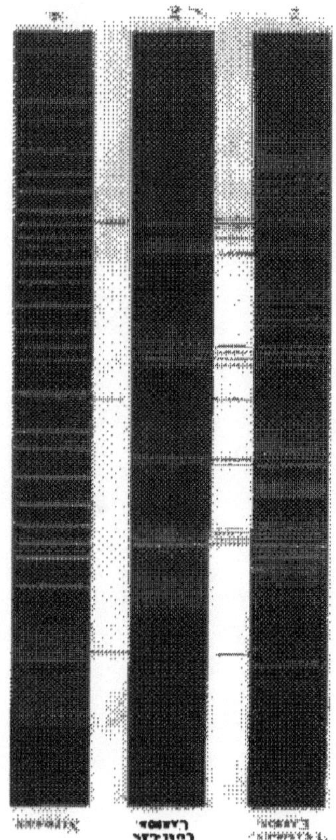

researches of three distinguished English men of science—Mr. Huggins, Dr. Miller, and Mr. Lockyer—to whom we are indebted for almost all our knowledge of celestial chemistry.

Although the moon and planets, shining by borrowed light, do not reveal to the spectroscope the nature of the material of which they are composed, like the sun and stars, yet something may be learned by an examination of the spectra of these bodies. You will remember that some of the dark lines in the solar spectrum are caused by absorption in our own atmosphere: now if an atmosphere of a similar kind exist round the moon or planets, the atmospheric absorption lines must appear more intense in the light reflected from these luminaries than they do in the light which passes through our air alone. With regard to the moon, the observations of Mr. Huggins and Dr. Miller have been negative. No signs of a lunar atmosphere presented themselves. A still more delicate means of ascertaining whether the moon possesses an atmosphere was employed by Mr. Huggins. On January 4th, 1865, he observed the spectrum of a star at the moment the dark edge of the moon passed over it. If an atmosphere existed in the moon, the observer would see the starlight by refraction after the occultation had occurred—just as the setting sun is visible to us after it has actually disappeared below the horizon. The variously coloured rays are, however, differently refrangible; and if any atmosphere existed round the moon, the red rays being least so would die out soonest, and the spectrum of the star would be seen progressively to diminish in intensity, beginning from the red end. Mr. Huggins observed nothing of this kind, all the rays of the stellar spectrum disappearing simultaneously: and the conclusion must be drawn that the moon is devoid of any appreciable atmosphere.

In the spectrum of Jupiter lines are seen which indicate the existence of an absorptive atmosphere about this planet. These lines plainly appeared when viewed simultaneously with the spectrum of the sky, which at the time of observation reflected the light of the setting sun. One strong band corresponds with some terrestrial atmospheric lines, and probably indicates the presence of vapours similar to those which float about the earth. Another band has no counterpart amongst the lines of absorption of our atmosphere, and tells us of some gas or vapour which does not exist in the earth's atmosphere. From observations upon Saturn it appears probable that aqueous vapour exists in the atmosphere of this planet, as well as in that of Jupiter. In Venus no intensifying of the atmospheric lines could be observed; but some remarkable groups of lines, corresponding to those seen when the sun is low, were noticed on the more refrangible side of the line "D," in the Mars spectrum; and these indicate the existence of matter similar to that occurring in our own atmosphere. The red colour which distinguishes this planet appears not to be caused by absorption in its atmosphere, as the light reflected from its polar regions is free from the ruddy tint peculiar to the other portions of the planet. Padre Secchi and M. Janssen have likewise made similar observations, and they also conclude that in all probability the vapour of water exists in the planetary atmospheres.

I must now pass on to the subject proper of this day's discourse, which is to consider the properties of the light from the fixed stars. The more we learn about this subject, the more I think we must be surprised at the accuracy of the observing powers of those philosophers who have given us this information. By means of this beautiful instrument (made by Mr. Browning, and a facsimile of the one used by Mr. Huggins, Fig. 78) we have been placed in possession of facts respecting the composition of the atmospheres, and the physical constitution of these stars, as accu-

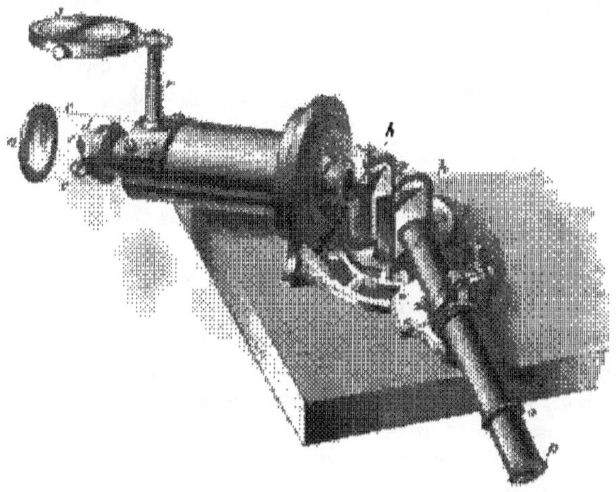

FIG. 78.

rate as the knowledge we possess concerning the composition of the solar atmosphere. It would be impossible for me to give you, even if time permitted, an accurate description of the method employed by Mr. Huggins. (See Appendix A.) Suffice it to say, that at the end of his telescope he has placed this spectroscope, containing two prisms (*h h*): and that, by very accurate adjustment, he is able to bring the image of the star on the slit of his spectroscope (d). You may imagine how difficult these observations are, when you remember that the light of the star emanates from a point,—that is to say, the star has no sensible magnitude; that the image of the star has to be kept steady upon a slit only the $\frac{1}{300}$ part of an inch in breadth; and, moreover, that the effect of the earth's motion has to be counteracted. When you add to this, that the amount of light which even the brightest stars give is excessively feeble, that this line of light must be still further weakened by being spread out by a cylindrical lens (a) into a band, and when you remember that in our climate on a few only of those nights in which the stars appear to the naked eye to shine brilliantly is the air steady enough to prevent the flickering and confusion of the spectra, fatal to these extremely delicate observations, I think you will easily understand how exceedingly difficult these researches must have been, and I am sure you will acknowledge the debt of gratitude which the world owes to those gentlemen who, by devoted labours, have brought the subject to this interesting issue.

In order to get a knowledge of the chemical composition of the stars, or to ascertain what chemical elements are present in them, it is necessary to use excessively delicate arrangements, by which not only the light from the star is allowed to pass through the prisms and to be received on the retina, but also that emitted by the various substances, the presence or absence of which in the stellar atmosphere it is desired to ascertain. These rays must pass together with the beam of starlight, or rather over or under the starlight, into the eyepiece, through the same prism, so that we may be able to compare the position of the dark lines in the stellar spectrum with that of the bright lines in the spectrum of the body under examination. For this purpose a very ingenious arrangement is attached to a part of the telescope-spectroscope. It consists of a moveable mirror (*f*), placed above the slit of the spectroscope, by means of which the light of the spark passing from the metallic poles, held between metal holders, is reflected by the small prism (*e*) placed on the slit into the optical

FIG. 79.

arrangement, and is received into the eye, the metal spectrum being ranged close above that derived from the star; so that the coincidence or otherwise of the two sets of lines can be accurately observed. In this way alone is it possible to arrive at any trustworthy conclusion respecting the composition of the stars, and the existence of certain metals in the stellar atmospheres. An improved and compact form of spark condenser, as manufactured by Mr. Browning, is shown in Fig. 79. It is of very simple construction, and may be employed either for burning the metals or for getting the spectra of gases. There are only two connexions needed, one at each end of the box; and thus the arrangements are much simplified.

The first result which we have to notice, then, is that the spectra of various stars differ very widely indeed from one another. As I mentioned to you, Fraunhofer in the year 1814 showed that the stellar spectra were not the same, and that they did not contain the same lines as the spectrum of the sun. I have here coloured drawings which will indicate to you, to begin with, the different nature of these stellar spectra: but these drawings do not pretend to give the exact positions of the various lines in the spectra, but only approximately to represent their general appearance. Here (see Nos. 1 and 2 on the Chromolith. Plate facing this lecture), for example, is a picture of the spectra of the two stars composing β Cygni, in each of which, as you see, the arrangement of the lines is totally different; and moreover the arrangement of the lines here is quite different from that of the lines in the solar spectrum.

Mr. Huggins specially describes the spectra of two particular stars, of which we have here an exact diagram (Fig. 80). The upper drawing represents the spectrum of Aldebaran, and the lower of Betelgeux, the star known as α in the constellation of Orion. This drawing is made on a similar plan to Kirchhoff's diagrams of the dark lines in the solar spectrum. The longer lines represent the dark bands in the stellar spectrum, the shorter ones beneath represent the bright lines of the metals with which the star spectrum was compared, the symbols of the elements thus examined being added. In the first place, then, the result at which we have arrived is that

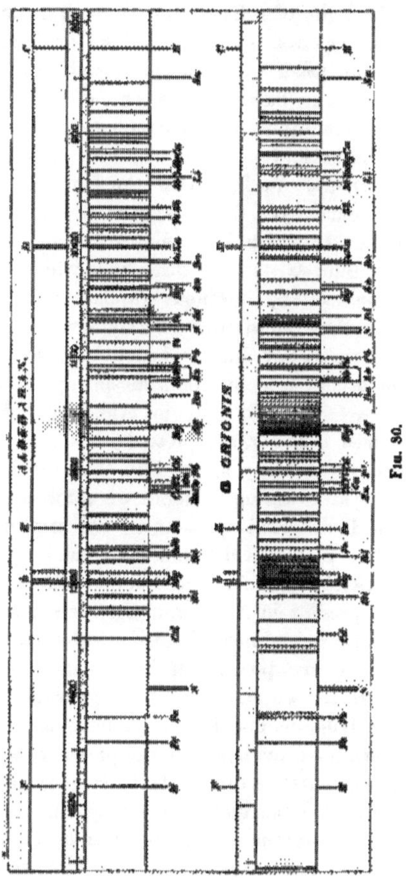

Fig. 30.

the constitution of the starlight, although not identical

with the light given off by the sun, is yet similar—that is to say, the light of a fixed star gives off a continuous spectrum, interspersed by dark shadows or bands; and hence the conclusion we come to is that the physical constitution of the fixed stars is similar to that of our sun, that their light also emanates from intensely white-hot matter, and passes through an atmosphere of absorbent vapours—in fact, that the stars are suns of different systems. We find, for instance, in these two particular stars to which I am now referring, the D line caused by sodium exists: the three lines which we know as *b* are produced by luminous vapour of magnesium. The lines of these substances exactly agree in position with the dark stellar lines; hence both sodium and magnesium are present in the atmosphere of these far distant stars. We also find in Aldebaran that two hydrogen lines, C and F, are present; but if we look at the spectrum of α Orionis, we find that the hydrogen lines C and F are wanting. Hence we come to the conclusion that hydrogen is present in the atmosphere of the sun and in that of Aldebaran, but that it is wanting in that of Betelgeux. And so I might show you that silver is not present in Aldebaran, nor seen in α Orionis, but that four bright lines of calcium, also seen in the sun's spectrum, are present in both stars. The lines observed in these two stars are at least seventy in number, and Mr. Huggins and Dr. Miller have found that in Aldebaran we have evidence of the presence of no less than nine elements: namely (1) hydrogen, giving the lines C and F; (2) the metal sodium, giving the double line D; (3) magnesium, giving the lines *b*; (4) calcium, giving four lines; (5) iron, giving four lines, and E; (6) bismuth, giving four lines (bismuth is not found in the sun); (7) tellurium, four lines; (8) antimony is also found, three lines; and (9) mercury, four lines. Thus the element tellurium, whose name implies a purely earthly origin, is found in the star, although it does not exist in the sun, and is very rare on this earth. There are only two stars—Betelgeux, to which I have just referred, and another star called β Pegasi—in which the hydrogen lines are wanting; all the other stars contain hydrogen.

We have, then, now arrived at a distinct understanding of the physical constitution of the fixed stars: they consist of a white-hot nucleus, giving off a continuous spectrum, surrounded by an incandescent atmosphere, containing the absorbent vapours of the particular metals. These results are interesting, as bearing on Laplace's nebular theory, because they show that the visible universe is mainly composed of the same elementary constituents, although certain of the stars differ from one another widely in their chemical constitution.

The next question to which the attention of the observers was directed was the different character of the light produced by the stars. It is well known that the stars are variously coloured: some shine with a bright white light, others with a yellow light, others with a blue light. Could spectrum analysis give any explanation of the variety of colours exhibited by these different stars? This is proved to be possible, as I shall show you by reference to some diagrams from Mr. Huggins' drawings, for most of what I have to say to-day will be the result of his investigations. Here we have, in the first place, the spectrum of a white star, of the star which we all well know as Sirius. In this coloured drawing (No. 3 on the Chromolith. facing the beginning of this lecture) we find a representation of what Mr. Huggins observed in the spectrum of Sirius: you will notice that we have a continuous spectrum with dark lines, and we find that these dark lines or shadows are interspersed pretty generally throughout the length of the spectrum, so that, when all the light enters the eye at once, it produces upon the retina the effect of white light.

We next take an orange-coloured star, known as a Herculis, which is a double star. Here (Chromolith. No. 4) we have a totally different spectrum, and the lines which are most marked in this spectrum exist in the green, blue, and deep red. The light is comparatively free from shadows in the yellow and orange portion; and hence, the light from the red, green, and blue portions of this star being weakened, the star shines with a yellow light. This, then, illustrates to us the explanation given by spectrum analysis of the cause of these differently coloured stars in the heavens. We have, however, yet to learn the nature of the substances which produce many of these dark bands in the stellar spectra, and cause the peculiar colour which the stars exhibit.

Another very interesting and well-known astronomical fact next attracts our attention, viz. the existence of certain twin or double stars. It appears that amongst these twin stars, which invariably differ in colour, the blue, green, and purple stars are faint telescopic stars, never found alone, but associated under the protection, as it were, of a brighter red or orange star. Does the same explanation which has been given of the variety of the differently coloured stars also apply to these double stars?

We have here (No. 2 on the Chromolith.) a diagram of the combined spectra of the two double stars existing in β Cygni—above, the orange star; below, the blue star. This one is orange because there are so many dark lines in the blue and red, whilst there are none at all in the orange portion of the spectrum. In the blue star, on the other hand, we have a vast number of very fine lines existing in the red and in the orange, and a much smaller number existing in the blue: hence the light of this star produces upon the retina the effect of blue light. Padre Secchi[1] observing under the clear skies of Rome has investigated the spectra of many hundred stars. He finds it possible to arrange all these stars in four groups, each characterised by a special form of spectrum. Group 1 contains the white stars, Sirius, α Lyræ, Vega, &c., whose spectra are especially characterised by four black lines, coincident with those of hydrogen. Group 2 consists of the yellow stars, having spectra intersected by numerous fine lines resembling those of our sun: in this group Secchi reckons Pollux, Capella, γ Aquilæ, and our sun. The third group contains the red and orange stars, α Orionis, α Herculis, β Pegasi, &c., the spectra of which are divided into eight or ten parallel columnar clusters of alternate dark and bright bands, increasing in intensity towards the red. Group 4 is made up of the small red stars, whose spectra are distinguished by a succession of three bright zones, increasing in intensity towards the violet. Out of 316 stars examined Secchi found that 164 belonged to the first type and 140 to the second, whilst the few remaining constituted the third and fourth classes.

A very interesting and remarkable observation was made in the month of May, 1866. All at once, in the constellation of the Northern Crown, a star which was

[1] Astronomische Nachrichten, Jan. 28, 1869.

entirely or almost entirely unknown, and which was at any rate a star of very small size, suddenly blazed out, and attained a magnitude almost equal to that of the largest stars seen in the heavens. The examination of the spectrum of this particular star naturally excited the liveliest interest, and Mr. Huggins and Dr. Miller were fortunate enough to be able to investigate at frequent intervals this very remarkable phenomenon by means of Mr. Huggins' spectroscope, and to their astonishment found that this star, of which I here show you a diagram (Chromolith. No. 6), differed altogether in its character from the ordinary stellar spectra, inasmuch as superposed on, or in addition to, the ordinary stellar spectrum which you see exhibited here (viz. one consisting of dark lines upon a bright ground), there were, in this particular star, bright lines. Now what do bright lines indicate? They indicate the presence of certain gaseous bodies; and the result of the examination of the position of these particular bright lines, which you see here, showed them to be coincident with the bright lines produced by hydrogen.

As this star made its appearance suddenly, so it soon gradually began to diminish in brilliancy, and at last died out, returning, as it were, to its original telescopic dimensions of about the tenth magnitude. How was this diminution of the brightness of the star to be explained? The cause of the diminution was revealed to us by the spectroscope, inasmuch as these bright lines were found to dwindle and fade away, and it was observed after a lapse of twelve days, when the star had diminished in brilliancy from the second to the eighth magnitude, that these bright lines became quite invisible. I had the good fortune to see through Mr. Huggins' telescope the very spectrum the drawing of which is now cast upon the screen. The lines when I happened to see them had, however, nearly faded away; but they were still visible. The conclusion to which we must come with regard to this violent outburst is that it was probably due to a rapid ignition of hydrogen, of a similar kind, though enormously larger than the sudden outbreaks of incandescent gases seen in the red prominences of our sun.[1]

An analogous increase of light has been observed in other stars;[2] and Padre Secchi, the Roman astronomer, has ascertained that several very small stars also exhibit bright bands, and therefore have a constitution similar to τ Coronæ: but he has not ascertained the accurate position of these lines; and it is therefore only in the case

[1] Mr. Baxendell's careful estimates of the varying brightness of this star (Manch. Proc. Nov. 27, 1866) led him to conclude that the intensity of its light on August 20th, when it reached its minimum, was only $\frac{1}{760}$th part of that emitted at its maximum on May 12th. From the recent observations of Lockyer and Janssen (see Lecture V.) we learn that the red prominences in the sun are also caused by glowing hydrogen, so that we have a new reason for believing that the sun may belong to the family of variable stars. The question at once suggests itself to the mind, Could a similar conflagration burst out in our system? Of the effects there can be no doubt. The intensity of the sun's rays being increased nearly eight hundredfold, our solid globe would be dissipated in vapour almost as soon as a drop of water in a furnace. The temperature in the sunlight would rise at once to that only attainable in the focus of the largest burning-glass, and all life on our planet would instantly cease. In thus speculating on such a possible termination to our terrestrial history, it must be well understood that the probability of such an event occurring is undoubtedly infinitely small, and that the researches of geologists do not lead us to suppose that any approach to such an occurrence has ever taken place in former geologic ages.

[2] See Appendix C. "On Variable Stars."

of the star examined by Mr. Huggins that we are really able to indicate the possible cause of the phenomenon.

From these observations you see that the stars possess chromospheres of ignited hydrogen, and you will not fail to draw the inference, already pointed out by astronomers on other grounds, that our sun belongs to the family of variable stars.

It is interesting to notice that the spectra of fixed stars contain, like the solar light, invisible chemically active rays. The spectrum of Sirius has been photographed by Mr. Huggins. The intensity of the light of this star is, according to the best measurements, the [illegible] part of that of the sun: and although probably not less in size than sixty of our suns, it is estimated to be at the enormous distance of more than 130,000,000,000,000 miles; and yet even this immense distance does not prevent us registering the chemical intensity of the rays which left Sirius twenty-one years ago (Miller): and Mr. Lockyer has recently shown that in the spectrum we have probably a means of determining the atmospheric pressure in the last layer of its chromosphere.

The next point to which Mr. Huggins directed his attention was the examination of those most interesting and singular astronomical phenomena, the nebulæ. The first nebula which Mr. Huggins examined with his spectroscope was one of that class of luminous bodies termed planetary nebulæ, in the constellation Draco. On the 20th August, 1864, Mr. Huggins turned his telescope on to this particular nebula. I am afraid I cannot give you any idea of the delicacy of such observations. Those, however, of my audience who have seen such a planetary nebula through a telescope will know that the light which those bodies give off is less than the light given off by perhaps even the smallest fixed star; and the difficulty of obtaining a spectrum and of examining the nature of this light is therefore exceedingly great.[1] What, however, was Mr. Huggins' astonishment, on bringing the image of this nebula on to the slit of his spectroscope, to observe that he no longer had to do with a class of bodies of the nature of stars!—that instead of having a band of light intersected by dark lines, indicating the physical constitution of the body to be that corresponding to the sun and stars, he found the light from

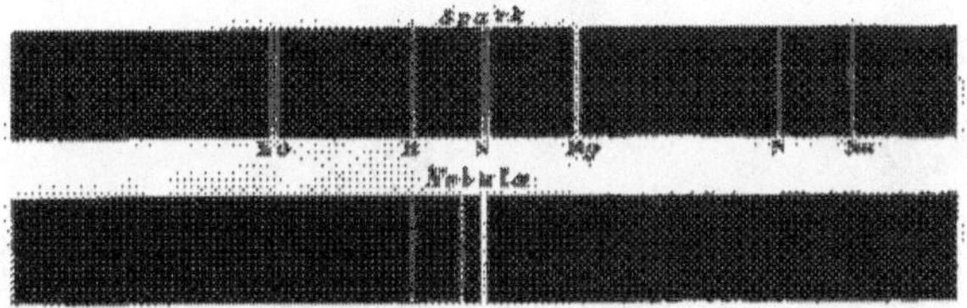

FIG. 81.

this nebula consisted simply of three isolated bright lines, of which we have here (Fig. 81, and in No. 7 of the Chromolith.) a very rough representation. If the spectrum of this nebula had been continuous, it would have been very difficult to see it. It was only because the light given off consisted of three bright lines that he was enabled to examine this spectrum at all. You will have already anticipated me in the conclusion that

[1] Mr. Huggins gives an idea of the extreme faintness of the more distant nebulæ. "The light of some of those visible in a moderately large instrument has been estimated to vary from $\frac{1}{1500}$ to $\frac{1}{20000}$ of the light of a single sperm candle consuming 158 grains of material per hour, viewed at a distance of a quarter of a mile; that is, such a candle a quarter of a mile off is 20,000 times more brilliant than the nebula!"

these most curious bodies do not consist of a white-hot nucleus, enveloped in an atmosphere, passing through which the light is absorbed, giving us dark lines; but, on the contrary, that these nebulæ are in the condition of luminous gases, and that it really is nebulous matter with which we have here to do.

The history of these nebulæ is one into which I cannot enter. You all know that the names of Herschel and of Rosse are associated with the most accurate and careful examination of these particular bodies, and that it is especially to the late Lord Rosse that we are indebted

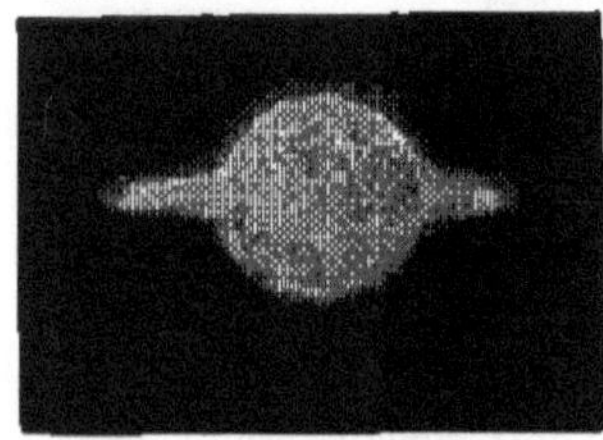

FIG. 82.

FIG. 83.

for the very careful examination, by means of his magnificent telescope, of these most singular bodies. It now became a matter of the very greatest interest to examine the character of the light given off by the other nebulæ. I will indicate to you the appearance of some of these nebulæ, though very roughly, by means of the drawings. The nebula in Aquarius is seen in Fig. 82. The drawing of this nebula gives you but a faint notion of its appearance in the telescope. I may also show you another nebula (Fig. 83) having a spiral form, and whose spectrum exhibits a fourth bright line. Mr. Huggins then found on examining the character of the lines which these nebulæ give off, that the spectrum was likewise distinguished by the same three distinct bright lines. The questions will occur to every one, Do all the nebulæ give similar spectra? and especially, Do those which the telescope had certainly resolved into a close aggregation of bright points give gaseous spectra?

Mr. Huggins has examined the spectra of about seventy nebulæ, and he finds that these can be divided into two great groups. One group (about one-third of the whole number) consists of the nebulæ giving spectra

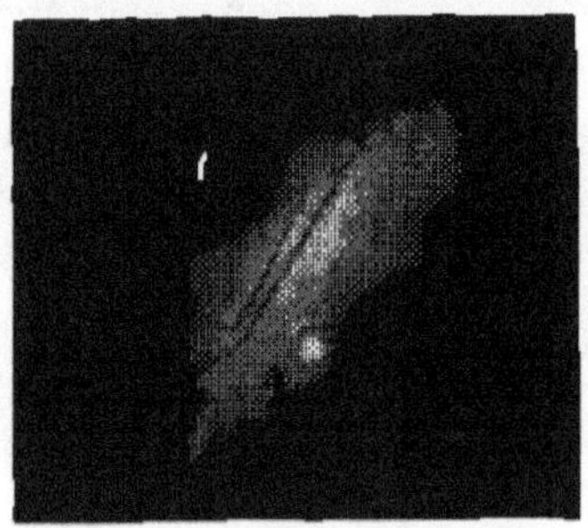

FIG. 84.

of three bright lines similar to those which I have shown you, or else containing only one or two of these bright lines. "Of these seventy nebulæ, about one-third belong to the class of gaseous bodies: the light of the remaining nebulæ and clusters becomes spread out by the prism into a spectrum which is apparently continuous." To the class of nebulæ giving continuous spectra the well-known nebula in Andromeda belongs. This singularly shaped body is visible to the naked eye (Fig. 84), and is not unfrequently mistaken for a comet. It was observed as early as the year 1612, by Simon Marius. The spectrum of this nebula, though apparently continuous, possesses some curious characteristics, the whole of the red and a portion of the orange being wanting, besides the brighter parts exhibiting an unequal and mottled appearance.

It next becomes a most important point to ascertain the chemical nature of the three bright lines in the spectra of the gaseous nebulæ. Mr. Huggins finds that the brightest of the lines of the nebula coincides with the

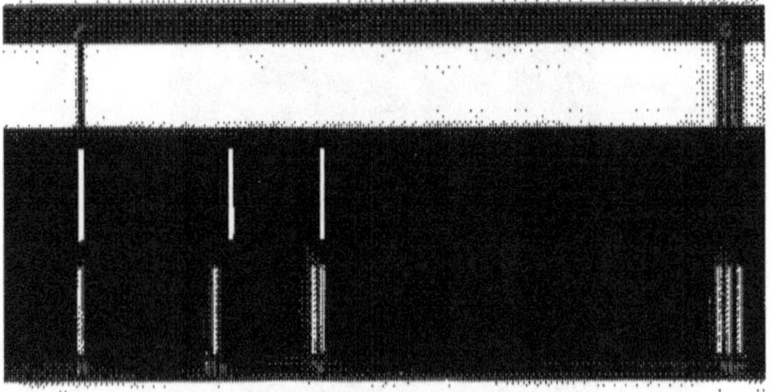

Fig. 85.

strongest of the lines which are peculiar to nitrogen, whilst the faintest of the lines was found to coincide with the green line (F) of hydrogen. The middle line of the three does not coincide with a line of any known element.

The upper part of this drawing (Fig. 85) is intended to represent a portion of the solar spectrum. Here you see the dark line F, due to hydrogen, and the lines formed by magnesium, corresponding with the letter *b*. Below are the lines corresponding with some of the bright lines of hydrogen, barium, nitrogen, and magnesium, whilst between them are the three lines observed in these nebulæ (Fig. 85). Now it may be asked, "How is it, if one of these three lines is due to hydrogen, and another to nitrogen, that the other well-known lines of these elements are not present in the spectra of the nebulæ? Can we come to the conclusion that nitrogen and hydrogen are contained in the nebulæ, when we only see two out of the many characteristic lines? Why do not the others appear?" With regard to this point, Mr. Huggins has recently shown that if the intensity of the light coming from glowing nitrogen be diminished to a certain point, only one line is seen, and if you diminish the intensity of the hydrogen spectrum, this one blue line (F) alone becomes visible.[1] We may therefore safely follow in Mr. Huggins' steps, and take all his conclusions as being the result not only of careful experimentation, but of philosophic caution, for in all these new and difficult subjects that is an absolute necessity. I think we may be well satisfied to adopt his decision, that in fact nitrogen and hydrogen do exist in the nebulæ, and that the cause of the non-appearance of the other lines is simply to be ascribed to the fact which I have already endeavoured to point out to you, that the light coming from these nebulæ is of such excessively slight intensity.

I am almost afraid to take up your time in exhibiting to you many of these diagrams; still I must not omit to show you one of the well-known nebula in the sword-handle of Orion (see Fig. 86), which was discovered by no less a personage than the astronomer Huyghens in 1656. I will read to you Sir John Herschel's description of this nebula. "The general aspect of the less luminous and

[1] This fact has since been observed by Padre Secchi, and Messrs. Lockyer and Frankland.

cirrous portion is simply nebulous and irresolvable; but the brighter portion immediately adjacent to the trapezium forming the square front of the head is shown with the eighteen-inch reflector broken up into masses, whose mottled and curdling light evidently indicates, by a sort of granular texture, its consisting of stars, and when examined under the great light of Lord Rosse's reflector, or the exquisite defining power of the great achromatic at Cambridge, U.S., is evidently perceived to consist of

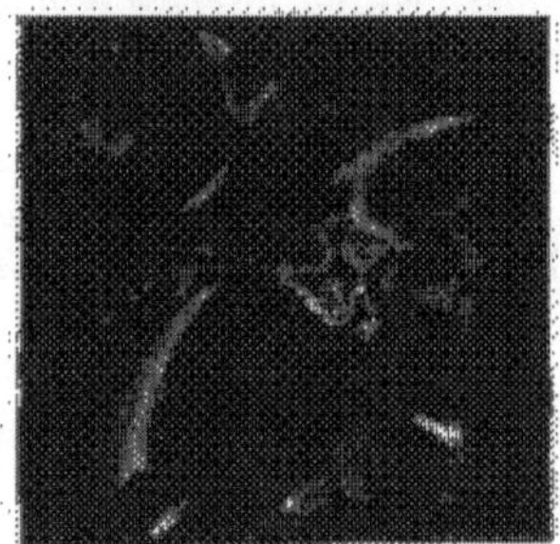

FIG. 80.

clustering stars. There can therefore be little doubt as to the whole consisting of stars too minute to be discerned individually, even with those powerful aids, but which become visible as points of light when closely adjacent in the more crowded parts."

It becomes a matter of the greatest interest to learn the conclusions to which the spectroscope leads us, concerning the nature of these resolvable portions of this nebula. Here you have Mr. Huggins' own words on this important subject. "The results of telescopic observation on this nebula seem to show that it is suitable for observation as a crucial test of the correctness of the usually received opinion, that the resolution of a nebula into bright stellar points is a certain and trustworthy indication that the nebula consists of discrete stars after the order of those which are bright to us. Would the brighter portions of the nebula adjacent to the trapezium, which have been resolved into stars, present the same spectrum as the fainter and outlying portions? In the brighter parts would the existence of closely aggregated stars be revealed to us by a continuous spectrum, in addition to that of the true gaseous matter?" The answer of the spectroscope comes to us in no doubtful tone. "The light from the brightest parts of the nebula near the trapezium was resolved by the prisms into three bright lines, in all respects similar to those of the gaseous nebula. . . . The whole of this great nebula, as far as lies within the power of my instrument, emits light which is identical in its characters; the light from one part differs from the light of another in intensity alone."

The conclusion is obvious, that the close association of points of light in a nebula can no longer be accepted as proof that the object consists of true stars. These luminous points, in some nebulæ at least, must be regarded as portions of matter, denser probably than the outlying parts of the great nebulous mass, but still gaseous. Another point of interest here presents itself with regard to the opinions entertained of the enormous distances of the nebulæ, founded upon the remoteness at which these supposed star-clusters must exist, as they cannot be resolved into stars by the most powerful telescopes. Such opinions it is clear cannot now be upheld, at least with respect to those nebulæ which have been proved to be gaseous.

Carrying on his observations still further, to well-known distant clusters of stars, a representation of which I will throw upon the screen, Mr. Huggins has found that even some of those which were supposed to be well-authenticated masses of stars do not really consist of stars, for the light given off by these clusters is also identical in character with the light given off by the true nebulæ. Hence we must be careful in drawing our conclusions respecting the existence of these bodies as groups of far-distant suns, because we find that the light which some of them give out is not the kind of light which such far-distant fixed stars must emit.

The true nature of comets is involved in even greater obscurity than that of the nebulæ. Mr. Huggins has examined some of these singular bodies, and the result of his observations leaves us even in greater uncertainty respecting their character. A small comet which made its appearance in 1866 and 1867 was first examined: the spectrum of this was a faint continuous one, on which *bright lines* were visible. Brorsen's comet was next examined: it is a recurring comet, having a period of rotation of 5½ years, and its spectrum consisted of three bright bands, the central one of which lay between F and *b*, and in addition a very faint continuous spectrum was seen. These observations settled the physical character of the comet: it consists of a mass of glowing gas, and is self-luminous, a portion only feebly reflecting the sunlight. From the drawing (Fig. 87) it will be seen that the light from Brorsen's comet differs from the light emitted by the nebulæ, inasmuch as the lines in the comet spectrum are not identical in position with the lines yielded by the nebulæ. Nor, in fact, are these particular lines—roughly represented here—identical in

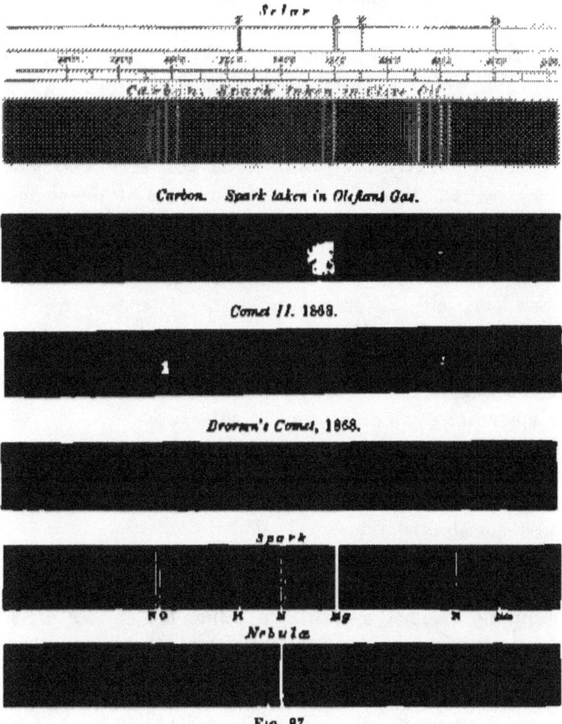

FIG. 87.

position with the bands of any known substance. (See Fig. 87.) At the bottom of the diagram the lines of the nebulæ are given; next we have the spark spectra of nitrogen, hydrogen, magnesium, and sodium; at the top we see the solar lines; and between these lie the particular lines of the comet: hence this comet contains something not found in the nebulæ, whose lines do not coincide with any substance known on earth, so far as examination has yet proceeded. So that we really do not yet know of what this comet consists. These observations, I am sure you will all admit, open out to us subjects of the deepest interest.

We are entirely at a loss to know how such a body as the comet can be self-luminous: the mass of the comet, I believe, is astronomically speaking inappreciable. We do not know whether there is as much matter in this comet as would fill this room, or as much as would fill one's hat; and this amount of matter is spread over an enormous space. The diameter of this comet has been determined for me by Mr. Baxendell, who tells me that it is about 60,700 miles—an immense space over which to spread so small an amount of substance.

How matter in this attenuated form can be kept up at the high temperature necessary for the gas to become incandescent is a subject on which we cannot at present even speculate.

Mr. Huggins has also published an account of his observations on the second comet of 1868, known as Winnecke's comet, a drawing of which is seen in Fig. 88. Spectroscopic investigation has shown that this comet contains luminous carbon, or carbon compounds: its spectrum, together with that of Brorsen's comet, is seen in Fig. 87. This gives that modification of the carbon spectrum which we obtain when the electric spark is passed through olefiant gas, the coincidence of the bright lines of Winnecke's comet with those of the spark taken in olefiant gas being clearly seen. In order to obtain an exact comparison of the lines of the comet with those of incandescent olefiant gas, the arrangement shown in Fig. 89 was employed. This consists of a glass bottle, *a*,

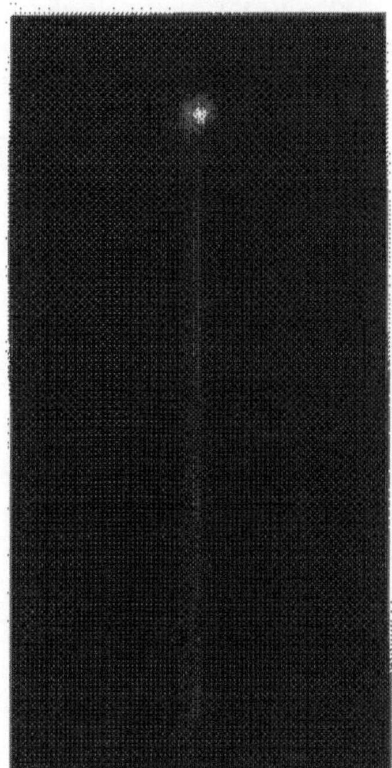

FIG. 89.

converted into a gasholder containing the olefiant gas. This was connected with the glass tube, *b*, through which the gas passed, and into which two platinum wires, *e* and *f*, had been soldered. This tube being then placed before the mirror of the spectroscope *c*, the light of the spark, passing through the gas in the tube by means of the wires, was reflected into the instrument *d*, and its spectrum was seen immediately below that of the comet. The two sets of bands were not only found to agree precisely in position, but they corresponded in their general characters, and in their relative brightness. Hence we can

FIG. 89.

scarcely doubt that carbon is really the cause of these bright lines; but whether the carbon is present in the free state in the comet, or in the state of combination, cannot be as yet definitively decided: nor can we explain how carbon can be transformed into the gaseous state, or heated so as to become luminous, unless indeed it be present in the form of a hydrocarbon which becomes ignited or enters into combination with some other constituent of the comet by the action of the sun's heat. That carbon is an element widely distributed throughout the universe we learn from the fact that it has been detected in considerable quantities in the extra-terrestrial matter of meteoric stones. This observation gains a further interest, as it appears probable that the orbits of many comets are identical with the paths of the recurring meteors. Hence an intimate connexion probably exists between the comets and falling stars, so that meteorites may perhaps consist of condensed cometary matter.

I have still to speak of another result of these interesting experiments of Mr. Huggins. Not only are we in a position thus to determine the constitution of the stars and of the nebulæ, but, strange as it may appear, we can actually, by these observations, get some ideas respecting the relative motions of these bodies and our earth. It is impossible in the time at my disposal to explain to you the mode in which philosophers or physicists have arrived at the conclusion, originally propounded by Doppler in 1841, that, when a luminous body is approaching another very rapidly, the kind of light which is received on the retina from that body moving at a very great speed differs in some respects from the light which the retina would receive were that body at rest. An illustration from sound may perhaps render this matter more plain. If in a railway train you listen to the whistle of the engine of another train, which is meeting you, you will notice that as the two trains approach the pitch of the note of the whistle alters.[1] This is because (owing to the sound being produced by

[1] An exact experiment of this kind was made in 1845 by Ballot of Utrecht, in which the alteration of tone for a given velocity was determined.

the vibration of the particles of the air), when the two trains are approaching each other, the waves of sound are, as it were, forced together and fall more rapidly upon the ear than they would do if the two trains were in a state of rest. The same thing happens with regard to light. If the one object which is luminous is approaching the retina very rapidly, the vibrations causing light will fall more frequently on the retina than if the bodies were at rest; and then the position of the dark lines will be shifted in the direction of the most refrangible rays; whilst, if the bodies were separating, the shifting would

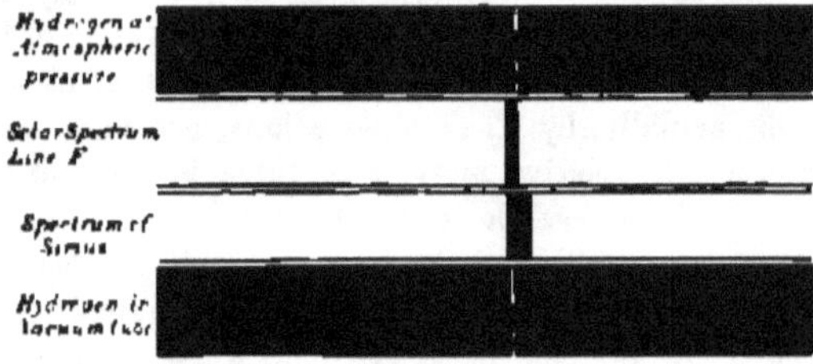

FIG. 90.

take place in the direction of the red or least refrangible rays. Mr. Huggins has actually found that in some of these stars there is a slight disturbance in the position of the hydrogen line F: he first most beautifully proved that it is really hydrogen which is present, and then he showed that there is a slight deviation observed between the hydrogen line and the line existing in the star; and hence he comes to the conclusion that there is motion of recession between the earth and star. Here you see a diagram (Fig. 90) showing the slight deviation which the line F exhibits in Sirius light. You see that the narrow line of hydrogen from the vacuum-tube does not coincide with the middle of the Sirius F line, but crosses it at a distance from the middle, which may be represented by saying that the want of coincidence is *apparently* equal to about one-third or one-fourth of the interval separating the two D lines. Hence the F line has been distinctly deviated towards the red rays, or the vibrations proceeding from the star must have been retarded in their passage; or, in other words, there is a motion of recession between our earth and Sirius of such a nature that the wave-length of the F ray has been increased by the 0·109 millionth part of a millimetre. The velocity of this recession can easily be calculated. Light travels at the rate of 185,000 miles per second; the wave-length of the F line is 486·5 millionths of a millimetre: now the velocity with which the two bodies move away from each other stands to the velocity of light in the same proportion as the observed difference of wave-length does to the wave-length of the particular ray; or—

$$486{\cdot}5 : 0{\cdot}109 :: 185000 : x = 41{\cdot}4.$$

Hence we conclude that this motion of recession between the earth and the star Sirius is 41·4 miles per second, or that if the earth were stationary, instead of moving in its orbit, as it did at the time of the experiment, away from Sirius, there would be a proper motion of recession of the star of twenty-nine miles per second.

Astronomers have, as you are aware, long ago shown by telescopic observation that the whole of our solar system is moving in space in the direction of the constellation Hercules. Such observations can, however, only be made when the motion to be noticed is at right angles to the line of vision, whilst the deductions which I have just described as being made with the spectroscope can only be arrived at when the motion is in the direction of the line of vision. We see, therefore, that by a happy combination of these two methods we are enabled to ascertain the actual rate of motion of the stars in space. Before we can, however, attain an exact knowledge of these motions, the delicacy of our spectroscopes must be much improved. This has already been partially accomplished by Zöllner in his reversing spectroscope, by means of which a displacement equal to the $\frac{1}{100}$ part of the distance between the D lines can be seen. With this instrument Zöllner expects not only to be able to measure the proper motion of the stars, but he hopes even to render visible the shifting in the Fraunhofer's lines which must occur from the rotation of the sun on its axis, and thus to distinguish between those which are really solar lines and those caused by absorption in our own atmosphere, which naturally cannot exhibit a shifting in position from this cause.

In the last lecture we learnt that the solar atmosphere, and especially the outlying portion, chiefly consisting of hydrogen gas, is constantly most violently agitated by storms of white-hot hydrogen, which blow with such fierceness that, compared with these, our most destructive tornadoes are mere summer breezes. We saw also that these storms give rise to the peculiar appearances which we observe on the solar surface, viz. the sun-spots and the faculæ. Lockyer has determined the velocity with which the glowing hydrogen rushes forwards or backwards in these storms by observing the peculiar alterations in breadth and position which the F line exhibits. These alterations in refrangibility can only be seen when the motion is one of approach or of recession towards the observer's eye; and hence upon the sun's disc we can only observe an upward or downward motion, whilst in the chromosphere or on the sun's limb we may obtain evidence of lateral or cyclonic movements, although we cannot there see any upward or downward currents. Now, in looking at the sun's surface through his spectroscope, Lockyer saw the line F sometimes appear as a *dark* line, and sometimes flash out as a *bright* line on a darker ground. In the former case the F line is seen to be bent and in several places shifted towards the red end of the spectrum, whilst at other points a displacement towards the violet end is noticed. Sometimes the line disappears altogether, and before it vanishes it is seen to swell or bulge out, or to terminate in a knotty bulbous form. Then it again becomes invisible as the slit passes over the faculæ, especially when they are near some small spots, and again is seen to swell out several times in the course of a few seconds, then to shift towards the violet, and lastly to appear as a straight bright line, without any thickening, over a small spot.

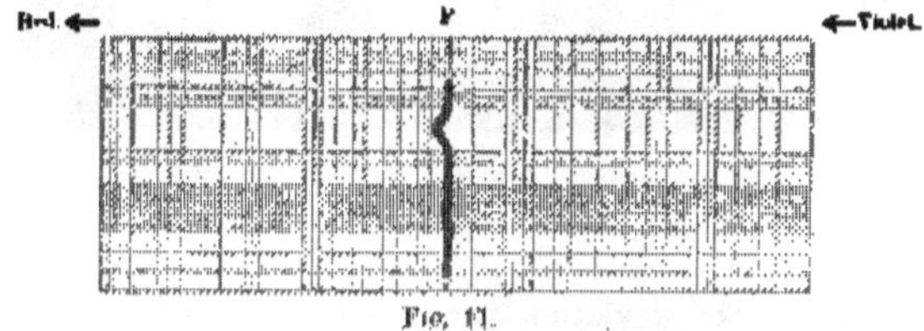

Fig. 91.
Deviation of the F line in a Spot-Spectrum (Lockyer).

These variations in the position and darkness or luminosity of the F line are attempted to be represented in Fig. 91, giving the appearance of a spot-spectrum: the dark horizontal lines show that a general absorption occurs where the spots are darkest. In the neighbourhood of one of these spots a facula is situated; and there the F line is seen to be bright and bulbous.

The other Fraunhofer's lines are, as you see, represented by vertical lines in the drawing (Fig. 91). These serve, according to Lockyer, as milestones[1] by which to measure the velocity, upwards or downwards, with which these eruptive masses of hydrogen are moving, for these shiftings and twistings of the F line are of course nothing else than rapid alterations in the wave-length of the F rays.

If, therefore, a shifting towards the violet is observed equal to the $\frac{1}{1000000}$ part of a millimetre (see No. 1, Fig. 92), this shows that the incandescent hydrogen is rushing upwards with a velocity of thirty-eight miles per second;

FIG. 92.

whilst a like deviation towards the red, as in No. 3 in the figure, proves that a downward current is blowing with an equal velocity. Sometimes the bright line is seen at the violet side of its normal position, whilst a

[1] Still alterations of wave-length have also been detected in the sodium, iron, and magnesium lines in a spot-spectrum.

dark line is found shifted towards the red: this proves that an *upward* rush of intensely-heated incandescent hydrogen occurs on the one side, and a *downward* rush of cool absorbing gas on the other.

The lateral motions near the limb are observed by means of the shifting of the bright lines in the spectrum of the chromosphere. The velocity of these cyclones is almost incredible. Lockyer observed such a circular storm on the 14th March, 1869 (No. 2, Fig. 92). The slit of his spectroscope was about $\frac{1}{500}$ of an inch in width; and as the diameter of the image of the sun cast by his telescope was only 0·94 of an inch, he could observe a strip on the sun's surface some 1,800 miles wide.

If we now assume that the solar circular storm spreads over a width of 1,500 miles and is moving at a fearful rate, we must be able by help of the spectroscope to distinguish between those portions which move towards us and those portions which are moving in the opposite direction. The drawing (Fig. 92) shows that this is indeed the case. When the slit was placed upon the centre of this cyclone, the bright line which, in fact, is a continuation of the dark F line, was seen to be shifted to a distance which corresponds to a velocity of forty miles per second. When the slit was directed towards the edge of the storm, it was clear that on the one side the current was moving towards us, and on the other side away from us, because the deviation in the first case was towards the violet, and in the second towards the red end of the spectrum. The great rapidity with which the prominences appear and disappear shows that the hydrogen gas of which these flames are composed is often in a state of the most violent eruption. From the observation of the F line in a prominence on May 12, 1869, the following conclusions were drawn: (1) a portion of this flame was in a position of rest as regards us, as is seen from the fact that near the photosphere the bright F line was exactly coincident with the ordinary dark line F of the solar spectrum; (2) the bright continuation of this line being displaced in a slanting direction towards the violet end of the spectrum shows that a portion of the hydrogen was moving towards the earth, but with different velocities, the upper layers moving more rapidly than the lower portions. Lockyer on this occasion saw the F line triple, and the greatest lateral displacement of the line corresponded to the almost incredible velocity of 120 miles per second!

These spectroscopic results become, if possible, still more interesting when they are combined with telescopic observations. On April 21, 1869, Lockyer was observing a spot near the sun's limb, and at 7h. 30m. A.M. a protuberance in full activity was seen in the field of view. The hydrogen lines were very bright, and, as the spectrum of the spot was visible at the same time, it was easy to see that the red flame was moving more quickly than the spot, and that the prominence was fed, so to speak, from the preceding edge of the spot. The violent eruption had torn up a portion of the upper layer of the photosphere beyond the usual limits of the chromosphere, and high up in the hydrogen flame floated a cloud of magnesium vapour! At 8h. 30m. there was comparative quiet, but after an hour had elapsed the action had commenced afresh, and the existence of a violent cyclone was plainly seen. On the same day, and nearly at the same time, a photograph of the sun was taken at Kew, in which the violent changes observed by Lockyer could be clearly traced, the limb of the sun being torn away exactly at the place where the spectroscope had shown a cyclone was situated.

Who could have dreamt ten years ago that we should so soon attain such an insight into the processes of creation? And yet, great though the results of spectrum analysis already are, they are but a tithe of the numerous questions which this branch of discovery has opened up, —questions of such number and magnitude, that many generations of men will pass away before they are all satisfactorily answered.

In conclusion, I may say that I feel it would be idle in me to attempt to add any words as to the importance and grandeur of the subjects which in these lectures I have so imperfectly brought before you. I leave the facts to speak for themselves, and I have only to thank you most heartily for the kind attention with which you have honoured me.

LECTURE VI.—APPENDIX A.

EXTRACT FROM A MEMOIR ON THE SPECTRA OF SOME OF THE FIXED STARS.[1]

BY W. HUGGINS, F.R.A.S., AND W. MILLER, M.D., LL.D., TREAS. AND V.P.R.S.[2]

§ I. *Introduction.*

1. The recent discovery by Kirchhoff of the connexion between the dark lines of the solar spectrum and the bright lines of terrestrial flames, so remarkable for the wide range of its application, has placed in the hands of the experimentalist a method of analysis which is not rendered less certain by the distance of the objects the light of which is to be subjected to examination. The great success of this method of analysis as applied by Kirchhoff to the determination of the nature of some of the constituents of the sun rendered it obvious that it would be an investigation of the highest interest, in its relation to our knowledge of the general plan and structure of the visible universe, to endeavour to apply this new method of analysis to the light which reaches the earth from the fixed stars. Hitherto the knowledge possessed by man of these immensely distant bodies has been almost confined to the fact that some of them, which observation shows to be united in systems, are composed of matter subjected to the same laws of gravitation as those which rule the members of the solar system. To this may be added the high probability that they must be self-luminous bodies, analogous to our sun, and probably in some cases even transcending it in brilliancy. Were they not self-luminous, it would

[1] Phil. Trans. 1864. [2] Professor of Chemistry, King's College, London.

be impossible for their light to reach us from the enormous distance at which the absence of sensible parallax in the case of most of them shows they must be placed from our system.

The investigation of the nature of the fixed stars by a prismatic analysis of the light which comes to us from them, however, is surrounded with no ordinary difficulties. The light of the bright stars, even when concentrated by an object-glass or speculum, is found to become feeble when subjected to the large amount of dispersion which is necessary to give certainty and value to the comparison of the dark lines of the stellar spectra with the bright lines of terrestrial matter. Another difficulty, greater because it is in its effect upon observation more injurious, and is altogether beyond the control of the experimentalist, presents itself in the ever-changing want of homogeneity of the earth's atmosphere through which the stellar light has to pass. This source of difficulty presses very heavily upon observers who have to work in a climate so unfavourable in this respect as our own. On any but the finest nights the numerous and closely approximated fine lines of the stellar spectra are seen so fitfully that no observations of value can be made. It is from this cause especially that we have found the inquiry, in which for more than two years and a quarter we have been engaged, more than usually toilsome; and indeed it has demanded a sacrifice of time very great when compared with the amount of information which we have been enabled to obtain.

2. Previously to January 1862, in which month we commenced these experiments, no results of any investigation undertaken with a similar purpose had been published. With other objects in view, two observers had described the spectra of a few of the brighter stars, viz. Fraunhofer in 1823,[1] and Donati, whose memoir, "Intorno alle Strie degli Spettri stellari," was published in the "Annali del Museo Fiorentino" for 1862.

Fraunhofer recognised the solar lines D, E, b, and F in the spectra of the moon, Venus, and Mars: he also found the line D in Capella, Betelgeux, Procyon, and Pollux;—in the two former he also mentions the presence of b. Castor and Sirius exhibited

[1] Gilbert's Annalen, vol. lxxiv. p. 374.

other lines. Donati's elaborate paper contains observations upon fifteen stars; but in no case has he given the positions of more than three or four bars, and the positions which he ascribes to the lines of the different spectra relatively to the solar spectrum do not accord with the results obtained either by Fraunhofer or by ourselves. As might have been anticipated from his well-known accuracy, we have not found any error in the positions of the lines indicated by Fraunhofer.

3. Early in 1862 we had succeeded in arranging a form of apparatus in which a few of the stronger lines in some of the brighter stars could be seen. The remeasuring of those already described by Fraunhofer and Donati, and even the determining the positions of a few similar lines in other stars, however, would have been of little value for our special object, which was to ascertain, if possible, the constituent elements of the different stars. We therefore devoted considerable time and attention to the perfecting of an apparatus which should possess sufficient dispersive and defining power to resolve such lines as D and *b* of the solar spectrum. Such an instrument would bring out the finer lines of the spectra of the stars, if in this respect they resembled the sun. It was necessary for our purpose that the apparatus should further be adapted to give accurate measures of the lines which should be observed, and that it should also be so constructed as to permit the spectra of the chemical elements to be observed in the instrument simultaneously with the spectra of the stars. In addition to this, it was needful that these two spectra should occupy such a position relatively to each other as to enable the observer to determine with certainty the coincidence or non-coincidence of the bright lines of the elements with the dark lines in the light from the star.

Before the end of the year 1862 we had succeeded in constructing an apparatus which fulfilled part of these conditions. With this some of the lines of the spectra of Aldebaran, α Orionis, and Sirius were measured; and from these measures, diagrams of these stars, in greater detail than had then been published, were laid before the Royal Society in February 1863. After the note was sent to the Society, we became acquainted with some similar observations on several other stars by Rutherfurd, in *Silliman's Journal* for 1863.[1] About the same time figures of a few stellar spectra were also published by Secchi.[2] In March 1863 the Astronomer Royal presented a diagram to the Royal Astronomical Society, in which are shown the positions of a few lines in sixteen stars.[3]

Since the date at which our note was sent to the Royal Society our apparatus has been much improved, and in its present form of construction it fulfils satisfactorily several of the conditions required.

§ II. *Description of the Apparatus, and Methods of Observation employed.*

4. This specially constructed spectrum apparatus is attached to the eye end of a refracting telescope of 8 inches aperture and 10 feet focal length, which is mounted equatorially in the observatory of Mr. Huggins at Upper Tulse Hill. The object-glass is a very fine one, by Alvan Clark of Cambridge, Massachusetts; the equatorial mounting is by Cooke of York; and the telescope is carried very smoothly by a clock motion.

As the linear spectrum of the point of light which a star forms at the focus of the object-glass is too narrow for the observation of the dark lines, it becomes necessary to spread out the image of the star; and to prevent loss of light, it is of importance that this enlargement should be in one direction only; so that the whole of the light received by the object-glass should be concentrated into a fine line of light as narrow as possible, and having a length not greater than will correspond to the breadth of the spectrum (when viewed in the apparatus), just sufficient to enable the eye to distinguish with ease the dark lines by which it may be crossed. No arrangement tried by us has been found more suitable to effect this enlargement in one direction than a cylindrical lens, which was first employed for

[1] Vol. xxxv. p. 71.
[2] Astronomische Nachrichten, No. 1405, March 3, 1863.
[3] Monthly Notices, Roy. Astron. Soc. vol. xxiii. p. 190.

this purpose by Fraunhofer. In the apparatus by which the spectra described in our "Note" of February 1863 were observed the cylindrical lens employed was plano-convex, of 0·5 inch focal length. This was placed within the focus of the object-glass, and immediately in front of the slit of the collimator.

The present form of the apparatus is represented in Fig. 78 (page 271), where the cylindrical lens is marked *a*. This is plano-convex, an inch square, and of about 14 inches focal length. The lens is mounted in an inner tube *b*, sliding within the tube *c*, by which the apparatus is adapted to the eye end of the telescope. The axial direction of the cylindrical surface is placed at *right angles* to the slit *d*, and the distance of the lens from the slit within the converging pencils from the object-glass is such as to give exactly the necessary breadth to the spectrum.

In consequence of the object-glass being over-corrected, the red and especially the violet pencils are less spread out than the pencils of intermediate refrangibility; so that the spectrum, instead of having a uniform breadth, becomes slightly narrower at the red end, and tapers off in a greater degree towards the more refrangible extremity.[1]

In front of the slit *d*, and over one-half of it, is placed a right-angled prism *e*, for the purpose of reflecting the light which it receives from the mirror *f*, through the slit. In the brass tube *c* are two holes: by one of these the light is allowed to pass from the mirror to the reflecting prism *e*; and by means of the other access to the milled head for regulating the width of the slit is permitted. Behind the slit, and at a distance equal

[1] The experiment was made of so placing the cylindrical lens that the axial direction of its convex cylindrical surface should be parallel with the direction of the slit. The line of light is in this case formed by the lens; and the length of this line, corresponding to the visible breadth of the spectrum, is equal to the diameter of the cone of rays from the object-glass when they fall upon the slit. With this arrangement, the spectrum appears to be spread out, in place of being contracted at the two extremities. Owing to the large amount of dispersion to which the light is subject, it was judged unadvisable to weaken still further the already feeble illumination of the extremities of the spectrum; and in the examination of the stellar spectra the position of the cylindrical lens with its axis at right angles to the slit, as mentioned in the text, was therefore adopted.

to its focal length, is placed an achromatic collimating lens, *g*, made by T. Ross: this has a diameter of 0·6 inch and a focal length of 4½ inches. These proportions are such that the lens receives the whole of the light which diverges from the linear image of the star, when this is brought exactly within the jaws of the slit.

A *plano-concave* cylindrical lens of about 14 inches negative focal length was also tried. The slight advantage which this possesses over the convex form is more than balanced by the inconvenience of the increased length given to the whole apparatus. The dispersing portion of the apparatus consists of two prisms, *h*, each having a refracting angle of about 60°: they were made by T. Ross, and are of very dense and homogeneous flint glass. The prisms are supported upon a suitable mounting, which permits them to be duly levelled and adjusted. Since the feebleness of the light from the stars limits the observations for the most part to the central and more luminous portions of the spectrum, the prisms have been adjusted to the angle of minimum deviation for the ray D. A cover of brass, *k*, encloses this part of the apparatus; and by this means the prisms are protected from accidental displacement and from dust.

The spectrum is viewed through a small achromatic telescope, *l*, furnished with an object-glass of 0·8 inch diameter and 6·75 inches focal length. This telescope has an adjustment for level at *m*. The axis of the telescope can be lowered and raised, and the tube can be also rotated around the vertical axis of support at *n*. At the focus of the object-glass are fixed two wires, crossing at an angle of 90°. These are viewed, together with the spectrum, by a positive eyepiece, *p*, giving a magnifying power of 5·7 diameters. As the eyes of the two observers do not possess the same focal distance, a spectacle lens, corresponding to the focal difference between the two, was fitted into a brass tube, which slipped easily over the eyepiece of the telescope, and was used or withdrawn as was necessary.

This telescope, when properly adjusted and clamped, is carried by a micrometer screw, *q*, which was constructed and fitted to the instrument by Cooke and Sons. The centre of motion about which it is carried is placed approximatively at the point of intersection of the red and the violet pencils from the last prism: consequently it falls within the last face of the prism nearest the small telescope. All the pencils therefore which emerge from the prism are, by the motion of the telescope, caused to fall nearly centrically upon its object-glass. The micrometer screw has 50 threads to an inch; and each revolution is read to the hundredth part, by the divisions engraved upon the head. This gives a scale of about 1,800 parts to the interval between the lines A and H of the solar spectrum. During the whole of the observations the same part of the screw has been used; and the measures being relative, the inequalities, if any, in the thread of this part of the screw do not affect the accuracy of the results. The eye lens for reading the divisions of the micrometer screw is shown at *s*.

The mirror *f* receives the light to be compared with that of the star spectrum, and reflects it upon the prism *e*, in front of the slit *d*. This light was usually obtained from the induction spark taken between electrodes of different metals, fragments of wires of which were held by a pair of small forceps attached to the insulating ebonite clamp *r*. Upon a moveable stand in the observatory was placed the induction coil, already described by one of us,[1] in the secondary circuit of which was inserted a Leyden jar, having 140 square inches of tinfoil upon each of its surfaces. The exciting battery, which, for the convenience of being always available, consisted of four cells of Smee's construction, with plates 6 inches by 3, was placed without the observatory. Wires, in connexion with this and the coil, were so arranged that the observer could make and break contact at pleasure without removing his eye from the small telescope. This was the more important, since, by tilting the mirror *f*, it is possible, within narrow limits, to alter the position of the spectrum of the metal relatively to that of the star. An arrangement is thus obtained which enables the observer to be assured of the perfect correspondence in relative position in the

[1] Phil. Trans. 1864, p. 141.

instrument of the stellar spectrum and the spectrum to be compared with it.

5. The satisfactory performance of this apparatus is proved by the very considerable dispersion and admirably sharp definition of the known lines in the spectra of the sun and metallic vapours. When it is directed to the sun, the line D is sufficiently divided to permit the line within it, marked in Kirchhoff's map as coincident with nickel, to be seen. The close groups of the metallic spectra are also well resolved.

When this improved apparatus was directed to the stars, a large number of fine lines was observed, in addition to those that had been previously seen. In the spectra of all the brighter stars which we have examined the dark lines appear to be as fine and as numerous as they are in the solar spectrum. The great breadth in the lines in the green and more refrangible parts of Sirius and some other stars, as seen in the less perfect form of apparatus which was first employed, and which band-like appearance was so marked as specially to distinguish them, has, to a very great extent, disappeared; and though these lines are still strong, they now appear, as compared with the strongest of the solar lines, by no means so abnormally broad as to require these stars to be placed in a class apart. No stars sufficiently bright to give a spectrum have been observed to be without lines. The stars admit of no such broad distinctness of classification. Star differs from star alone in the grouping and arrangement of the numerous fine lines by which their spectra are crossed.

6. For the convenience of reference and comparison, a few of the more characteristic lines of twenty-nine of the elements were measured with the instrument. These were laid down to scale, in order to serve as a chart, for the purpose of suggesting, by a comparison with the lines measured in the star, those elements the coincidence of the lines of which with stellar lines was probable.

For the purpose of ensuring perfect accuracy in relative position in the instrument between the star spectrum and the spectrum to be observed simultaneously with it, the following general method of observing was adopted:—The flame of a small lamp of alcohol, saturated with chloride of sodium, was placed centrally before the object-glass of the telescope, so as to furnish a sodium spectrum. The sodium spectrum was then obtained by the induction spark, and the mirror *f* was so adjusted that the components of the doubled line D, which is well divided in the instrument, should be severally coincident in the two spectra. The lamp was then removed, and the telescope directed to the sun, when Fraunhofer's line D was satisfactorily observed to coincide perfectly with that of sodium in the induction spark. Having thus ascertained that the sodium lines coincided in the instrument with the solar lines D, it was of importance to have assurance from experiment that the other parts of the solar spectrum would also accurately agree in position with those corresponding to them in the spectrum of comparison. When electrodes of magnesium were employed, the components of the triple group characteristic of this metal severally coincided with the corresponding lines of the group *b*, C and F also agreed exactly in position with the lines of hydrogen. The stronger of the Fraunhofer lines were measured in the spectra of the moon and of Venus, and these measures were found to be accordant with those of the same lines taken in the solar spectrum.

Before commencing the examination of the spectrum of a star, the alcohol-lamp was again placed before the object-glass of the telescope, and the correct adjustment of the apparatus obtained with certainty. The first observation was whether the star contained a double line coincident with the sodium line D. When the presence of such a line had been satisfactorily determined, we considered it sufficient in subsequent observations of the same star to commence by ascertaining the exact agreement in position of this known stellar line with the sodium line D.

Since from flexure of the parts of the spectrum apparatus the absolute reading of the micrometer might vary when the telescope was directed to stars differing greatly in altitude, the measure of the line in the star which was known to be coincident with that of sodium was always taken at the commencement and at the end of each set of measures. The distances of the other lines from this line, and not the readings of the micrometer, were then finally registered as the measures of their position; and these form the numbers given in the Tables, from which the diagrams of the star spectra have been laid down.

The very close approximation, not unfrequently the identity, of the measures obtained for the same line on different occasions, as well as the very exact agreement of the lines laid down from these measures with the stellar lines subsequently determined by a direct comparison with metallic lines, the positions of which are known, have given the authors great confidence in the minute accuracy of the numbers and drawings which they have now the honour of laying before the Society.

APPENDIX B.

ON THE SPECTRUM OF MARS, WITH SOME REMARKS ON THE COLOUR OF THAT PLANET.

BY WILLIAM HUGGINS, ESQ., F.R.S.

On several occasions during the late opposition of Mars I made observations of the spectrum of the solar light reflected from that planet.

The spectroscope which I employed was the same as that of which a description has appeared in my former papers.[1] Two

[1] "On the Spectra of some of the Fixed Stars." (Phil. Trans. 1864, p. 415.) During my prismatic researches I have tried, and used occasionally, several other arrangements for applying the prism to the telescope. Some of these instruments are fitted with compound prisms, which give direct vision. I have not found any apparatus equal in delicacy and in accuracy to that which is referred to in the text.

instruments were used, one of which is furnished with a single prism of dense glass, which has a refracting angle of 60°. The other instrument has two similar prisms.

In a paper "On the Spectra of some of the Fixed Stars," by myself and Dr. W. A. Miller, we state that on one occasion several strong lines of absorption were seen in the more refrangible parts of the spectrum of Mars.

During the recent more favourable opportunities of viewing Mars I again saw groups of lines in the blue and indigo parts of the spectrum. However, the faintness of this portion of the spectrum, when the slit was made sufficiently narrow for the distinct observation of the lines of Fraunhofer, did not permit me to measure with accuracy the position of the lines which I saw. For this reason I was unable to determine whether these lines are those which occur in this part of the solar spectrum, or whether any of them are new lines due to an absorption which the light suffers by reflection from the planet.

I have confirmed our former observation, that several strong lines exist in the red portion of the spectrum. Fraunhofer's C was distinctly seen, and its identity determined by satisfactory measures with the micrometer of the spectrum apparatus. From this line the spectrum, as far as it can be traced towards the less refrangible end, is crossed by dark lines. One strong line was satisfactorily determined by the micrometer to be situated from C at about one-fourth of the distance from C to B. As a similar line is not found in this position in the solar spectrum, the line in the spectrum of Mars may be accepted as an indication of absorption by the planet, and probably by the atmosphere which surrounds it. The other lines in the red may be identical, at least in part, with B and A, and the adjacent lines, of the solar spectrum.

On February 14 faint lines were seen on both sides of Fraunhofer's D. The lines on the more refrangible side of D were stronger than the less refrangible lines. These lines occupy positions in the spectrum apparently coincident with groups of lines which make their appearance when the sun's light traverses the lower strata of the atmosphere, and which are therefore supposed to be produced by the absorption of gases or vapours existing in our atmosphere. The lines in the spectrum of Mars probably indicate the existence of similar matter in the planet's atmosphere. I suspected that these lines were most distinct in the light from the margin in the planet's disc; but this observation was to some extent uncertain. That these lines were not produced by the portion of the earth's atmosphere through which the light of Mars had passed was shown by the absence of similar lines in the spectrum of the moon, which at the time of observation had a smaller altitude than Mars.

I observed also the spectra of the darker portions of the planet's disc. The spectrum of the dark zone beneath the southern polar spot appeared as a dusky band when compared with the spectra of the adjoining brighter parts of the planet. This fainter spectrum appeared to possess a uniform depth of shade throughout its length. This observation would indicate that the material which forms the darker parts of the planet's surface absorbs all the rays of the spectrum equally. These portions should be therefore neutral, or nearly so, in colour.

I do not now regard the ruddy colour of Mars to be due to an *elective* absorption; that is, an absorption of certain rays only so as to produce dark lines in the spectrum.

Further, it does not appear to be probable that the ruddy tint which distinguishes *Mars* has its origin in the planet's atmosphere, for the light reflected from the polar regions is free from colour, though this light has traversed a longer column of atmosphere than the light from the central parts of the disc. It is in the central parts of the disc that the colour is most marked. If indeed the colour be produced by the planet's atmosphere, it must be referred to peculiar conditions of it which exist only in connexion with particular portions of the planetary surface. The evidence we possess at present appears to support the opinion that the planet's distinctive colour has its origin in the material of which some parts of its surface are composed. Mr. Lockyer's observation, that the colour is most intense when the planet's atmosphere is free from clouds, obviously admits of an interpretation in accordance with this view.

This opinion appears to receive support from the photometric observations of Seidel and Zöllner, some of the results of which I will briefly state.

These observations show that Mars resembles the moon in the anomalous amount of variation of the light reflected from it as it increases and decreases in phase; also in the greater brilliancy of the marginal portions of its disc. Further, Zöllner has found that the albedo of Mars, that is, the mean reflective power of the different parts of its disc, is not more than about one-half greater than that of the lunar surface. Now these optical characters are in accordance with telescopic observation, that in the case of Mars the light is reflected almost entirely from the true surface of the planet. Jupiter and Saturn, the light from which has evidently come from an envelope of clouds, are, on the contrary, less bright at the margin than at the central part of the disc. These planets have an albedo, severally, about four and three times greater than that of the moon.[1]

The anomalous degradation in the brightness of the moon at the phases on either side of the full, as well as the greater brilliancy of the limb, may be accounted for by the supposition of inequalities on its surface, and also by a partly regular reflective property of its superficial rocks. Zöllner has shown that if these phenomena be assumed empirically to be due to inequalities, then the angle of mean elevation of these inequalities must be taken as 52°. On the same hypothesis the more rapid changes of Mars would require an angle of 76°.[2]

It appears to be highly probable that the conditions of surface which give rise to these phenomena are common to the moon and to Mars. The considerations referred to in a former paragraph suggest that these superficial conditions represent peculiarities which exist at the true surface of the planet. In this connexion it is of importance to remark that the

[1] Photometrische Untersuchungen, von Dr. J. C. Zollner; Leipsig, 1865.
[2] Ibid. pp. 118, 128.

darker parts of the disc of Mars gradually disappear, and the coloured portions lose their distinctive ruddy tint as they approach the limb.

The observations of Sir John Herschel[1] and Professor G. Bond[2] give a mean reflective power to the moon's surface, similar to that from a "grey, weathered sandstone rock." Zöllner has confirmed this statement. According to him,—

The albedo of the	Moon	= ·1736	of the incident light.	
,,	Mars	= ·2672	,,	,,
,,	Jupiter	= ·6238	,,	,,
,,	Saturn	= ·4981	,,	,,
,,	White Paper	= ·700	,,	,,
,,	White Sandstone	= ·237	,,	,,

From this table it appears that Mars takes in for its own use ·7328 of the energy which it receives as light. Jupiter's cloudy atmosphere, nearly as brilliant as white paper, rejects more than six-tenths of the light which falls upon it. Therefore, less than four-tenths of the light which this distant planet receives is alone available for the purposes of its economy.

The photographic researches of Mr. De la Rue and others show that the rays of high refrangibility, which are specially powerful in producing chemical action, are similarly affected.[3] At present we know nothing of the reflective power of the planets for those rays of slower vibration which we call heat.

[1] Outlines of Astronomy, p. 272.

[2] "On the Light of the Moon and Jupiter," Memoirs Amer. Academy, vol. iii. p. 212. In the same Memoir Prof. G. Bond estimates the albedo of Jupiter to be greater than unity. This estimate would require the admission that Jupiter shines in part by native light. (Ibid. p. 284.)

[3] Prof. G. Bond states that "the Moon, if the constitution of its surface resembled that of Jupiter, would photograph in *one-fourteenth* of the time it actually requires." (Ibid. p. 223.)

APPENDIX C.

ON THE OCCURRENCE OF BRIGHT LINES IN STELLAR SPECTRA, AND ON THE SPECTRA OF VARIABLE STARS.

"The spectrum of γ Cassiopeiæ appears to be in some respects at least analogous to that of τ Coronæ. In addition to the bright line near the boundary of the green and blue observed by Father Secchi, there is a line of equal brilliancy in the red, and some dark lines of absorption. The two bright lines are narrow and defined, but not very brilliant. Micrometrical measures made by Mr. Huggins of these lines show that they are doubtless coincident in position with Fraunhofer's C and F, and with two of the bright lines of luminous hydrogen. In these stars part of the light must be emitted by gas intensely heated, though not necessarily in a state of combustion. The nearly uniform light of γ Cassiopeiæ suggests that the luminous hydrogen of this star forms a normal part of its photosphere."—*Notices, Royal Astronomical Society,* vol. xxvii. p. 131.

"*Mira Ceti,* which gives a spectrum apparently identical, or nearly so, with α Orionis, was examined when at its maximum brilliancy, and on several subsequent occasions, after it had commenced its downward course. At the time the star was waning in brightness there was thought to be an appearance of greater intensity in several of the groups, but a continued series of observations is desirable before any opinion is hazarded as to the cause of the variation in brightness which has procured for this object the title of 'Wonderful.' At Mr. Baxendell's request the variable ρ Coronæ was examined when at its maximum, but without any successful result. . . . Mr. Huggins has confirmed the observation of MM. Wolf and Raget so far as to the presence of bright lines in the three small stars described by them. He has not determined the positions of these lines."—Ibid. vol. xxviii. p. 87.

APPENDIX D.

FURTHER OBSERVATIONS ON THE SPECTRA OF SOME OF THE STARS AND NEBULÆ, WITH AN ATTEMPT TO DETERMINE THEREFROM WHETHER THESE BODIES ARE MOVING TOWARDS OR FROM THE EARTH; ALSO OBSERVATIONS ON THE SPECTRA OF THE SUN AND OF COMET II. 1868.[1]

BY WILLIAM HUGGINS, ESQ. F.R.S.

§ I. *Introduction.*

In a paper "On the Spectra of some of the Fixed Stars,"[2] by myself and Dr. W. A. Miller, Treas. R.S., we gave an account of the method by which we had succeeded during the years 1862 and 1863 in making trustworthy simultaneous comparisons of the bright lines of terrestrial substances with the dark lines in the spectra of some of the fixed stars. We were at the time fully aware that these direct comparisons were not only of value for the more immediate purpose for which they had been undertaken, namely, to obtain information of the chemical constitution of the investing atmospheres of the stars, but that they might also possibly serve to tell us something of the motions of the stars relatively to our system. If the stars were moving towards or from the earth, their motion, compounded with the earth's motion, would alter to an observer on the earth the refrangibility of the light emitted by them, and consequently the lines of terrestrial substances would no longer coincide in position in the spectrum with the dark lines produced by the absorption of the vapours of the same substances existing in the stars.

The apparatus employed by us was furnished with two prisms of dense flint glass, each with a refracting angle of 60°, and permitted the comparisons to be made with so much accuracy

[1] Phil. Trans. 1868, p. 529. [2] Ibid. 1864, p. 413.

that the displacement of a line, or of a group of lines, to an amount smaller even than the interval which separates the components of Fraunhofer's D, would have been easily detected. We were therefore in possession of the information that none of the stars, the lines in the spectra of which we had compared with sufficient care, were moving in the direction of the visual ray with a velocity so great, relatively to that of light, as to shift a line through an interval corresponding to a difference of wave-length equal to that which separates the components of D. To produce an alteration of refrangibility of this amount a velocity of about 169 miles per second would be required. The following stars, with some others, were observed with the requisite accuracy:—Aldebaran, α Orionis, β Pegasi, Sirius, α Lyræ, Capella, Arcturus, Pollux, Castor.

It appeared premature at the time to refer to these negative results, as it did not seem to be probable that the stars were moving with velocities sufficiently great to cause a change of refrangibility which could be detected with our instrument. The insufficiency of our apparatus for this very delicate investigation does not, however, diminish the trustworthiness of the results we obtained respecting the chemical constitution of the stars, as the evidence for the existence or otherwise of a terrestrial substance was made to rest upon the coincidence, or want of coincidence, in general character as well as position of *several lines*, and not upon that of a single line.

According to the undulatory theory, light is propagated with equal velocity in all directions, whether the luminous body be at rest or in motion. The change of refrangibility is therefore to be looked for from the diminished or increased distance the light would have to traverse if the luminous object and the observer had a rapid motion towards or from each other. The great relative velocity of light to the known planetary velocities and to the probable motions of the few stars of which the parallax is known, showed that any alterations of position which might be expected from this cause in the lines of the stellar spectra would not exceed a fraction of the interval between the double line D, for that part of the spectrum.

I have devoted much time to the construction and trial of various forms of apparatus with which I hoped to accomplish the detection of so small an amount of change of refrangibility. The difficulties of this investigation I have found to be very great, and it is only after some years that I have succeeded in obtaining a few results which I hope will be acceptable to the Royal Society.

The subject of the influence of the motions of the heavenly bodies on the index of refraction of light had already, at the time of the publication of our paper in 1864, occupied the attention of Mr. J. C. Maxwell, F.R.S., who had made some experiments in an analogous direction. In the spring of last year, at my request, Mr. Maxwell sent to me a statement of his views and of the experiments which he had made. I have his permission to enrich this communication with the clear statement of the subject which is contained in his letter, dated June 10, 1867. . . . .

In 1841 Doppler showed that since the impression which is received by the eye or the ear does not depend upon the intrinsic strength and period of the waves of light and of sound, but is determined by the interval of time in which they fall upon the organ of the observer, it follows that the colour and intensity of an impression of light, and the pitch and strength of a sound, will be altered by a motion of the source of the light or of the sound, or by a motion of the observer, towards or from each other.[1]

Doppler endeavoured by this consideration to account for the remarkable differences of colour which some of the binary stars present, and for some other phenomena of the heavenly bodies. That Doppler was not correct in making this application of his theory is obvious from the consideration that, even if a star could be conceived to be moving with a velocity sufficient to alter its colour sensibly to the eye, still no change of colour would be perceived, for the reason that beyond the visible spectrum, at both extremities, there exists a store of invisible waves which would be at the same time exalted or degraded into visibility to take the place of the waves which had been raised or lowered in refrangibility by the star's motion. No change of colour,

[1] "Ueber das farbige Licht der Doppelsterne und einiger anderer Gestirne des Himmels." (Bohm. Gesell. Abh. ii. 1841-42, S. 465.)

therefore, could take place until the whole of those invisible waves of force had been expended, which would only be the case when the relative motion of the source of light and the observer was several times greater than that of light.

In 1845 Ballot published a series of acoustic experiments which support Doppler's theory in the case of sound. In the same paper Ballot advances several objections to Doppler's application of his theory to the colours of the stars.[1]

This paper was followed by several papers by Doppler in reply to the objections which were brought against his conclusions.[2]

In 1847 two memoirs were published by Sestini on the colours of the stars in connexion with Doppler's theory.[3]

More recently, in 1866, Klinkerfues[4] published a memoir on the influence of the motion of a source of light upon the refrangibility of its rays, and described therein a series of observations from which he deduces certain amounts of motion, in the case of some of the objects observed by him.

The method employed by Klinkerfues has been critically discussed by Dr. Sohncke.[5]

It may be sufficient to state that, as Klinkerfues employs an achromatic prism, it does not seem possible, by his method of observing, to obtain any information of the motion of the stars; for in such a prism the difference of period of the luminous waves would be as far as possible annulled. It is, however, conceivable that his observations of the light when travelling from E. to W., and from W. to E., might show a difference in the two cases, arising from the earth's motion through the ether.

[1] "Akustische Versuche auf der Niederländischen Eisenbahn nebst gelegentlichen Bemerkungen zur Theorie des Hrn. Prof. Doppler;" Pogg. Ann. B. lxvi. S. 321.

[2] See Pogg. Ann. B. lxxxi. S. 270, and B. lxxxvi. S. 371.

[3] "Memoria sopra i Colori delle Stelle del Catalogo de Baily osservati dal P. Band Sestini." Roma, 1847.

[4] "Fernere Mittheilungen über den Einfluss der Bewegung der Lichtquelle auf die Brechbarkeit eines Strahls." Von W. Klinkerfues, Nachr. K. G. der W. zu Göttingen, No. 4, S. 33.

[5] "Ueber den Einfluss der Bewegung der Lichtquelle auf die Brechung, kritische Bemerkungen zu der Entdeckung des Hrn. Prof. Klinkerfues." Von Hrn. Dr. Sohncke, Astron. Nachr. No. 1646.

Father Secchi has quite recently called attention to this subject.[1] In his paper he states that he has not been able to detect any change of refrangibility in the case of certain stars, of an amount equal to the difference between the components of the double line D. These results are in accordance with those obtained by myself and Dr. Miller in 1868, so far as they refer to the stars which had been examined by us . . . .

§ II. *Description of Apparatus.*

All the experiments were made with my refractor by Alvan Clark, of 8 inches aperture and 10 feet focal length, which is mounted equatorially, and carried very smoothly by a clock motion. As even on nights of unusual steadiness the lines in the spectra of the stars are necessarily, for several reasons, more difficult of minute discrimination of position than are those of the solar spectrum, it is important that the apparatus employed should give an ample amount of dispersion relatively to the degree of minuteness of observation which it is proposed to attempt.

In 1866 I constructed a spectroscope for the special objects of research described in this paper, which was furnished with three prisms of 60° of very dense flint glass. The solar lines were seen with great distinctness. I found, however, that, in order to obtain a separation of the lines sufficient for my purpose, an eyepiece magnifying ten or twelve diameters was necessary. Under these circumstances the stellar lines were not seen in the continued steady manner which is necessary for the trustworthy determination of the minute differences of position which were to be observed. After devoting to these observations the most favourable nights which occurred during a period of some months, I found that if success was to be obtained it would probably be with an apparatus in which a larger number of prisms and a smaller magnifying power were employed.

The inconvenience arising from the pencils, after passing through the prisms, crossing those from the collimator, when

[1] *Comptes Rendus*, 2 Mars, 1868, p. 398.

more than three or four prisms are employed, and also, in part, the circumstance that I had in my possession two very fine direct-vision prisms on Amici's principle, which had been made for me by Hofmann of Paris, induced me to attempt to combine in one instrument several simple prisms with one or two compound prisms which give direct vision. An instrument constructed in this way, as will be seen from the following description, possesses several not unimportant advantages.[1]

a is an adjustable slit; *b* an achromatic collimating lens of 4·5 inches focal length; *c* represents the small telescope with which the spectrum is viewed. The train of prisms consists of two compound prisms, *d* and *e*, and three simple prisms, *f*, *g*, *h*.

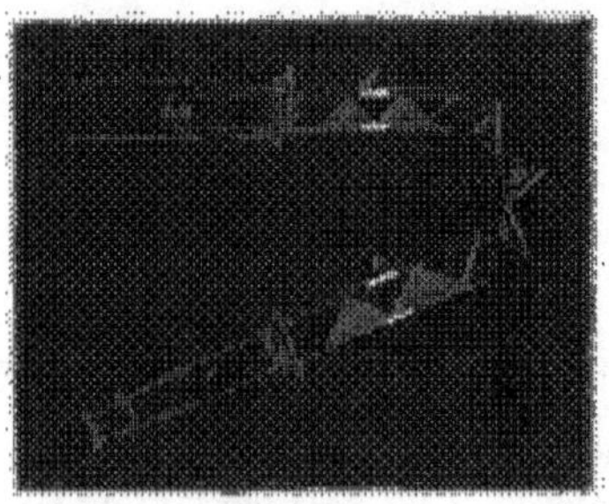

FIG. 93.

Each of the compound prisms contains five prisms, cemented together with Canada balsam. The shaded portions of the diagram represent the position of the two prisms of very dense flint glass in each compound prism. The compound prism marked *e* is much larger than the other, and is permanently connected with the telescope *c*, with which it moves. These compound prisms, which were made specially to my order by Hofmann, are of great perfection, and produced severally a dispersion fully equal to two prisms of ordinary dense flint

[1] An apparatus in many respects superior to the one here described has been constructed since.—October 1868.

glass. The prisms *f* and *g* were cut for me from a very fine piece of dense glass of Guinand by Messrs. Simms, and have each a refracting angle of 60°. The prism *h* was made by Mr. Browning from the dense flint glass manufactured by Messrs. Chance: this prism has a refracting angle of 45°. The great excellence of all these prisms is shown by the very great sharpness of definition of the bright lines of the metals when the induction spark is taken before the slit, even when considerable magnifying power is employed on the small telescope with which the spectrum is viewed. The instrument is provided with a second collimator, of which the object-glass has a focal length of 18 inches.

The compound prism *e* is so fixed that it can be removed at pleasure, when the total dispersive power of the instrument is reduced from about six and a half prisms of 60° to about four and a half prisms of 60°. The facility of being able to reduce the power of the instrument has been found to be of much service for the observation of faint objects, and also on nights when the state of the atmosphere was not very favourable.

The telescope with which the spectrum is viewed is carried by a micrometer screw, which, however, has not been employed for taking measures of the spectra, but only for the purpose of setting the telescope to the part of the spectrum which it is intended to observe. This precaution is absolutely necessary when nebulæ are observed which emit light of two or three refrangibilities only.

For the purpose of the simultaneous comparisons of the light of the heavenly bodies with the lines of the terrestrial elements, the slit was provided, in the usual way, with a small prism placed over one half of it, which received the light reflected upon it from a small mirror placed opposite the electrodes. The plan of observation formerly employed, and which is described in the paper "On the Spectra of some of the Fixed Stars," was adopted to ensure perfect accuracy of relative position in the instrument between the star spectrum and the spectrum to be compared with it, since it is possible, by tilting the mirror, to alter within narrow limits the position of the spectrum of the terrestrial substance relatively to that of the star. Before commencing an observation, a small alcohol-lamp, in the wick of which bicarbonate of soda was placed, was fixed before the object-glass of the telescope, and then the mirror and the electrodes were so adjusted that the components of the double line D were exactly coincident in both spectra.

This plan was soon found to be very inconvenient, and even in some degree untrustworthy for the more delicate comparisons which were now attempted. An unobserved accidental displacement of the spark, or of the mirror, might cause the two spectra to differ in position by an amount equal to the whole extent of want of coincidence which it was proposed to seek for in this investigation. The observations of many nights have been rejected, from the uncertainty as to the possible existence of an accidental displacement.

Another inconvenience, so great as even to seem to diminish the hope of ultimate success, was found to arise from the difficulty of bringing the lower margin of the star spectrum into actual contact with the upper margin of the spectrum of the light reflected into the instrument. The lines in the spectra of the stars are not, on ordinary nights, so steady and distinct as are those of the solar spectrum. Under these difficult circumstances, it is very desirable, as an assistance to the eye in its judgment of the absolute identity or otherwise of the position of lines, that the bright lines of comparison should not merely meet the dark lines in the star spectrum, but that they should overlap them to a small extent. When the two spectra are so arranged as to be in contact, the eye is found to be influenced to some extent by the apparent straightness or otherwise of the compound line formed by the coincident, or nearly coincident, lines in the two spectra. Owing to the unavoidable shortness of the collimator, the lines in a broad spectrum are slightly curved. From this cause the determination of the identity of lines in spectra which are in contact merely is rendered more difficult, and it may be less trustworthy.

The difficulties of observation which have been referred to

were in the first instance sought to be overcome by placing the spark before the object-glass of the telescope. In some respects this method appears to be unexceptionable, but there are disadvantages connected with it. The bright lines, under these circumstances, extend across the star spectrum, and make the simultaneous observation of dark lines, which are coincident, or nearly so, with them, very difficult. When the spark is taken between electrodes, the consequent disturbance of the air in front of the object-glass is unfavourable to good definition. An important disadvantage arises from the great diminution in the brightness of the spark from the distance (10 feet) at which it is placed from the slit; since in consequence of its nearness to the object-glass the divergence of the light from it is diminished in a small degree only by that lens. It is obvious that, by means of a lens of short focal length placed between the spark and the object-glass, the light from the spark might be rendered parallel, or even convergent; but the adjustments of such a lens, so that the pencils transmitted by it should coincide accurately in direction with the optical axis of the telescope, would be very troublesome. When two Leyden jars, connected as one jar, were interposed, and the spark was taken in air between platinum points, there was visible in the spectroscope only the brightest of the lines of the air spectrum, namely, the double line belonging to nitrogen, which corresponds to the principal line in the spectra of the gaseous nebulæ. When a vacuum-tube containing hydrogen at a low tension was placed before the object-glass, the line corresponding to F was seen with sufficient distinctness, but the line in the red was visible with difficulty. Some observations, however, have been made with the spark arranged before the object-glass.

The following arrangement for admitting the light from the spark appeared to me to be free from the objections which have been referred to, and to be in all respects adapted to meet the requirements of the case. In place of the small prism, two pieces of silvered glass were securely fixed before the slit at an angle of 45°. In a direction at right angles to that of the slit an opening of about 1/50 inch was left between the pieces of glass for the passage of the pencils from the object-glass. By means of this arrangement the spectrum of a star is seen accompanied by two spectra of comparison, one appearing above and the other below it. As the reflecting surfaces are about 0·5 inch from the slit, and the rays from the spark are divergent, the light reflected from the pieces of glass will have encroached upon the pencils from the object-glass by the time they reach the slit, and the upper and lower spectra of comparison will appear to overlap to a small extent the spectrum formed by the light from the object-glass. This condition of things is of great assistance to the eye in forming a judgment as to the absolute coincidence or otherwise of lines. For the purpose of avoiding some inconveniences which would arise from glass of the ordinary thickness, pieces of the thin glass used for the covers of microscopic objects were carefully selected, and these were silvered by floating them upon the surface of a silvering solution. In order to ensure that the induction spark should always preserve the same position relatively to the mirror, a piece of sheet gutta-percha was fixed above the silvered glass; in the plate of gutta-percha, at the proper place, a small hole was made of about $\frac{1}{16}$ inch in diameter. The ebonite clamp containing the electrodes is so fixed as to permit the point of separation of these to be adjusted exactly over the small hole in the gutta-percha. The adjustment of the parts of the apparatus was made by closing the end of the adapting tube, by which the apparatus is attached to the telescope, with a diaphragm with a small central hole, before which a spirit-lamp was placed. When the lines from the induction spark, in the two spectra of comparison, were seen to overlap exactly, for a short distance, the lines of sodium from the light of the lamp, the adjustment was considered perfect. The accuracy of adjustment has been confirmed by the exact coincidence of the three lines of magnesium with the component lines of *b* in the spectrum of the moon.

In some cases the spectra produced by the spark are inconveniently bright for comparison with those of the stars and nebulæ. If the spark is reduced in power below a certain point, many of the lines are not then well developed. The plan, therefore, was adopted of diminishing the brightness of the spectrum by a wedge of neutral-tint glass, which can be moved at pleasure between the plate of gutta-percha and the silvered mirror.

Two eyepieces were employed with the apparatus; the one magnifying four diameters, and the other six diameters. . . . .

§ III. *Observations of Nebulæ.*

For the greater convenience of reference and of comparison, the spectrum of 37 II. IV. Draconis from my paper "On the Spectra of some of the Nebulæ"[1] has been added (Fig. 85, p. 266). The spectrum of this nebula may be taken as characteristic, in its general features, of the spectra of all the nebulæ which do not give a continuous spectrum. At present I have determined satisfactorily the general characters of the spectra of about seventy nebulæ. This number forms but a part of the much larger list of nebulæ which I have examined, but in the case of many of these objects their light was found to be too feeble for a satisfactory analysis. Of the seventy nebulæ about one-third give a spectrum of bright lines. The proportion which is indicated by this examination, of the nebulæ which give a spectrum of bright lines to those of which the spectrum is continuous (namely, as one to two), is probably higher than would result from a wider observation of the objects contained in such catalogues as those of Sir John Herschel and Dr. D'Arrest, since many of the objects which I examined were specially selected, on account of the probability (which was suggested by their form or colour) that they were gaseous in constitution.

All the differences which I have hitherto observed between the spectra of the gaseous nebulæ may be regarded as modifications only of the typical form of spectrum which is represented in the diagram, since they consist of differences of relative intensity, of the deficiency of one or two lines, or of the presence of one or two additional lines. It is worthy of remark that, so far as the nebulæ have been examined, the brightest of the three

[1] Phil. Trans. 1864, p. 438.

lines, which agrees in position in the spectrum with the brightest of the lines of the spectrum of nitrogen, is present in *all* the nebulæ which give a spectrum indicative of gaseity. It is a suggestive fact that should not be overlooked, that in no nebula which has a spectrum of bright lines has any additional line been observed on the less refrangible and brighter side of the line common to all the gaseous nebulæ.

The faint continuous spectrum, which in some cases is also seen, has been traced in certain nebulæ, by its breadth, to a distinct brighter portion of the nebula which it is convenient still to distinguish by the term "nucleus," though at present we know nothing of the true relation of the bright points of the nebulæ to the more diffused surrounding portions.

It must not be forgotten that when gases are rendered luminous there may usually be detected a faintly luminous continuous spectrum. In the case of several of the nebulæ, such as the annular nebula of Lyra and the Dumb-bell nebula, no existence of even a faint continuous spectrum has been yet certainly detected.

The determination of the position in the spectrum of the three bright lines was obtained by simultaneous comparison with the lines of hydrogen, nitrogen, and barium. The instrument which I employed had two prisms, each with a refracting angle of 60°, and the positions of the lines were trustworthy within the limits of about the breadth of the double line D.

The objects which I proposed to myself, in attempting a re-examination of some of the nebulæ with the large instrument described in this paper, were to determine, first, whether any of the nebulæ were possessed of a motion which could be detected by a change of refrangibility; secondly, whether the coincidence which had been observed of the first and the third line with a line of hydrogen and a line of nitrogen would be found to hold good when subjected to the test of a spreading out of the spectrum three or four times greater than that under which the former observations were made. It would not, it seemed, be difficult, in the case of the detection of a want of coincidence, to separate the effects of the two distinct sources referred to, from both of which equally a minute difference of refrangibility between the nebular lines and those of terrestrial substances might arise. The probability is very great indeed that, in all the nebulæ which give the kind of spectrum of which I am speaking, the two lines referred to are to be attributed to the same two substances, and that therefore, in all these nebulæ, they were originally of the same degree of refrangibility. On the other hand, it is not to be supposed that nebulæ situated in different positions in the heavens would have a similar motion relatively to the earth. An examination of several nebulæ would therefore show to which of these causes any observed want of coincidence was to be attributed.

The great Nebula in Orion.—In my description of this nebula[1] I stated that the light from all the parts of this strangely diversified object, which were bright enough to be observed with my instrument, was resolved into three bright lines similar to those represented in the diagram.

On the present occasion I applied myself in the first place to as careful a comparison as possible of the brightest line with the corresponding line of the spectrum of nitrogen.

My first observations were made with the light from the induction spark taken in pure nitrogen sealed in a tube at a tension a little less than that of the atmosphere, which was reflected into the instrument, as in my former series of observations, by means of a mirror and a small prism. The precaution was taken to verify the accuracy of the position of the spectrum of comparison relatively to that of the nebula, by placing a small lamp before the object-glass in the way already described.

The coincidence of the line in the nebula with the brightest of the lines of nitrogen, though now subjected to a much more severe trial, appeared as perfect as it did in my former observations. I expected that I might discover a duplicity in the line in the nebula corresponding to the two component lines of the line of nitrogen; but I was not able, after long and careful scrutiny, to see the line double. The line in the nebula was narrower than the double line of nitrogen: this latter may have appeared,

[1] Proc. Roy. Soc. vol. xiv. p. 39.

broader in consequence of irradiation, and it was much brighter than the line in the nebula.

The following observations are suggestive in connexion with the point under consideration. Electrodes of platinum were placed before the object-glass in the direction of a diameter, so that the spark was as nearly as possible before the centre of the lens. The spark was taken in air. I expected to find the spectrum faint, for the reasons which have been stated in a previous paragraph; but I was surprised to find that only one line was visible in the large spectroscope when adapted to the eye end of the telescope. This line was the one which agrees in position with the line in the nebula; so that under these circumstances the spectrum of nitrogen appeared precisely similar to the spectra of those nebulæ of which the light is apparently monochromatic. This resemblance was made more complete by the faintness of the line; from which cause it appeared much narrower, and the separate existence of its two components could no longer be detected. When this line was observed simultaneously with that in the nebula, it was found to appear but a very little broader than that line. When the battery circuit was completed, the line from the spark coincided so accurately in position with the nebular line that the effect to the eye was as if a sudden increase of brightness in the line of the nebula had taken place. In order to make this observation, and to compare the relative appearance of the lines, the telescope was moved so that the light from the nebula occupied the lower half only of the slit. The line of the spark was now seen to be a very little broader than the line of the nebula, and appeared as a continuation of it in an unbroken straight line. These observations were repeated many times on several nights.

An apparent want of coincidence, which would be represented by 0·02 division of the head of the micrometer screw, would be about the smallest difference that could be observed under the circumstances under which these observations were made. At the part of the spectrum where this line of nitrogen occurs the angular interval measured by ·02 division of the micrometer corresponds to a difference of wave-length of ·0460 millionth of a millimetre.

At the time the comparisons were made the earth was receding from the part of the heavens in which the nebula is situated by about half its orbital velocity. If the velocity of light be taken at 185,000 miles per second, and the wave-length of the nitrogen line at 500·80 millionths of a millimetre, the effect of half the orbital motion would be to degrade the refrangibility of the line by 0·023, an alteration of wave-length which would correspond to about 0·01 of the large micrometer head, an interval too small to be detected.

We learn from these observations, that if the line be emitted by nitrogen, the nebula is not receding from us with a velocity greater than ten miles per second; for this motion, added to that of the earth's orbital velocity, would have caused a want of coincidence that could be observed. Further, that if the nebula be approaching our system, its velocity may be as much as twenty miles or twenty-five miles per second; for part of its motion of approach would be masked by the effect of the motion of the earth in the contrary direction.

The double line in the nitrogen spectrum does not consist of sharply defined lines, but each component is nebulous, and remains of a greater width than the image of the slit.[1] The breadth of these lines appears to be connected with the conditions of tension and of temperature of the gas. Plücker[2] states that when an induction spark of great heating power is employed the lines expand so as to unite and form an undivided band. Even when the duplicity exists, the eye ceases to have the power to distinguish the component lines, if the intensity of the light be greatly diminished.

Though I have been unable to detect duplicity in the corresponding line in the nebula, it might possibly be found to be double if seen under more favourable conditions: I incline to the belief that it is not double.[3]

[1] Secchi states that with his direct spectroscope this line in the annular nebula in Lyra appears double. As the image of the nebula is viewed directly, after elongation by the cylindrical lens, and without a slit, it is probable that the two lines may correspond to the two sides of the elongated annulus of the nebula.

[2] Phil. Trans. 1863, p. 13.

[3] "On the Spectra of the Chemical Elements," Phil. Trans. 1864, p. 111.

In my Tables of the lines of the air[1] I estimated the brightness of each of the components of the double line in the spectrum of nitrogen at 10, and the components of the double line next in brightness in the orange at 7 and 5, and those of a third double line on the less refrangible side of D at 6 and 4. It was with reference to these two double lines next in apparent brilliancy that I wrote,[2] in speaking of the line in the nebula, "If, however, this line were due to nitrogen, we ought to see other lines as well; for there are specially two strong double lines in the spectrum of nitrogen, one at least of which, if they existed in the light of the nebulæ, would be easily visible."

As the disappearance of the whole spectrum of nitrogen, with the exception of the one double line was unexpected,—though, indeed, in accordance with my previous estimations, I examined the spectrum of nitrogen with a spectroscope furnished with one prism with a refracting angle of 60°, in which the whole of the spectrum from C to G is included in the field of view,—I then moved between the eye and the little telescope of the spectroscope a wedge of neutral-tint glass corrected for refraction by an inverted similar wedge of crown glass, and which I had found to be sensibly equal in absorbing power on the different parts of the visible spectrum. As the darker part of the wedge was brought before the eye, the two groups in the orange were quite extinguished, while the lines in the green still remained of considerable brightness. The line which under these circumstances remained longest visible, next to the brightest line, was one more refrangible at 2660 of the scale of my map. This observation was made with a narrow slit. When the induction spark was looked at from a distance of some feet with a direct-vision prism held close to the eye, I was surprised to observe that the double line in the orange appeared to me to be the brightest in the spectrum; and when the neutral-tint wedge was interposed, this line in the orange remained alone visible, all the other lines being extinguished.

When, however, in place of the simple prism a small direct-vision spectroscope provided with a slit was employed, I found

[1] Phil. Trans. 1864, p. 141. [2] Ibid. p. 443.

it to be possible, by receding from the spark, to find a position in which the double line in the green, with which the line in the nebula coincides, was alone visible, and the spectrum of the spark in nitrogen resembled that of a monochromatic nebula.

It is obvious that, if the spectrum of hydrogen were reduced in intensity, the line in the blue, which corresponds to that in the nebula, would remain visible after the line in the red and the lines more refrangible than F had become too feeble to affect the eye.

It therefore becomes a question of much interest whether the one, two, or three, or four lines seen in the spectra of these nebulæ represent the whole of the light emitted by these bodies, or whether these lines are the strongest lines only of their spectra, which, by reason of their greater intensity, have succeeded in reaching the earth. Since these nebulæ are bodies which have a sensible diameter, and in all probability present a continuous luminous surface, or nearly so, we cannot suppose that any lines have been extinguished by the effect of the distance of these objects from us.

If we had evidence that the other lines which present themselves in the spectra of nitrogen and hydrogen were quenched on their way to us, we should have to consider their disappearance as an indication of a power of extinction residing in cosmical space, similar to that which was suggested from theoretical considerations by Chéseaux, and was afterwards supported on other grounds by Olbers and the elder Struve. Further, as the lines which we see in the nebulæ are precisely those which experiment shows would longest resist extinction, at least so far as respects their power of producing an impression on our visual organs, we might conclude that this absorptive property of space is not *elective* in its action on light, but is of the character of a *general* absorption acting equally, or nearly so, on light of every degree of refrangibility. Whatever may be the true state of the case, the result of this re-examination of the spectrum of this nebula appears to give increased probability to the suggestion that followed from my former observations, namely,—that the substances hydrogen and nitrogen are the principal constituents of the nebulæ of the class under consideration.

I now pass to observations of the third line of the nebular spectrum, the one which I found to coincide with the line of hydrogen which corresponds to Fraunhofer's F. The substance in the nebulæ which is indicated by this line appears to be subject to much greater variation in relative brilliancy, or to be more affected by the conditions under which it emits light; for while the brightest line is always present, the line of which I am speaking seems to be wholly wanting in some nebulæ, and to be of different degrees of relative brightness in some other nebulæ.

In the nebula of Orion this line is relatively stronger than in 37 H. IV. Draconis, and some other nebulæ. I have suspected that the relative brightness of this line varies slightly in different parts of this nebula. It may be estimated perhaps in the nebula of Orion at about the brightness of the second line. The second line suffers in apparent brilliancy from its nearness to the brightest line, and may, without due regard to this circumstance, be estimated as brighter than the third line.

In order to compare the position of the line with that of the corresponding line in the spectrum of hydrogen, I employed a vacuum-tube containing hydrogen at a very small tension, which was placed before the object-glass of the telescope. Under these conditions the line appears narrow when the slit is narrow, without any sensible nebulosity at the edges. The character of the line is altered, as has been shown by Plücker, when hydrogen at the atmospheric pressure is employed: the line then expands into a nebulous band of considerable width, even with a very narrow slit. Such a condition of the line is obviously unsuitable for the delicate comparisons which it was proposed to attempt.

The narrow, sharply-defined line of hydrogen, when the vacuum-tube was before the slit, was observed to coincide perfectly in position with the third line of the nebula. This observation, which shows the coincidence of these lines with an accuracy three or four times as great as my former observations, increases in the same ratio the probability that the line in the nebula is really due to luminous hydrogen.

I suspect that, although the third line in this nebula may impress the eye as strongly as the second line, yet it is not so narrow and well defined as that line. If this suspicion be correct, this condition of the line might indicate that the hydrogen exists at a rather greater tension than that in the so-called vacuum-tubes, but that it is not nearly so dense as would correspond to the atmospheric pressure at the surface of the earth. As, however, the character of the lines of hydrogen is also greatly modified by temperature, it is not possible to reason with any certainty as to the state of things in this distant object, the light of which we have now under examination.

I am still unable to find any terrestrial line which corresponds to the middle line. I have made the additional observation that the line in the nebula is in a very slight degree less refrangible than the line of oxygen at 2060 of the scale of my map. It is in a rather larger degree less refrangible than the strong line of barium at 2075 of my scale.

Several other nebulæ have been observed with the large spectroscope: I prefer, however, to re-examine these objects before I publish any observations of them.

§ IV. *Observations of Stars.*

The chief difficulties which I have had to encounter have arisen from the unsteadiness of our atmosphere. There is sufficient light from stars of the first and second magnitude for the large spectroscope described in this paper, and, so far as the adjustments of the instrument are concerned, the lines in the spectra of the stars would be well defined. Unless, however, the air is very steady, the lines are seen too fitfully to permit of any certainty in the determination of coincidences of the degree of delicacy which is attempted in the present investigation. I have passed hours in the attempt to determine the position of a single line, and have then not considered that the numerous observations which I had obtained were possessed, even collectively, of sufficient weight to establish with any certainty the coincidence of the line with the one compared with it.

I prefer, therefore, to reject a large number of observations which appear unsatisfactory from this cause, and to give in this place a very few of the most trustworthy of the observations which I have made.

Sirius.—The brilliant light of this star and the great intensity of the four strong lines of its spectrum make it especially suitable for such an examination. The low altitude of this star in our latitude limits the period in which it can be successfully observed to about one hour on each side of the meridian.

I have confined myself to comparisons of the strong line in the position of F with the corresponding line of the spectrum of hydrogen. My first trials were made with hydrogen at the ordinary atmospheric pressure: the width of the band of hydrogen, under these circumstances, was greater than the line of Sirius. This line in Sirius, from some cause, is narrower relatively to the length of the spectrum, when considerable dispersion and a narrow slit are employed, than when the image of the star, rendered linear by a cylindrical lens, is observed with a single prism.[1] (See Fig. 90, p. 296.)

When the large spectroscope was employed, I estimated the breadth of the line to be about equal to that of the double line D. In Kirchhoff's map the line F of the solar spectrum is represented as a little more than one-fourth of the interval separating the lines D. When the spectroscope attached to the telescope was directed to the moon, the line F appeared even narrower than it is represented in Kirchhoff's map; I estimated it at about one-sixth of the apparent breadth of the corresponding line in the spectrum of Sirius. The character of the line agrees precisely with the line of hydrogen under certain conditions of tension and temperature.

As it was obviously impossible to determine with the required accuracy the coincidence of the line of Sirius when the much broader band of hydrogen at the ordinary pressure was compared with it, I employed a vacuum-tube fixed before the object-glass. In all these observations the slit used was as narrow as possible. The air at the time of the present observations was more favourable than usual, and the line in Sirius was seen with great distinctness. The line from the spark appeared in comparison

[1] See Phil. Trans. 1864, p. 42.

very narrow, not more than about one-fifth of the width of the line of Sirius. When the battery circuit was completed, the line of hydrogen could be seen distinctly upon the dark line of Sirius. The observation of the comparison of the lines was made many times, and I am certain that the narrow line of hydrogen, though it appeared projected upon the dark line in Sirius, did not coincide with the middle of the line, but crossed it at a distance from the middle, which may be represented by saying that the want of coincidence was *apparently* equal to about one-third or one-fourth of the interval separating the components of the double line D. I was unable to measure directly the distance between the centre of the line of hydrogen and that of the line in the spectrum of Sirius, but several very careful estimations by means of the micrometer give a value for that distance of 0·040 of the micrometer head. This value is probably not in error by so much as its eighth part.

Comparisons on many other nights were also made, sometimes with the vacuum-tube before the object-glass, and sometimes with the vacuum-tube placed over the small hole in the gutta-percha plate. On all these occasions the numerous comparisons which were made gave for the line in Sirius a very slightly lower refrangibility than that of the line of hydrogen, but on no one occasion was the air steady enough for a satisfactory determination of the amount of difference of refrangibility.

I have not been able to detect any probable source of error in this result, and it may therefore, I believe, be received as representing a relative motion of recession between Sirius and the earth.

The probability that the substance in Sirius by which this line is produced is really hydrogen is strengthened almost to certainty by the consideration that there is a strong line in the red part of the spectrum which is also coincident with a strong line of hydrogen. There is a third line more refrangible than F, which appears to coincide with the line of hydrogen in that part of the spectrum.

As the line in Sirius is more expanded than that of the vacuum-tube, it seemed of importance to have proof from experiment that this line of hydrogen, when it becomes broad, expands equally in both directions. I made the comparisons of the narrow line of the vacuum-tube with the more expanded band which appears when denser hydrogen is employed. For this purpose the intersection of the wires of the eyepiece was brought, as nearly as could be estimated, upon the middle of the expanded line which is produced by dense hydrogen. The vacuum-tube was then arranged before the slit, when the narrow line which it gives was observed to fall exactly upon the point of intersection of the wire. Under these terrestrial conditions the expansion of the line may be considered to take place to an equal amount in both directions. There is very great probability that a similar equal expansion takes place under the conditions which determine the absorption of light by this gas in the atmosphere of Sirius, for the reason that the nebulosity at the edges of the line in the spectrum of that star is sensibly equal on both sides.

I made some attempts to compare the strong line at C with the corresponding line of hydrogen; but when the large spectroscope was employed, though the lines could be seen with tolerable distinctness, they were not bright enough to admit of a trustworthy determination of their relative position. When one of the compound prisms was removed, the lines were much more easily seen, but under these circumstances the amount of dispersion was insufficient for my present purpose.

The lines of Sirius which, in conjunction with Dr. Miller, I had compared with those of iron, magnesium, and sodium, are not sufficiently well seen in our latitude for comparison when a powerful train of prisms is employed, such as is necessary for this special inquiry.

From these observations it may, I think, be concluded that the substance in Sirius which produces the strong lines is really hydrogen, as was stated by Dr. Miller and myself in our former paper. Further, that the aggregate result of the motions of the star and the earth in space, at the time when the observations were made, was to degrade the refrangibility of the line in Sirius by an amount corresponding to 0·040 of the micrometer screw. Now the value of the wave-lengths of 0·01 division of the micrometer at the position of F is 0·02725 millionth of a millimetre.[1] The total degradation of refrangibility observed amounts to 0·109 millionth of a millimetre. If the velocity of light be taken at 185,000[2] miles per second, and the wave-length of F at 486·50 millionths of a millimetre (Ångström's value is 486·52, Ditscheiner's 486·49), the observed alteration in period of the line in Sirius will indicate a motion of recession existing between the earth and the star of 41·4 miles per second.

Of this motion a part is due to the earth's motion in space. As the earth moves round the sun in the plane of the ecliptic, it is changing the direction of its motion at every instant. There are two positions, separated by 180°, where the effect of the earth's motion is a maximum; namely, when it is moving in the direction of the visual ray, either towards or from the star. At two other positions in its orbit, at 90° from the former positions, the earth's motion is at right angles to the direction of the light from the star, and therefore has no influence on its refrangibility.

The effect of the earth's motion will be greatest upon the light of a star situated in the plane of the ecliptic, and will decrease as the star's latitude increases, until, with respect to a star situated at the pole of the ecliptic, the earth's motion during the whole of its annual course will be perpendicular to the direction of the light coming to us from it, and will be therefore without influence on its period.

[1] The value in wave-lengths of the divisions of the micrometer for different parts of the spectrum was determined by the aid of the tables of the wave-lengths corresponding to every tenth line of Kirchhoff's map by Dr. Wolcott Gibbs (Silliman's Journal, vol. xliii. January 1867). A paper on the same subject by the Astronomer Royal, presented to the Royal Society, is not yet in print. The Astronomer Royal's paper is contained in the Philosophical Transactions for 1868, Part I. p. 29. The wave-lengths computed by him differ slightly from those assigned to Kirchhoff's numbers by Dr. Gibbs at the part of the spectrum under consideration in the text. The difference is due in part to the employment, by the Astronomer Royal, of Ditscheiner's later measures. These give for F the higher value of 486·37.—October 1868.

[2] The new determination of the value of the solar parallax by observations of Mars requires that the usually received velocity of light, 192,000 miles per second, should be reduced by about the one twenty-seventh part. The velocity, when diminished in this ratio, agrees nearly with the result obtained by Foucault from direct experiment.

That part of the earth's resolved motion which is in the direction of the visual ray, and which has alone to be considered in this investigation, may be obtained from the following formula:

$$\text{Earth's motion towards star} = r.\cos\lambda.\sin(l - l'),$$

where r is the earth's velocity, l the earth's longitude, l' the star's longitude, and λ the star's latitude.

At the time when the estimate of the amount of alteration of period of the line in Sirius was made the earth was moving from the star with a velocity of about twelve miles per second.

There remains unaccounted for a motion of recession from the earth amounting to 29·4 miles per second, which we appear to be entitled to attribute to Sirius.

It may be not unnecessary to state that the solar motion in space, if accepted as a fact, will not materially affect this result, since, according to M. Otto Struve's calculations, the advance of the sun in space takes place with a velocity but little greater than one-fourth of the earth's motion in its orbit. If the apex of the solar motion be situated in Hercules, nearly the whole of it will be from Sirius, and will therefore diminish the velocity to be ascribed to that star.

It is interesting, in connexion with the motion of Sirius deduced from these prismatic observations, to refer to the remarkable inequalities which occur in the rather large proper motion of that star. In 1851 M. Peters[1] showed that the variable part of the proper motion of Sirius in right ascension might be represented by supposing that Sirius revolves in an elliptic orbit, round some centre of gravity without itself, in a period of 50·093 years. This hypothesis has acquired new interest, and seems indeed to have received confirmation from direct observation by Alvan Clark's discovery of a small companion to Sirius.

Professor Safford[2] and Dr. Auwers[3] have investigated the

[1] Astron. Nachrichten, No. 748.

[2] Proceedings of the American Academy, vol. vi.; also Astron. Notices, Ann Arbor, No. 28; Monthly Notices, vol. xxii. p. 145.

[3] Astron. Nachrichten, No. 1506; Monthly Notices, vol. xxii. p. 148, and vol. xxv. p. 39.

periodical variations of the proper motion of Sirius in declination, and they have found that these variations, equally with those in right ascension, would be reconcilable with an elliptic orbital motion round a centre not in Sirius. The close coincidence of the observed positions of the new satellite with those required by theory seems to show that it may be the hypothetical body suggested by Peters, though we must then suppose it to have a much greater mass relatively to Sirius than that which its light would indicate.

At the present time the proper motion of Sirius in declination is less than its average amount by nearly the whole of that part of it which is variable. May not this smaller apparent motion be interpreted as showing that a part of the motion of the star is *now in the direction of the visual ray?* This circumstance is of much interest in connexion with the result arrived at in this paper.

Independently of the considerations connected with the variable part of the star's proper motion, it must not be forgotten that the whole of the motion which can be directly observed by us is only that portion of its real motion which is at right angles to the visual ray. Now it is precisely the other portion of it, which we could scarcely hope to learn from ordinary observations, which is revealed to us by prismatic investigations. By combining the results of both methods of research we may perhaps expect to obtain some knowledge of the real motions of the brighter stars and nebulæ.

It seems therefore desirable to compare with the result obtained by the prism the motion of Sirius which corresponds to its assumed constant proper motion. The values adopted by Mr. Main,[1] and inserted by the Astronomer Royal in the Greenwich "Seven-year Catalogue," are — 0"·035 in R.A. and + 1"·24 in N.P.D.

The parallax of Sirius from the observations of Henderson, corrected by Bessel, = 0"·150. A recent investigation by Mr. C. Abbe[2] gives for the parallax the larger value of 0"·27. If the radius of the earth's orbit be taken at its new value of

[1] Memoirs of the Royal Astronomical Society, vol. xix.

[2] Monthly Notices of the Royal Astronomical Society, vol. xxviii. p. 2

91,600,000 miles, the assumed annual constant proper motion in N.P.D. of 1"·24 would indicate, with the parallax of Henderson, a velocity of Sirius of twenty-four miles nearly per second; with the larger parallax of Mr. Abbe, a velocity of 43·2 miles per second. It may be that in the case of Sirius we have two distinct motions; one peculiar to the star, and a second motion which it may share in common with a system of which it may form a part.

Observations and comparisons, similar to those on Sirius, have been made on a Canis Minoris, Castor, Betelgeux, Aldebaran, and some other stars. I reserve for the present the results which I have obtained, as I desire to submit these objects to a re-examination. It is seldom that the air is sufficiently favourable for the successful prosecution of this very delicate research.

* * * * * * *

§ VI. *Observations of Comet II. 1868.*

On June 13 a comet was discovered by Dr. Winnecke, and also independently the same night by M. Becquet, Assistant Astronomer at the Observatory of Marseilles.

I was prevented by buildings existing near my observatory from making observations of this comet before June 22. On that evening the comet was much brighter than Brorsen's comet, a description of the spectrum of which I recently presented to the Royal Society,[1] and it gave a spectrum sufficiently distinct for measurement and comparison with the spectra of terrestrial substances.

Telescopic Appearance of the Comet.—A representation of the comet as it appeared on June 22 at 11 P.M. is given in Fig. 89, p. 293. The comet consisted of a nearly circular coma, which became rather suddenly brighter towards the centre, where there was a nearly round spot of light. The diameter of the coma, including the exterior faint nebulosity, was about 6'20". The tail, which was traced for more than a degree, was

[1] Proc. Roy. Soc. vol. xvi. p. 386.

sharply defined on the following edge, but faded so gradually away on the opposite side that no limit could be perceived. No connexion was traced between the tail and the brighter central part of the coma. The circular form of the coma was uninterrupted on the side of the tail, which appeared as an extension of the faint nebulosity which formed the extreme margin of the coma.

The bright roundish spot of light in the centre, when examined with eyepieces magnifying from 200 to 600 diameters, presented merely a nebulous light without a defined form.

Spectrum of the Comet.—When a spectroscope furnished with two prisms of 60° was applied to the telescope, the light of the comet was resolved into three very broad bright bands, which are represented in the diagram. (Fig. 87.)

In the two more refrangible of these bands the light was brightest at the less refrangible end, and gradually diminished towards the other limit of the bands. This gradation of light was not uniform in the middle and brightest band, which continued of nearly equal brilliancy for about one-third of its breadth from the less refrangible end. This band appeared to be commenced at its brightest side by a bright line.

The least refrangible of the three bands did not exhibit a similar marked gradation of brightness. This band, though of nearly uniform brilliancy throughout, was perhaps brightest about the middle of its breadth.

These characters, which are peculiar to the light emitted by the cometary matter, must be distinguished from some appearances which the bands assumed in consequence of the mode of distribution of the light in the coma of the comet. The two more refrangible bands became narrower towards their most refrangible side, as well as diminished in brightness. This appearance was obviously not due to any dissimilarity of the light in the parts of the coma, but to the circumstance that, as the light of the coma became brighter towards the centre, it was emitted by a smaller area of the cometary matter. The strong light of the central spot could be traced the whole breadth of the band; but the light surrounding this spot, in proportion as it became fainter and broader, was seen for a shorter distance, so that the light from the faintest parts near the margin of the coma was visible only at the brightest side of the band. Since in the least refrangible band a similar gradation of light did not take place, this band appeared of nearly the same width throughout.

The increasing brightness of the coma up to the brilliant spot in the centre showed itself in this band as a bright axial line fading off gradually in both directions.

On this evening I took repeated measures of the positions of these bands with the micrometer attached to the spectroscope. These measures give the following numbers for the commencement and termination of the three bands on the scale adopted in the diagram:—

First band { 1094. 1198. Second band { 1298. 1440. Third band { 1580. 1700.

I could not resolve the bands into lines. When the slit was made narrower, the bands became smaller both in breadth and length, from the invisibility of the fainter portions. I suspected, however, the presence of two or three bright lines in the bright central part of the middle band near its less refrangible limit. This part would consist chiefly of light from the bright central spot.

As has been stated, the middle band commences probably with a bright line; for the limit of the band is here abrupt and distinct. On the contrary, the exact point of commencement and termination of the other bands could not be observed with certainty.

I could perceive no other bands, nor light of any kind beyond the three bands, in the parts of the spectrum towards the red and the violet.

When the marginal portions of the coma were brought upon the slit, the three bands of light could still be traced. When, however, the spectrum became very faint, it appeared to me to become continuous; but the light was then so very feeble that it could not be traced beyond the three bands towards the violet or the red.

On this evening I observed the spectrum of the comet in a

larger spectroscope, which gives a dispersion equal to about five prisms. In this instrument the middle band was well seen. It retained its nebulous, unresolved character, and the abrupt commencement, as if by a bright line, already mentioned, was distinctly seen.

For convenience of comparison, the spectrum of Brorsen's comet, and that of the gaseous nebulæ, have been added to the diagram, Fig. 87. The spectrum of Brorsen's comet consisted of three bright bands and a faint continuous spectrum. These bands appeared, as represented in the diagram, narrower than those of the comet now under examination. It is not possible to say to what extent this circumstance may be due to the much feebler light of this comet. Though the bands of Brorsen's comet fall within the limits of position occupied by the broad bands of Comet II., they do not correspond to the brightest parts of these bands. In the middle band I suspected two bright lines which appeared shorter than the band, and may be due to the nucleus. Brorsen's comet differed from the two small comets which I had previously examined[1] in the much smaller relative proportion of the light which forms a continuous spectrum. In Brorsen's comet the bright middle part of the coma seemed to emit light similar to that of the nucleus; in the other comets the coma appeared to give a continuous spectrum. The three comets resembled each other in the circumstance that the light of the central part was emitted by the cometary matter, while the surrounding nebulosity reflected solar light.

It will be seen in the diagram that the bands of Brorsen's comet and those of Comet II. occupy positions in the spectrum widely removed from those in which the lines of the nebulæ occur. The spectra of gaseous nebulæ consist of true lines, which become narrow as the slit is made narrower.

The following day I carefully considered these observations of the comet, with the hope of a possible identification of its spectrum with that of some terrestrial substance. The spectrum

[1] Comet I. 1866, Proceedings, vol. xv. p. 5; and Comet 1867, Monthly Notices of Royal Astronomical Society, vol. xxvii. p. 288.

of the comet appeared to me to resemble some of the forms of the spectrum of carbon which I had observed and carefully measured in 1864. On comparing the spectrum of the comet with the diagram of these spectra of carbon, I was much interested to perceive that the positions of the bands in the spectrum, as well as their general characters and relative brightness, agreed exactly with the spectrum of carbon when the spark is taken in olefiant gas.

These observations on the spectrum of carbon were undertaken in continuation of my researches "On the Spectra of the Chemical Elements."[1] I have not presented them to the Royal Society, as they are not so complete as I hope to make them.

Though the essential features of the spectrum of carbon remained unchanged in all the experiments, certain modifications were observed when the spectrum was obtained under different conditions. One of those modifications, which was referred to in my paper "On the Spectra of the Chemical Elements,"[2] may be mentioned here. One of the strongest of the lines of carbon is a line in the red a little less refrangible than the hydrogen line, which corresponds to Fraunhofer's C. Now this line is not seen when the carbon is subjected to the induction spark in the presence of hydrogen. Two of the other modifications of the spectrum of carbon are given in Fig. 87. The first spectrum represents the appearance of the spectrum of carbon when the induction spark, with Leyden jars intercalated, was taken between the points of wires of platinum sealed in glass tubes, and placed almost in contact in olive-oil. In this spectrum are seen the principal strong lines which distinguish carbon. The shading of fine lines which accompanies the strong lines cannot be accurately represented, on account of the small size of the diagram. A spectrum essentially the same is produced when the spark is taken in a current of cyanogen. It may be mentioned that when the heating power of the spark was reduced below a certain limit, though the decomposition of the oil still took place, the carbon was not volatilized, and the spectrum was continuous.

[1] Phil. Trans. 1864, p. 139. [2] Ibid. p. 145.

The third spectrum in the diagram represents the modification of this typical spectrum when the induction spark is taken in a current of olefiant gas. The highly-heated vapour of carbon emits light of the same refrangibilities as in the case of the oil; but the separate strong lines, with a similar power of spark, were no longer to be distinguished. The shading, when the carbon was obtained from the olefiant gas, was not composed of numerous fine lines, but appeared as an unresolved nebulous light.

Of course in all these experiments the lines of the other elements present were also seen, but they were known, and could therefore be disregarded.

In the case of the spark in olefiant gas, the three bands in the diagram constitute the whole spectrum, with the exception of a faint band in the more refrangible part of the spectrum.

It was with the spectrum of carbon, as thus obtained, that the spectrum of the comet appeared to agree. It seemed, therefore, to be of much importance that the spectrum of the spark in olefiant gas should be compared directly in the spectroscope with the spectrum of the comet. The comparison of the gas with the comet was made the same evening, June 23.

My friend Dr. William Allen Miller visited the observatory on this evening, and kindly took part in the following observations.

The general arrangements of the apparatus with which the comparison was made is shown in the following diagram (Fig. 94.)

A glass bottle converted into a gasholder, *a*, contained the olefiant gas. This was connected by means of a flexible tube *b*, into which were soldered two platinum wires. The part of the tube in front of the points of the wires had been cut away, and the surfaces carefully ground. A small plate of glass closed the opening, being held in its place by a band of vulcanized india-rubber. This tube was arranged in its proper position, before the small mirror of the spectroscope *c*, by which the light of the spark was reflected into the instrument, and its spectrum was seen immediately beneath the spectrum of the comet. The spectroscope employed was furnished with two prisms of 60°.

The brightest end of the middle band of the cometic spectrum was seen to be coincident with the commencement of the corresponding band in the spectrum of the spark. As this limit of the band was well defined in both spectra, the coincidence could be satisfactorily observed up to the power of the spectroscope,

FIG. 91.

and may be considered to be determined within about the distance which separates the components of the double line D. As the limits of the other bands were less distinctly seen, the same amount of certainty of exact coincidence could not be obtained. We considered these bands to agree precisely in position with the bands corresponding to them in the spectrum of the spark.

The apparent identity of the spectrum of the comet with that of carbon rests not only on the coincidence of position in the spectrum of the bands, but also upon the very remarkable resemblance of the corresponding bands in their general characters, and in their relative brightness. This is very noticeable in the middle band, where the gradation of brightness is not uniform. This band in both spectra remained of nearly equal brightness for the same proportion of its length.

On a subsequent evening, June 25, I repeated these comparisons, when the former observations were fully confirmed in every particular. On this evening I compared the brightest band with that of carbon in the larger spectroscope, which gives a dispersion of about five prisms.

The remarkably close resemblance of the spectrum of the comet to the spectrum of carbon necessarily suggests the identity of the substances by which in both cases the light was emitted.

It may be well to state that some phosphorescent and fluorescent bodies give discontinuous spectra, in which the light is restricted to certain ranges of refrangibility. There are, however, several considerations which seem to oppose the idea that the light of comets can be of a phosphorescent character. Phosphorescent bodies are usually so highly reflective that the phosphorescence emitted by them is not seen so long as they are exposed to light. This comet was still in the full glare of the sun, and yet the continuous spectrum corresponding to reflected solar light was of extreme feebleness compared with the three bright bands which we have under consideration. The phenomenon of phosphorescence seems to be restricted to bodies in the solid state, a condition which is not apparently in accordance with certain phenomena which have been observed in large comets, such as the outflow of the matter of the nucleus and the formation of successive envelopes.

There are, indeed, some phenomena of fluorescence, such as that of a nearly transparent liquid becoming an object of some brightness by means of the property which it possesses of absorbing the nearly invisible rays of the spectrum, and dispersing them in a degraded and much more luminous form, which are less obviously inconsistent with cometary phenomena than are those of phosphorescence.

The violent commotions and internal changes which we witness in comets when near the sun seem, however, to connect the great brightness which they then assume more closely with that part of the solar force we call heat. There is also to be considered the fact of the polarized condition of the light of the tail and some parts of the comæ of comets, which shows that a part of their light is reflected.

The observations of the spectrum of Comet II. contained in this paper, which show that its light was identical with that emitted by highly-heated vapour of carbon, appear to be almost decisive of the nature of cometary light. The great fixity of carbon seems indeed to raise some difficulty in the way of accepting the apparently obvious inference of these prismatic observations. Some comets have approached the sun sufficiently near to acquire a temperature high enough to convert even carbon into vapour.[1] Indeed for these comets a body of great fixity seems to be necessary. In the case of comets which have been submitted to a less fierce glare of solar heat, it may be suggested that this supposed difficulty is one of degree only; for we do not know of any conditions under which even a gas, permanent at the temperature of the earth, could maintain sufficient heat to emit light, a state of things which appears to exist permanently in the case of the gaseous nebulæ.

If the substance of the comet be taken to be pure carbon, it would appear probable that the nucleus had been condensed from the gaseous state in which it existed at some former period. It would therefore probably consist of carbon in a state of excessively minute division. In such a form it would be able to take in nearly the whole of the sun's energy, and thus acquire more speedily a temperature high enough for its conversion into vapour. In the liquid or gaseous state, or in a continuous solid state, this substance appears, from Dr. Tyndall's researches, to be diathermanous. Still, under the most favourable of known conditions, the solar heat, to which the majority of comets are subjected, would seem to be inadequate to the production of luminous vapour of carbon.

[1] The comet of 1843 "approached the luminous surface of the sun within about a seventh part of the sun's radius. The heat to which the comet was subjected (a glare as that of 47,000 suns, such as we experience the warmth of) surpassed that in the focus of Parker's great lens in the proportion of 24½ to 1 without, or 3½ to 1 with, the concentrating lens. Yet that lens so used melted cornelian, agate, and rock-crystal."—SIR JOHN HERSCHEL, *Outlines of Astronomy*, 7th edit. p. 401.

It should be stated that olefiant gas when burnt in air may give a similar spectrum of shaded bands. If the gas be ignited at the orifice of the tube from which it issues, the flame is brilliantly white, and gives a continuous spectrum. When a jet of air is directed through the flame, it becomes less luminous, and of a greenish-blue colour. The spectrum is now no longer continuous, but exhibits the bands distinctive of carbon. Under these circumstances, for obvious reasons, the bright lines of the hydrogen spectrum are not seen. In this way a spectrum resembling that of the comet may be obtained, with the difference that the fourth more refrangible band, which was not seen in the cometic spectrum, is stronger relatively to the other bands than is the case when the spark is taken in olefiant gas. If we were to conceive the comet to consist of a compound of carbon and hydrogen, we should diminish in some degree the necessity for the excessively high temperature which pure carbon appears to require for its conversion into luminous vapour: but other difficulties would arise in connexion with the decomposition we must then suppose to take place; for we have no evidence, I believe, that olefiant gas or any other known compound of carbon can furnish this peculiar spectrum of shaded bands without undergoing decomposition. If, indeed, it were allowable to suppose a state of combustion, with oxygen or some other element, set up by the solar heat, we should have an explanation of a possible source of a degree of heat sufficient to render the cometary matter luminous, and which the sun's heat would be directly inadequate to produce.

There is one observation made by Bunsen which appears to stand as an exception to the rule that only bodies in the gaseous state give, when luminous, discontinuous spectra. Bunsen discovered that solid erbia, when heated to incandescence, gives a spectrum containing bright bands. It is therefore conceivable, though all the evidence we possess from experience is opposed to the supposition, that carbon might exist in some form in which it would possess a similar power of giving a discontinuous spectrum without volatilization. There is the further objection to this hypothesis, that the telescopic phenomena observed in comets appear to show that vaporization does usually take place. However this may be, a state of gas appears to accord with the very small power of reflexion which the matter of the coma of this comet possessed, as was shown by the great faintness of the continuous spectrum.

A remarkable circumstance connected with comets is the great transparency of the bright cometary matter. The most remarkable instance is that of Miss Mitchell's comet in 1847, which passed centrally over a star of the fifth magnitude. "The star's light appeared in no way enfeebled: yet such a star would be completely obliterated by a moderate fog extending a few yards from the surface of the earth."[1] We do not know what amount of transparency is possessed by the vapour of carbon, but the absence of a continuous spectrum seems to show that, as it existed in the comet, it was almost perfectly transparent. The light of a star would suffer, therefore, only that kind and degree of absorption which corresponds with its power of radiation, as shown by its spectrum of bright lines. As these occur in the brightest part of the spectrum, we should expect a noticeable diminution of the star's light, if it were not for the luminous condition of the gas, in consequence of which it would give back to the beam light of precisely the same refrangibilities as it had taken, and so enable the part of the field occupied by the image of the star to appear of its original brightness, or nearly so. This state of things would not prevent an apparent diminution of the star's light from the effect upon the eye of the brightness of the surrounding field. In the case of the tails of comets, the great transparency observed is more probably to be referred to the widely-scattered condition of the minute particles of the cometary matter.

I may be permitted to repeat here a paragraph from my paper on the Spectrum of Comet I. 1866.[2]

"Terrestrial phenomena would suggest that the parts of a comet which are bright by reflecting the sun's light are probably in the condition of fog or cloud.

[1] Outlines of Astronomy, p. 373.
[2] Proc. Roy. Soc. vol. xv. p. 5.

"We know from observation, that the comæ and tails of comets are formed from the matter contained in the nucleus.[1]

"The usual order of the phenomena which attend the formation of a tail appears to be that, as the comet approaches the sun, material is thrown off, at intervals, from the nucleus in the direction towards the sun. This material is not at once driven into the tail, but usually forms in front of the nucleus a dense luminous cloud, into which for a time the bright matter of the nucleus continues to stream. In this way a succession of envelopes may be formed, the material of which afterwards is dissipated in a direction opposite to the sun, and forms the tail. Between these envelopes dark spaces are usually seen.

"If the matter of the nucleus is capable of forming by condensation a cloudlike mass, there must be an intermediate state in which the matter ceases to be self-luminous, but yet retains its gaseous state, and reflects but little light. Such a non-luminous and transparent condition of the cometary matter may possibly be represented by some at least of the dark spaces which, in some comets, separate the cloudlike envelopes from the nucleus and from each other."

Now considerable differences of colour have been remarked in the different parts of some comets. The spectrum of this comet would show that its colour was bluish green. Sir W. Herschel described the head of the comet of 1811 to be of a greenish or bluish-green colour, while the central point appeared to be of a pale ruddy tint. The representations of Halley's comet at its appearance in 1835, by the elder Struve, are coloured bluish

[1] The head of Halley's comet in 1835 in a telescope of great power "exhibited the appearance of jets as it were of flame, or rather of luminous smoke, like a gas fan-light. These varied from day to day, as if wavering backwards and forwards, as if they were thrown out of particular parts of the internal nucleus or kernel, which shifted round, or to or fro, by their recoil, like a squib not held fast. The bright smoke of these jets, however, never seemed to be able to get far out towards the sun, but always to be driven back and forced into the tail, as if by the action of a violent wind setting against them (always *from* the sun), so to make it clear that the tail is neither more nor less than the accumulation of this sort of luminous vapour darted off, in the *first instance towards the sun*, as it were something raised up and, as it were, exploded by the sun's heat out of the kernel, and then immediately and forcibly turned back and *repelled from* the sun."—SIR JOHN HERSCHEL, *Familiar Lectures on Scientific Subjects*, p. 115.

green, and the nucleus on October 9 is coloured reddish-yellow. He describes the nucleus on that day thus:—"Der Kern zeigte sich wie eine kleine, etwas ins gelbliche spielende, glühende Kohle von länglicher Form."[1] Dr. Winnecke describes similar colours in the bright Comet of 1862:—"Die Farbe des Strahls erscheint mir gelbröthlich; die des umgebenden Nebels (vielleicht aus Contrast) mattbläulich." "Die Farbe der Ausströmung erscheint mir gelblich; die Coma hat bläuliches Licht."[2]

Now carbon, if incandescent in the solid state, or reflecting, when in a condition of minute division, the light of the sun, would afford a light which, in comparison with that emitted by the luminous vapour of carbon, would appear as yellowish or approaching to red.

The views of comets presented in this paper do not, however, afford any clue to the great mystery which surrounds the enormous rapidity with which the tail is often projected to immense distances. There are not any known properties peculiar to carbon, even when in a condition of extremely minute division, which would help to a solution of the enigma of the violent repulsive power from the sun which appears to be exerted upon cometary matter shortly after its expulsion from the nucleus, and upon matter in this condition only. It may be that this apparent repulsion takes place at the time of the condensation of the gaseous matter of the coma into the excessively minute solid particles of which the tail probably consists. There is a phenomenon occasionally seen which must not be passed without notice, namely, the formation of faint narrow rays of light, or secondary tails, which start off usually from the brightest side of the principal tail, not far from the head. Sir John Herschel[3] considers that "they clearly indicate an analysis of the cometic matter by the sun's repulsive action, the matter of the secondary tails being darted off with incomparably greater velocity (indicating an incomparably greater intensity of repulsive energy)

[1] Beobachtungen des Halleyschen Cometen, S. 41.

[2] Mémoires de l'Académie Impériale des Sciences de St. Pétersbourg, tome vii. No. 7.

[3] Familiar Lectures on Scientific Subjects, p. 120.

than that which went to form the primary one." The important differences which exist between the spectrum of Brorsen's comet and that of Comet II. 1868 appear to show that comets may vary in their constitution. If the phenomena of the secondary tails were observed in a comet which, like Comet II. 1868, appears to consist of carbon, the analytical action supposed by Sir John Herschel might be to separate between particles of carbon in different conditions, or possibly in a state of more or less subdivision. The enormous extent of space, sometimes a hundred millions of miles in length, over which a comparatively minute portion of cometary matter is in this way diffused, would suggest that we have in this phenomenon a remarkable instance of the extreme division of matter. Perhaps it would be too bold a speculation to suggest that, under the circumstances which attend the condensation of the gaseous matter into discrete solid particles, the division may be pushed to its utmost limit, or nearly so. If we could conceive the separate atoms to be removed beyond the sphere of their mutual attraction of cohesion, it might be that they would be affected by the sun's energy in a way altogether different from that of which we have been hitherto the witnesses upon the earth.

Though comets may differ in their constitution, reference may be permitted to the periodical meteors, which have been shown to move in orbits identical with those of some comets. If these consist of carbon, we might have some explanation of the appearances presented by these meteors, though their light is doubtless greatly modified by that of the air rendered luminous by their passage, as well as by the degree of temperature to which they are raised. Carbon is abundantly present in some meteorites, but we have no certain evidence at present that the periodical meteors belong to this class of celestial bodies.

APPENDIX E.

PRELIMINARY NOTE OF RESEARCHES ON GASEOUS SPECTRA IN RELATION TO THE PHYSICAL CONSTITUTION OF THE SUN.

BY E. FRANKLAND, F.R.S., AND J. NORMAN LOCKYER.[1]

"WITH reference to the orange line of the chromosphere, we have failed to detect any line in the hydrogen spectrum in the place indicated, *i.e.* near the line D. With regard to the thickening of the F line, we may remark that we have convinced ourselves that this widening out is due to pressure, and not appreciably, if at all, to temperature *per se*.

"Having determined that the phenomena presented by the F line were phenomena depending on and indicating varying pressures, we were in a position to determine the atmospheric pressure operating in a prominence in which the red and green lines are nearly of equal width, and in the chromosphere, through which the green line gradually expands as the sun is approached. With regard to the higher prominences, we have ample evidence that the gaseous medium of which they are composed exists in a condition of *excessive* tenuity, and that at the lower surface of the chromosphere itself the pressure is very far below the pressure of the earth's atmosphere. . . . .

"We believe that the determination of the above-mentioned facts leads us necessarily to several important modifications of the received theory of the physical condition of our central luminary. According to Kirchhoff's theory the photosphere itself is either solid or liquid, and is surrounded by an atmosphere composed of gases and the vapours of the substances incandescent in the photosphere. We find, however, instead of this compound atmosphere, one which gives us nearly, or at all events mainly, the spectrum of hydrogen; and the tenuity of this incandescent atmosphere is such that it is extremely improbable that any considerable atmosphere, such as the corona

[1] Proc. Roy. Soc. Feb. 11, 1869.

has been imagined to indicate, lies outside it: with regard to the photosphere itself, so far from being either a solid surface or a liquid ocean, that it is cloudy or gaseous follows both from our observations and experiments.

"1. The gaseous condition of the photosphere is quite consistent with its continuous spectrum.

"2. The spectrum of the photosphere contains bright lines when the limb is observed, indicating probably an outer shell of gaseous matter.

"3. The sun-spot is a region of greater absorption.

"4. Occasionally photospheric matter appears to be injected into the chromosphere.

"May not these facts indicate that the absorption to which the reversal of the spectrum and Fraunhofer's lines are due takes place in the photosphere itself, or extremely near to it, instead of in an extensive outer absorbing atmosphere?"

ADDENDUM.

LOCKYER'S SPECTROSCOPIC OBSERVATIONS ON THE SUN.[1]

"Since the date on which the foregoing paper was written I have obtained additional evidence on the points referred to. I beg therefore to be permitted to make the following additions to it.

"The possibility of our being able to determine the velocity of movements of uprush and downrush taking place in the chromosphere depends upon the alterations of wave-length observed.

"It is clear, therefore, that a mere uprush or downrush at the sun's limb will not affect the wave-length, but that if we have at the limb cyclones, or backward and forward movements, the wave-length will be altered; so that we may have:—

"1. An alteration of wave-length near the centre of the disc caused by upward or downward movements.

[1] April 29, 1869.

"2. An alteration of wave-length close to the limb, caused by backward or forward movements. . . . .

"On April 21st I was enabled to extend my former observations. The spot under observation was very near the limb, so near that its spectrum and that of the chromosphere were both visible in the field of view.

"The spot-spectrum was very narrow, owing to the foreshortening of the spot; but the spectrum of the chromosphere showed me that the whole adjacent limb was covered with prominences of various heights all blended together. Further, the prominences seemed fed, so to speak, from apparently the preceding edge of the spot; for both C, F, and the line near D, *were magnificently bright on the sun itself*, the latter especially striking me with its thickness and brilliancy.

"In the prominences, C and F were observed to be strangely gnarled, knotty, and irregular; and I thought at once that some 'injection' must be taking place. I was not mistaken. On turning to the magnesium lines, I saw them far above the spectrum of the limb, and unconnected with it. A portion of the upper layer of the photosphere had in fact been lifted up beyond the usual limits of the chromosphere, and was there floating cloudlike. The vapour of sodium was also present in the chromosphere, though not so high as the magnesium, or unconnected with the spectrum of the limb: and, as I expected, with such a tremendous uplifting force, I saw the iron lines (for the first time) in the spectrum of the chromosphere. My observations commenced at 7.30 A.M.; by 8.30 there was comparative quiet; at 9.30 the action had commenced afresh. There was now a single prominence.

"The changes in the F line are seen better than those of any other line. On the 17th the following changes were noted:—

"I. It often stopped short of one of the small spots, swelling out prior to disappearance.

"II. It was invisible in a facula between two small spots.

"III. *It was changed into a bright line, and widened out on both sides two or three times in the very small spots.*

"IV. Once I observed it to become bright *near* a spot, and to expand over it on both sides. . . . .

"I observed it in all gradations of darkness; when the bright and dark lines were alongside, the latter was always the least refrangible."

RESEARCHES ON GASEOUS SPECTRA IN RELATION TO THE PHYSICAL CONSTITUTION OF THE SUN, STARS, AND NEBULÆ.

BY E. FRANKLAND, F.R.S., AND J. N. LOCKYER.[1]

"I. The Fraunhofer line on the solar spectrum, named *h* by Ångström, which is due to the absorption of hydrogen, is not visible in the tubes we employ with low battery and Leyden jar power: it may be looked upon therefore as an indication of relatively high temperature. As the line in question has been reversed by one of us in the spectrum of the chromosphere, it follows that the chromosphere, when cool enough to absorb, is still of a relatively high temperature.

"II. Under certain conditions of temperature and pressure the very complicated spectrum of hydrogen is reduced in our instrument to *one line in the green* corresponding to F in the solar spectrum.

"III. The equally complicated spectrum of nitrogen is similarly reducible to one bright line in the green, with traces of other more refrangible faint lines.

"By removing the tubes from the slit the combined spectra (II. and III.) were reduced to two bright lines.[2] By reducing the temperature all spectroscopic evidence of the nitrogen vanished, and by increasing it many new nitrogen lines made their appearance; the hydrogen lines always remaining visible.

"These latter observations bear on the observations of the nebulæ, especially on the conclusions of Mr. Huggins."

[1] Proc. Roy. Soc. xvii. p. 453 (June 10, 1869).

[2] This observation had previously been made by Huggins [H. E. R.]. See Appendix D.

APPENDIX F.

TABLES OF THE DARK LINES OF THE SOLAR SPECTRUM, SHOWING THE COINCIDENCES WITH THE BRIGHT LINES OF THE SPECTRA OF MANY METALS, AS GIVEN IN PROFESSOR KIRCHHOFF'S DRAWINGS.[1]

PLATE III. STRIP 1.

381.7	1c		447.0	2a		483.3	4d	509.9	2b	565.0	2c
384.1	2c		448.4	1b		484.1	2d	510.9	1a	566.0	2c
385.9	2d		452.6	2c		485.1	3d	512.9	2b	566.9	2b
387.5	3d		453.0	1b		486.2	6e	513.6	3b	567.4	3b
388.9	4d		434.4	1b		486.8	2c	517.1	2b	(568.6	2b
390.4	4e		460.0	1c	from	(488.2	1	519.3	2b	(	1
392.1	5e		461.0	1b		(488.8	5a	521.6	1b	569.2	2b
393.6	6e		462.2	2b		489.6	6e	529.4	1b	(	3e
395.0	6e		463.3	2a		(491.2	3e	530.4	1c	(570.0	2b
396.2	5e		466.0	1b		(491.5	5b	532.8	1b	570.6	3b
397.4	4e		466.5	2c		491.9	4c	536.9	2b	572.2	1b
398.4	4d		467.0	1b		493.1	2c	537.3	1b	572.9	3c
399.2	4d		468.1	2e		494.1	3b	540.6	3b	573.6	1b
399.8	4d		470.0	2b		(495.4	1e	541.1	2c	574.4	2d
400.4	3d		470.5	3c		(495.7	2b	542.0	1a	575.1	2d
(401.9	4c		470.9	2b		497.2	1b	543.6	4b	576.6	3d
(402.4	3		(472.4	2e		497.5	2a	544.6	3d	578.1	3d
402.8	4		(472.7	3c		498.4	4c	547.0	4c	579.6	3e
403.2	5		(473.8	4d		499.0	5b	547.9	2b	581.1	3e
	6		(	1		499.9	5d	549.0	3e	582.5	4e
405.0	5		474.7	3b		500.8	3d	551.2	3c	583.8	4f
405.6	4	from	(475.7	2		(501.8	2c	552.5	3c	585.0	4e
(406.2	3		(476.4	1b		(502.0	5b	(553.8	1c	586.2	3e
(406.8	3e		(	2		502.6	5e	(554.0	3b	587.0	2b
408.5	1d		477.0	5b		503.8	6d	554.6	2b	587.9	3b
423.7	2b		(	2		504.3	5b	557.0	1a	589.0	3b
426.6	2b		(477.8	4b		505.1	6c	557.7	2b	589.4	3b
433.8	2c		(479.1	2c		(506.2	2b	558.1	1b	589.9	3b
437.0	2b		(	1		(506.4	5b	559.7	1c	590.3	3b
442.8	2d		(480.1	6c		(506.6	2b	561.5	1b	590.7	3b
444.6	2e		(480.4	4d		(507.4	5e	562.5	3b	591.1	3b
445.8	2b		481.2	4c		508.2	3b	563.0	2c	591.5	4b
446.1	2b		482.1	2d		509.1	3b	564.1	4c	591.9	4b

[1] See Plates facing Lecture V.

Plate III. Strip 1 (*continued*).

592.3	3b	602.8	1a	638.4	1b		669.5	2b		690.9	1a	
(592.7	6a	606.0	1b	639.8	1b		678.6	1b		692.1	2a	
(593.1	4g	608.3	1a	641.0	2b	*Ca*	681.4	1a	from	693.4	1	
595.0	1a	612.4	1b	645.3	1b		682.8	1b		694.1	6e	*Air*
596.6	1a	613.4	1a	648.1	1b		683.1	2a	to	694.1	1	
597.4	1b	623.4	1b	654.3	2b		685.3	1b		698.1	2a	
601.2	1a	626.1	1b	659.3	2a		689.8	2b		700.0	2a	
601.8	1b	631.4	1b	665.7	2a							

Plate III. Strip 2.

	690.9	1a		729.0	2b	*Ca*	798.1	3a	
	692.1	2a		731.7	5b	*Ca*	799.5	1a	*Fe*
from	693.4	1		734.0	1d		799.8	2b	
	694.1	6a	*Air*	736.9	3b	*Ca*	800.3	2b	
to	694.8	1		740.9	3b	*Ca, Cd*	801.2	1a	
	698.1	2a		743.7	2b		801.5	1a	
	700.0	2a		744.3	4b		802.7	1b	
	701.1	2b		748.1	4b		803.8	2a	
	702.1	2a		748.7	3b		805.8	1b	
	702.6	1b		750.1	1a		807.4	2b	
	705.5	2a		751.0	1b		808.2	2c	
	705.9	2a		752.3	4b		808.7	1c	
	707.5	1b		753.8	3b	*Sr*	809.5	3b	*Au*
	708.6	2b		756.9	5b	*Fe*	809.9	2d	
	710.5	2a		759.3	3b		812.7	1a	
	711.4	3c		764.2	1a		813.1	2a	
	713.0	2a		771.8	1a	*Zn*	815.0	4b	
	713.2	1b		773.1	2b		816.8	2b	
	714.1	1c		774.8	2b		818.0	3c	
	717.8	2b	*Ca*	778.3	1b	*(Ru, Ir)*	819.0	4b	
from	718.7	2	*Ba*	779.5	1b		820.1	4b	
	719.6	3a		781.9	3b		820.9	4b	
	720.1	2c	*Ca*	783.1	4b		823.5	1a	
	721.1	2b	*Fe*	783.8	3b		824.0	4b	
	823.7	2c		786.8	1a		824.9	1d	
	724.2	1b		788.9	3b		826.4	3a	
	725.1	1b	*Air*	791.0	1d		827.6	1a	
	726.7	3c		791.4	3b		828.0	2a	
	727.8	1c		792.0	2d		830.2	3b	
	728.0	2a		794.5	1d		831.6	4c	*Fe*

PLATE III. STRIP 2 (*continued*).

831.7	1*b*		896.1	1*a*		970.5	1*b*	
836.5	2*b*		896.7	1*b*		971.5	2*c*	
838.2	1*b*		898.9	1*a*		972.1	1*b*	
838.6	2*b*		899.1	1*a*		973.1	3*a*	
839.2	2*b*		900.2	1*a*		973.5	3*a*	
845.7	2*b*		901.4	1*a*		974.3	2*a*	
849.7	3*c*	*Fe*	901.6	1*a*		975.0	2*a*	
851.2	1*a*		902.4	1*a*		976.8	3*a*	
851.8	1*a*		903.1	1*a*		977.4	2*a*	
855.0	2*a*		903.6	1*a*		977.7	2*a*	
856.8	2*a*		904.6	1*a*		979.1	1*b*	
857.5	2*a*		906.1	2*c*		980.8	1*a*	
858.3	2*a*		912.1	3*b*	*Fe*	981.2	3*b*	
859.7	3*a*		916.3	2*b*		982.0	1*a*	
860.2	3*d*	*Ca*	923.0	2*b*		982.3	2*a*	
861.6	2*a*		929.5	2*b*		983.0	3*c*	
862.2	1*a*		931.3	4*b*	*Fe*	984.5	1*c*	
863.2	2*c*		932.5	4*b*		986.3	1*a*	
863.9	5*b*	*Ca*	933.3	4*c*		986.7	2*c*	
864.4	1*d*		935.1	4*b*		987.4	1*b*	
866.2	2*b*		936.7	4*b*		988.9	2*a*	
867.1	2*b*		937.4	1*b*		989.2	2*a*	
867.6	1*a*		940.1	3*b*		989.6	2*a*	
869.2	2*b*		940.4	2*b*		990.8	2*a*	
870.9	1*b*		943.4	3*b*		991.2	1*a*	
871.4	2*b*		946.6	3*b*		991.9	3*b*	*Fe*
872.5	1*b*		947.0	1*a*		992.4	1*a*	
874.0	1*b*		949.4	1*b*		993.9	1*b*	
874.3	4*b*	*Ba*	949.8	1*b*		994.3	1*b*	
876.5	4*a*		951.7	1*c*		995.0	1*a*	
877.0	4*c*	*Fe*	952.9	3*b*		997.2	2*b*	
879.8	1*b*		954.3	3*b*		998.1	1*a*	
880.9	1*a*		954.8	3*b*		998.9	1*a*	
881.6	2*a*		958.8	3*b*		999.2	1*a*	
882.6	1*a*		959.6	3*b*		1000.0	1*a*	
883.2	1*b*		961.9	1*a*		1000.4	1*a*	
884.9	4*b*	*Cu, Co*	963.7	1*c*		1001.4	1*a*	
887.7	2*a*	*Ni*	964.4	1*c*		1002.8	6*b*	*Na*
890.2	1*b*	*Ba*	968.7	2*a*		1005.0	2*b*	*Ni*
891.7	2*a*	*Ni*	969.0	2*a*		1006.8	6*b*	*Na*
894.9	2*c*	*Ca, Li*	969.6	3*a*				

PLATE III. STRIP 3.

1000.0	1a		1122.6	2a		1189.3	3b	
1000.4	1a		1128.3	2b		1190.1	2b	
1001.4	1a		1130.9	2b		1193.1	3a	
1002.8	6b	Na	1133.1	3c		1199.6	2d	
1005.0	2b	Ni	1133.9	3c		1200.6	4b	Fe
1006.8	6b	Na	1135.1	4d		1201.0	2a	
1011.2	3a		1135.9	2c		1203.5	2c	
1023.0	1a		1137.0	2b		1204.2	2c	
1025.5	3a		1137.8	3b		1204.9	2d	
1027.7	2a		1141.3	2c		1206.1	1c	
1029.3	3c	Ca, Ni	1143.6	2c		1207.3	5g	Fe
1031.8	2a	Ba	1146.2	1b		1217.8	5d	Fe, Ca
1032.8	1a		1147.2	1b		1219.2	8c	Ca
1035.3	1a		1148.6	1b		1220.1	2c	
1058.0	2b		1149.4	1b		1221.6	5d	Ca
1063.0	2b		1151.1	4b		1224.7	5d	Ca
1065.0	2b		1152.5	2b		1225.3	1b	
1066.0	1a		1154.2	2b		1226.6	2d	
1067.0	2b		(1155.7	3b		1228.3	2d	Ca
1070.5	2b		(1155.9	2c		1229.6	4c	Ca
1073.5	1a		1158.3	2a		1230.5	2	
1074.2	1a		1160.9	2a		1231.3	5d	Fe
1075.5	3a		1165.2	1a		1232.8	2b	
1077.5	1a		1165.7	1a		1235.0	3d	Ca
from 1078.9	1		1167.0	1d		1237.8	2c	
to 1079.7			1168.3	1a		1239.9	4a	Fe
1080.3	1a		1169.4	1a		1242.6	6c	Fe
1080.9	1a		1170.6	2c		1245.6	4d	Fe
1081.8	2b	Cu	1174.2	5d		1247.4	3b	
1083.0	2a	Ba	1175.0	2a		1248.6	3d	
1087.5	2a		1176.6	3c		1250.4	3c	
1089.6	2a		1177.0	2a		1251.1	2b	
1096.1	3c	Fe	1177.3	1a		1253.3	2b	
1096.8	1a		1177.6	1a		1255.2	2b	
1097.8	1a		1178.6	1a		1257.5	3c	
1100.4	1a		1179.0	1a		1258.5	2b	
1102.1	3b		1179.4	1a		1264.4	1a	
1102.9	3a		1179.8	1a		1264.9	2a	
1103.3	2b		1180.2	1a		1267.3	3a	
1104.1	2b		1183.4	2a		1268.0	3a	
1107.1	2c		1184.8	3a		1271.9	1a	
1111.4	1a		1186.8	2a		1272.4	1a	
1119.0	2a		1187.1	2a		1274.2	3b	Ba

PLATE III. STRIP 3 (*continued*).

1274.7	3*a*	*Sr*	1289.7	2*c*		1299.7	2*c*	
1276.2	2*a*		1291.9	3*c*		1302.0	2*c*	
1276.7	1*a*		1293.8	3*c*		1303.5	5*c*	
1280.0	6*d*		1294.5	3*c*		1306.7	5*c*	
1281.3	3*c*		1295.6	1*a*		1315.0	4*c*	
1282.6	2*c*		1296.3	2*c*		1315.7	2*b*	
1285.3	2*c*		1297.5	1*a*		1319.0	3*c*	*Co*
1287.5	1*c*	*Ba*	1298.9	5*c*				

PLATE III. STRIP 4.

1315.0	4*c*		1372.6	5*b*	*Fe*	1427.5	3*b*	
1315.7	2*b*		1374.8	1*c*		1428.2	5*b*	*Fe*
1319.0	3*c*	*Co*	1375.8	2*a*		1430.1	5*b*	
1320.6	4*c*	*Sr*	1377.4	1*a*		1431.2	1*b*	
1321.1	3*b*		1379.0	1*a*		1438.9	4*c*	*Co*
1323.3	2*b*		1380.5	4*c*	*Fe*	1440.2	1*b*	*Co*
1324.0	2*b*		1384.7	4*c*	*Fe*	1443.1	2*b*	
1324.8	4*d*	*Ni*	1385.7	5*b*	*Cr*	1443.5	2*b*	*Ca*
1325.3	2*d*		1386.3	2*b*		1444.4	4*b*	
1327.7	4*b*		1387.4	2*b*		1446.7	4*c*	
1328.7	2*b*		1389.4	6*c*	*Fe*	1448.7	2*a*	*Co*
1330.4	3*b*		1390.9	5*d*	*Fe*	1449.4	1*a*	*Co*
1333.3	1*a*		1394.2	4*c*		1450.8	5*c*	*Fe*
1334.0	4*b*		1395.3	1*c*		1451.8	5*b*	*Fe*
1336.3	1*b*		1396.4	2*c*		1453.7	1*a*	
1337.0	4*d*	*Fe*	1397.5	5*c*	*Fe*	1454.7	3*b*	
1337.8	1*b*		1400.2	3*b*		1456.6	1*a*	
1338.5	1*b*		1401.6	4*c*	*Fe*	1458.6	3*c*	
1343.5	6*c*	*Fe*	1403.1	3*c*		1461.5	2*c*	
1351.1	5*d*	*Fe*	1404.1	1*b*		1462.2	2*c*	
1352.7	5*b*	*Fe*	1405.2	3*b*		1462.8	5*c*	*Fe*
1356.5	1*a*		1410.5	4*c*	*Fe*	1463.3	5*c*	*Fe*
1360.9	1*a*		1412.5	2*b*		1464.8	1*a*	
1361.6	1*a*		1414.0	2*b*		1465.3	1*a*	
1362.9	5*b*	*Fe*	1415.8	2*b*		1466.8	5*c*	*Fe*
1364.3	1*a*		1419.4	2*b*		1468.8	2*b*	
1364.7	1*a*		1421.5	6*c*	*Fe*	1469.6	1*b*	
1367.0	6*d*	*Fe*	1423.0	5*b*	*Fe*	1473.9	5*b*	*Fe*
1371.4	1*b*	*Ba*	1423.5	2*b*		1475.3	1*a*	
1372.1	1*b*		1425.4	5*b*	*Fe*	1476.8	1*a*	

PLATE III. STRIP 4 (*continued*).

1477.5	1a		1527.7	5c	*Fe, Co*	1579.4	2a	
1483.0	4b		1528.7	5c	*Ca*	1580.1	2a	
1487.7	5b	*Fe*	1530.2	4c	?	1588.3	1g	*Cu*
1489.2	2c		1531.2	4c		1589.1	3b	
1489.9	1a		1532.5	4b	*Ca*	1590.7	3b	
(1491.2	1c		1533.1	4b	*Ca*	1592.3	3b	
1491.6	3c		(1541.4	1g		1598.9	2b	
1492.4	4b		1541.9	3b		(1601.4	6b	*Cr*
1493.1	4b		1543.7	2a		1601.7	3d	
1494.5	1a		1545.5	2a		1604.4	5b	*Cr*
1495.9	1a		1547.2	3a		1606.4	5b	*Cr*
1497.3	1a	*Cu*	1547.7	2a		1609.2	5b	
1501.3	2b		1551.0	2a		1611.3	1c	
1504.8	1a		1551.6	2a		1613.9	3b	
1505.3	1a		1555.6	2a		1615.6	2b	
1505.7	2a		1557.3	3a		1616.6	1b	
1506.3	5c	*Fe*	1561.0	1a		1617.4	2b	
1508.6	5b	*Fe*	1564.2	1a		1618.2	3b	
1510.3	2c	*Co*	1566.5	2b	*Co*	1618.9	4b	
1515.5	1d		1567.5	2b		1621.5	1b	
1516.5	4c		1569.6	5c	*Fe*	1622.3	5c	*Fe*
1519.0	4d		1573.5	5a		1623.4	5b	*Fe*
1522.7	6c	*Fe, Ca*	1575.4	1b		1627.2	5b	*Ca*
1523.7	6c	*Fe*	(1577.2	5c	*Fe*	1628.2	1b	
1525.0	1b	*Co*	1577.8	3c				

PLATE IV. STRIP 1.

1621.5	1b		(1648.4	4e		1670.3	1a	
1622.3	5c	*Fe*	1648.8	6f	*Mg*	1671.5	3b	
1623.4	5b	*Fe*	(1649.2	4e		1672.2	4a	*Ni*
1627.2	5b	*Ca*	1650.3	6b	*Fe*	1673.7	4a	
1628.2	1b		1653.7	6b	*Fe, Ni*	1674.7	3c	*Cu*
(1631.5	1b		1654.0	4c		(1676.2	2d	
1633.5	4g		1655.6	6e	*Fe, Mg*	1676.5	4b	
(1634.1	6g	*Mg*	1655.9	4d		1677.9	4c	
1634.7	4g		1657.1	5b		1681.6	4c	
1638.7	1b		(1658.3	2b		1684.0	4a	*Ni*
1642.1	1b		to 1659.4	1		1684.4	1b	
1643.0	1b	*Ni*	1662.8	5b	*Fe*	1685.9	2a	
1647.3	5a		1667.4	3a		1686.3	2a	

Plate IV. Strip 1 (*continued*).

1689.5	5c		1784.4	1b		1868.4	5b	
1690.0	5b	*Ni*	1785.0	4b		1869.5	1c	*Ni*
1691.0	5b		1787.7	2c		1870.6	3a	
1693.8	6c	*Fe*	1788.7	3b		1872.4	5b	
1696.5	3c		1793.8	4b		1873.4	6b	
1697.0	3c	*Ni*	1795.4	1a		1874.2	2a	
1701.8	5c	*Fe*	1796.0	3a		1874.8	2a	
(1704.6	2c		1797.8	1a		1875.8	2c	
(1704.9	3b		1799.0	4c		1876.5	6b	
(1707.6	2c		1799.6	3b		1884.3	6b	
(1707.9	3b		1806.4	2b		1885.8	6b	
1710.7	5a		1818.7	5b		1886.4	6b	
1712.2	3b		1821.4	5b		1889.5	1g	
1713.4	5b		1822.6	3a		1891.0	3b	
1715.2	4b		1823.2	2a		1892.5	5b	
1717.9	4b		1823.6	2a		1893.8	1b	
1719.4	1c		1828.6	1b		1894.8	3b	
1720.9	1a		1830.1	3b		1896.2	4b	
1727.3	3b	*Ni*	1832.8	2a	*Ca*	1897.9	1c	
1733.6	5b		1833.4	6c		1900.0	1c	
1734.6	3b		1834.3	6c		1904.5	4b	
1737.7	5d		1835.9	3b		1905.1	2c	
1741.0	4b	*Cu*	1836.7	3c		1908.5	5d	
1742.7	1a		1837.5	3c		1911.9	3c	
1743.1	1a		1841.0	4b		1916.2	1d	
1744.6	2a		1841.6	4b		1917.5	4b	
1748.9	3c	*Ni*	1842.2	4b	*Ni*	1917.9	4b	
1749.6	2d	*Ni*	1848.9	2c		1919.8	4b	
1750.4	5c		1851.0	1c		1920.2	4b	
1752.0	2b		1853.2	3b		1921.1	4b	*Ni*
1752.8	4c		1854.0	2b		1922.0	4b	
1762.0	3c		1854.9	4c		1922.4	4b	
1771.5	3c		1856.9	1c		1923.5	4b	
1772.5	3c		1857.9	2b		1925.8	4b	*Ni*
1774.0	2b		1860.4	2b		1928.0	4b	
1775.8	3b	*Ni*	1861.3	3c		1931.2	1c	
1776.5	3c	*Ni*	1862.3	2b		1932.5	1c	
1777.5	3c		1864.9	3b		1936.2	3c	
1778.5	3c		1867.1	5d	*Fe*	1939.5	2c	
1782.7	3b							

PLATE IV. STRIP 2.

1931.2	1c		2008.6	1b		2080.0	6g	
1932.5	1c		2000.8	2b		2080.5	4c	
1936.2	3c		2013.9	2a		2082.0	6a	*Fe*
1939.5	2c		2014.3	2a		2084.6	2b	
1940.6	2c		2015.7 to 16.9	1		2086.0 to 86.9	1	
1941.5	3b		2017.7	2b		2086.9	3b	*Ni*
1943.5	2c			1		2087.6	1a	
1944.5	3b		2018.5	2b		2089.7	1a	
1947.6	4c		2019.5	2a		2090.9	1a	
1949.4	1c		2021.2	1g		2094.0	2b	
1953.6	2b		2024.9	1a		2096.8	1b	
1960.8	6b		2025.7	4a	*Ni*	2098.8	1a	
	4	*Fe*	2026.8	4b		2099.8	2a	
1961.2	6b		2031.1	2c	*Ba*	2100.4	1a	
1964.3	2c		2035.4	1b		2102.6	4a	
1966.2	2b		2039.6	1b		2103.3	4b	
1966.7	2b		2041.3	6c	*Fe*	2104.0	4a	
1970.1	3b		2042.2	6b	*Fe*	2105.1	4b	
1974.7	4b		2044.5	5b		2107.0	1a	
1975.7	2d		2045.0	5b		2107.4	2a	
1979.2	3c		2047.0	3d		2109.1	2b	
1982.8	5a		2047.8	3b		2111.1	3b	
1983.3	5a		2049.3	3a		2112.7	3b	*Ni*
1983.8	5a		2049.7	3a		2115.0	3a	*Ni*
1984.5	4b		2051.3	3c		2115.4	3a	
1985.8	4b		2053.0	4b		2119.8	1b	
1986.9	2a		2053.7	4c		2121.2	4b	
1987.5	3a	*Ni*	2058.0	6c	*Fe, Ca*	2121.9	5c	
1989.5	6c	*Ba*	2060.0	2b		2124.3	1b	
1990.4	5b		2060.6	5a		2125.1	2b	
1991.8	1b		2061.0	1a		2127.7	3b	
1994.1	5b		2064.7	2c	*Ni*	2132.3	2a	*Co*
1996.9	2a		2066.2	5c	*Fe*	2132.7	1a	
1997.5	2a		2067.1	5c	*Fe*	2133.8	2a	
1999.6	2c		2067.8	3b		2134.3	1a	
2000.6	5a		2068.8	3b		2136.0	5a	*Zn*
2001.6	5c	*Fe*	2070.6	1b		2138.0	2g	
2003.2	3b		2071.3	1b	*Co*	2138.4	4a	
2003.7	1a		2073.5	3b	*Ni*	2139.5	4a	
2004.9	2d		2074.6	2b		2140.4	4a	
2005.2	6d	*Fe*	2076.5	1b		2141.0	2a	
2007.2	6c	*Fe*	2077.3	2b		2142.4	5a	
2008.1	1b	*Ni*	2079.5	4c		2144.6	4a	

PLATE IV. STRIP 2 (*continued*).

2146.9	3*a*		2187.9	5*a*		2219.8	3*b*	
2147.4	4*a*		2188.5	5*a*		2221.3	1*a*	
2148.5	4*a*		2190.1	5*b*		2221.7	1*a*	
2148.9	3*a*		(2191.9	3*c*		2222.3	5*c*	
2150.1	3*a*		2192.3	5*b*		2223.5	3*c*	
2150.5	3*a*		2193.3	5*a*		2225.4	2*b*	
2157.0	3*a*	*Co*, 4*a*	2195.7	2*b*		2226.2	4*b*	
2157.4	5*a*		2197.1	2*b*		2227.6	2*a*	
2159.0	1*c*		2197.7	2*b*		2228.6	2*a*	
2160.6	5*a*		2198.8	4*a*		2229.1	1*a*	
2160.9	4*a*		2199.2	3*a*	*Ni*	2230.7	4*a*	
2161.7	4*a*		2201.1	2*b*		2231.2	2*a*	
2162.6	3*a*		2201.9	5*c*		2232.3	1*a*	
2163.7	4*a*		2203.3	2*a*		(2233.7	5*c*	
2164.0	4*a*	*Ni*	2203.8	1*a*		2234.0	2*c*	
2167.5	6*b*		2205.1	1*b*		2237.4	1*b*	
2171.5	3*b*	*Co*	2206.4	1*a*	*Co*	2238.7	1*b*	
2172.3	2*a*		2206.7	1*a*		2240.0	3*b*	*Zn*
2175.7	2*b*		2209.1	4*c*		2241.4	2*b*	
2176.4	1*b*		2211.7	4*b*		2245.1	3*b*	
2179.0	5*b*		2213.4	4*b*		2246.2	1*b*	
2181.2	3*c*		2215.1	1*b*		2248.2	3*c*	
2184.9	5*b*		2216.7	3*b*		(2249.7	6*a*	*Ni*
2186.5	3*b*		2217.5	3*b*		2250.0	3*d*	
2187.1	5*a*		2218.3	3*c*				

PLATE IV. STRIP 3.

2240.0	3*b*	*Zn*	2259.4	4*c*			2276.4	4*c*
2241.4	2*b*		2261.4	1*b*			2279.8	2*a*
2245.1	3*b*		2262.1	2*a*			2280.7	2*a*
2246.2	1*b*		2263.4	2*a*			2282.0	1*a*
2248.2	3*c*		2264.3	6*d*			2282.3	1*b*
2249.7	6*b*	*Ni*	2266.2	2*a*			2283.6	2*a*
(2250.0	3*d*		2266.6	2*a*			2284.9	2*b*
2253.4	4*b*		2268.0	3*a*			2286.1	2*b*
2256.2	2*b*		2269.1	3*a*		from	(2288.1	2*a*
2257.1	1*d*		2269.9	3*a*			2289.1	1
2257.6	2*b*		2270.2	3*a*			2289.5	2*b*
2258.5	2*c*		2274.2	1*d*			2290.4	1*b*

Plate IV, Strip 3 (continued).

	2291.8	2g	Zn	2361.0	1d	2416.0	3d	
	2293.1	2a		2362.2	1c	2416.3	5b	
		1		2362.6	4b	2418.0	3b	
	2293.6	3b		2364.0	4b	2419.3	5b	Co
	2294.5	2b	Cd	2365.9	2b	2420.6	2b	
	2301.7	4c		2366.8	1b	2422.3	6d	
	2302.9	3b		2367.7	2b	2423.8	3c	
	2305.3	3d		2369.7	2b	2424.4	4b	
	2306.8	4c		2371.4	2b	2426.5	4b	
	2307.8	1b		2371.6	4b	2428.4	1a	
	2308.2	5b		2372.4	4b	2429.5	3b	
	2309.0	5c		2374.2	3b	2431.9	2b	
to	2310.4	1		2375.0	2b	2432.4	1b	
	2310.9	2e		2375.6	4b	2435.3	2b	
	2312.5	3b		2376.1	1b	2435.5	5c	
	2313.7	3b		2379.0	6c	2435.7	2b	
	2314.3	3b		2381.6	6c	2436.5	5a	
	2316.0	2b		2386.1	3b	2438.5	1a	
	2316.6	1b		2386.6	2a	2439.4	2b	
	2322.0	2b		2388.7	2c	2440.0	1a	
	2323.0	2b		2389.7	2c	2441.8	2a	
	2325.3	6d		2390.7	5a	2442.4	1a	
	2328.3	5b		2391.2	1b	2443.9	5a	
	2329.5	5b	Cu	2393.1	5b	2444.2	5a	
	2332.8	2b		2394.4	4a	2445.3	1c	
	2333.0	5b		2395.8	1f	2446.0	5b	
	2334.1	2d	Ni	2396.1	3b	2452.1	2c	
	2335.0	5b		2396.7	2a	2454.1	4b	
	2336.2	2d			1	2457.5	4b	
	2336.8	5b		2397.4	2a	2457.9	4b	
	2339.9	4b		2399.6	3a	2458.6	3a	
	2342.5	1d		2399.9	3a	2459.5	2b	
from	2343.7	1		2402.2	3b	2460.4	1c	
	2345.1	2d		2403.2	3b	2461.2	6b	Ba
	2346.7	4b		2404.9	2b	2463.4	4b	
	2347.3	4b		2406.2	2b	2466.0	3a	
	2349.4	1b		2406.6	6c	2467.3	3c	
	2349.9	2b		2407.2	1b	2467.6	5c	
	2351.4	1c		2408.2	1b	2467.9	3c	
	2352.2	2b		2409.0	1b	2468.7	3a	
	2354.1	6c		2410.2	4b	2470.1	1a	
	2357.4	5a		2412.8	3b	2471.2	2b	
	2358.4	5b		2414.7	2b	2471.4	4a	

PLATE IV. STRIP 3 (continued).

2472.9	4a		2499.0	3b		2537.1	5c	
2473.8	2c		2499.8	3b		2538.0	1b	
2474.6	4b		2500.3	4c		2538.3	2a	
2475.5	1c		(2502.2	4c	} Ba	2540.5	2g	Pt
2477.4	2a		2502.4	1b		2543.5	4c	
2477.8	2a		2505.6	4d		2544.5	2d	
2478.7	2a		2509.1	2d		2545.4	1e	
2479.7	2a		2512.1	1e		2547.2	6c	
2480.1	2a		2512.5	2a		2547.7	2b	
2481.1	1a		2513.2	2b		2548.4	1c	
2482.1	1a		2513.5	1b		2549.7	1b	
2482.4	1c		2517.0	3b		2550.1	1b	
2486.6	5b		(2518.2	2c		(2551.2	1b	
2487.0	5b		2518.4	3a		2551.4	3a	
2488.2	4b		2520.9	3a		(2552.4	3a	
2489.4	5d		2522.3	1a		2552.6	1b	
(2490.5	5a		2525.0	2a		2553.6	3a	
2490.8	3d		2525.4	1b		2554.0	3a	
2493.0	3a		2527.0	4a		(2554.9	3a	
(2493.6	5a	Co	2532.0	2b		2555.1	2c	
2493.9	3f		2535.5	2b		2556.3	2c	
2495.8	5b		2535.9	2b		2559.9	3b	
2497.2	6d		2536.6	1b				

PLATE IV. STRIP 4.

2550.1	1b	2565.0	6c	2585.4	5b	2597.7	3b
(2551.2	1b	2565.9	2b	2587.9	3a	2598.5	1b
2551.4	3a	2566.3	3d	2588.5	5b	(2599.4	3c
(2552.4	3a	2567.8	3b	2589.7	1b	2599.7	5b
2552.6	1b	2568.4	2b	2591.3	4a	2600.6	2a
2553.6	3a	2574.4	5c	2591.7	2c	2601.0	2c
2554.0	3a	2579.3	3d	2593.0	1c	2602.1	4b
2554.9	3a	2581.0	1a	(2594.9	2b	2602.9	1a
2555.1	2c	2581.5	1a		1	2603.6	2b
2556.3	2c	2582.0	2a	2595.4	4a	2604.0	1a
2559.0	3b	2582.4	2a	2595.9	4a	2604.8	4b
2562.1	4b	2582.8	1a	(	1		
2564.0	3b	2584.0	3e	2596.4	2c		

Plate IV. Strip 4 (*continued*).

2605.8	3b	
	2	*Ca*
2606.6	3c	
2607.1	3c	
2608.2	1c	
2608.6	1b	
2609.0	1a	
2610.2	1a	
2612.7	3b	
2613.6	2c	
2614.1	3c	
2616.5	2b	
2619.1	5b	
2619.9	3a	
2620.3	3a	
2622.3	4b	
2624.1	1b	
2625.2	3a	
2625.9	4a	
2626.3	2a	
2627.0	5b	
2627.9	2a	
2628.9	1c	
2629.7	1b	
2630.5	1a	
2633.6	2c	
2634.4	1d	
2635.8	3b	
2636.4	2c	
2637.4	4b	
2638.5	4c	*Ca*
2638.8	5a	
2639.6	1c	
2640.6	2c	
2641.6	3c	
2642.5	2a	
2643.2	1a	
2643.5	1a	
2645.6	4b	
2646.2	2g	
2650.5	5b	(*La, Di*)
2650.7	3a	

	2652.9	1d	
	2653.3	5b	*Ca*
from	2656.7	1	
	2657.2	3b	
	2658.6	1d	
	2664.9	3a	
	2665.9	3b	
	2666.7	1b	
	2667.6	3a	
	2668.0	1b	
	2669.4	3b	
	2670.0	6c	
	2673.8	1a	*Fe*
	2674.3	2a	
	2675.8	3c	
	2676.5	2a	
	2677.2	1a	
	2678.4	1a	
	2679.0	2a	
	2680.0	6b	
	2680.2	3b	
	2681.2	5a	
	2683.1	4b	
	2686.0	3c	
	2686.4	6f	
	2686.8	3a	*Fe*
	2688.4	2c	
	2690.8	5b	
	2691.1	3c	
	2692.3	3c	
	2693.5	4c	
from	2695.2	1	
to	2696.8		
	2698.2	1f	
from	2699.[illegible]	1	
	2700.7	2a	
	2702.1	3b	
	2702.3	4a	
	2702.5	3b	
	2703.5	3a	
from	2703.8	1	
to	2704.9		

2707.4	1f	
2707.7	3a	
2708.0	1b	
2709.0	2b	
2710.6	3a	
2710.9	1g	
2711.9	1a	
2712.8	4a	
2713.3	3a	
2714.3	2a	
2715.3	2b	
2716.1	1d	
2718.5	3g	
2719.0	4c	
	1	
2720.3	2	
2720.8		
2721.6	6	*Fe*
2722.8	3	
2725.5	2d	
2725.8	3c	
2726.8	2a	
2728.0	4b	
2728.4	1b	
2729.8	2a	
2730.7	1b	
2731.6	3c	
2732.4	1c	
2733.7	3b	
2734.1	3b	
	1	
2735.7	3b	
2736.5	3b	
2736.9	3b	
2737.4	1a	
2737.8	2a	
2739.3	3c	
2739.9	1b	
2741.3	3d	
2741.7	3b	
2743.8	1f	
2744.1	4c	
2744.3	1	

PLATE IV. STRIP 4 (*continued*).

Wave-length		
2746.8	} 1	
(2747.2	}	
(2747.6	3*a*	
2748.0	4*c*	
2749.8	3*c*	
2750.6	3*a*	
2754.5	2*c*	
2755.1	1*b*	
2755.8	2*b*	
2756.5	1*c*	
2757.2	1*c*	
2759.4	1*a*	
2760.1	2*a*	
2760.6	2*d*	
2762.0	4*c*	
2763.8	3*f*	
2767.2	1*d*	
2768.2	2*a*	
2768.5	1*a*	
2770.0	2*b*	
2770.8	2*b*	
2771.0	5*c*	
2773.4	4*c*	
(2775.7	6*c*	
(2776.0	4*c*	
2777.3	3*a*	
	} 2	
(2777.8		
	} 1	
2778.5		
2781.2	2*b*	
2782.2	1*b*	
2782.9	3*b*	
2783.9	1*b*	
(2784.8	1*c*	
(2785.1	2*c*	
(2788.8	1*b*	
(2789.1	3*c*	
2790.5	1*c*	
2791.1	3*b*	
2793.0	} 1	
2794.0	} 2	
2795.7		

Wave-length		
2796.7	} 6	
	2	
(2797.6	3*b*	
	2	
(2798.0	3*b*	
	1	
(2798.9	2*c*	
	1	
(2799.5	2*c*	
	1	
(2800.1	3*b*	
	1	
(2800.7	3*b*	
	1	
2801.4	4*d*	
2804.5	1*b*	
2805.4	1*b*	
2806.9	1*c*	
2807.2	2*a*	
2808.6	1*b*	
(2808.8	2*a*	
(2809.0	1*b*	
2810.8	2*b*	
2811.7	2*a*	
2812.0	2*a*	
(2812.5	2*a*	
(2812.8	1*c*	
2814.1	1*b*	
2817.7	3*c*	
2819.2	3*b*	
2819.6	2*b*	
2820.6	} 2	
2821.0	} 3	
2821.6	} 6	
2822.3		
	3	*Fe*
(2823.4	4*c*	
2824.2	3*a*	
	2	
(2825.0	4*c*	
	3	
(2825.9	4*b*	
	3	
(2826.5	4*c*	

Wave-length		
2828.9	3*b*	
2830.7	3*g*	
2834.2	5*c*	*Ca*
2837.7	1*g*	
(2841.4	5*b*	
(2841.7	4*c*	
(2843.0	3*d*	
(2843.3	4*a*	
2844.0	3*b*	
(2845.3	4*f*	
	2	
(2846.1	3*c*	
	2	
(2846.9	4*c*	
	1	
(2847.7	4*a*	
	2	
(2848.0	4*a*	
	2	
(2848.4	3*b*	
	2	
(2848.9	3*b*	
	2	
(2849.3	3*b*	
	2	
(2849.8	3*b*	
	2	
(2850.2	3*b*	
	2	
(2850.7	3*b*	
	2	
(2851.1	3*b*	
	2	
(2851.6	3*b*	
	2	
(2852.0	4*a*	
	2	
(2852.3	4*a*	
	1	
2853.1	} 3	
2853.6	} 4	
2854.1	} 6	*Fe*
2854.7		*Ca*

Plate IV. Strip 4 (*continued*).

2855.2	4		(2863.1	3b		2869.7	5c	*Cu*
2855.7	3		2863.6				4	
2856.9	4d			4		2871.2		
from 2857.9	3	*Sr*	2864.2	5b		2872.2	4d	
2858.5	4a			2			1	
2858.9	2		2864.7	4b	*Ca*	2873.4	2b	
	3			2			1	
2859.4	1		2865.3	4c		2873.9	2b	
2860.2				1			1	
2860.9	2		2866.3	5b		2874.3	3b	
	1			3			1	
(2861.7	4b		2867.1			2874.7	2b	
2861.9	3b			2			1	
	1		2868.1	4c		2875.2	4c	
				3				

POSITIONS OF THE LINES OF CERIUM, LANTHANIUM, DIDYMIUM, PALLADIUM, PLATINUM, IRIDIUM, AND RUTHENIUM, ON PLATES III. & IV. (KIRCHHOFF).

(The lines marked with an asterisk appear to coincide with dark lines in the Solar Spectrum.)

	Ce		from	1364.5	1		*1303.4	2		1279.1	1
	1190.1	1	to	1365.2			1317.6	1	from	1400.0	2
	1249.9	1		1431.9	1		1345.4	1	*to	1400.7	
	1256.7	1		1471.1	1	from	1486.8	2		1430.1	1
	1329.1	2	from	1518.6	1	*to	1489.2			1447.0	1
	1332.4	2	*to	1519.4			*1622.3	1		1477.0	1
	1336.2	1		1536.0	1		*1623.3	1		1495.2	3
	1383.0	2		1541.4	1		1716.6	2		1540.0	1
	1401.7	2		1548.9	2		1728.8	2	from	1566.5	2
	*1438.9	3		*1567.5	1	from	1894.5	2	to	1567.1	
	1460.9	1		1709.2	2	*to	1895.2			1601.4	1
	1517.9	3					1903.0	1	from	1660.0	3
from	1571.0	1		*La*			1940.2	1	to	1660.7	
to	1572.4		from	1411.6	2	from	1988.6	1		1732.9	2
	1573.0	2	*to	1412.8		to	1989.5			1801.9	1
	1623.1	1		1416.8	2		2003.8	2		2002.0	2
from	1629.2	2		1431.0	1		2004.7	2		2123.6	2
to	1630.1			1606.8	2		2031.0	2		2162.0	2
	1683.1	1		1627.9	2		2081.0	2			
	1725.5	1		1634.8	2		2121.4	1		*Pt*	
	*1777.5	2		2136.8	1		2208.2	2		1325.7	1
from	1782.4	1					2214.5	2	from	1488.2	3
to	1784.5			*(La, Di)*			2217.5	2	to	1489.0	
	1938.8	2		1025.0	1					1576.8	1
	2052.3	1		1064.5	1		*Pd*		from	1806.1	2
	2221.5	1		1066.1	1		1114.7	1	*to	1806.9	
				1071.1	1		*1146.2	2		2057.0	1
				1075.6	1		1164.9	2			
	Di			1077.0	1		1185.0	1		*(Ru, Ir)*	
	1225.0	2		1092.1	2		1261.6	2		1348.3	2
	1230.0	1		*1302.0	1		1269.0	2		*1489.9	1

ATMOSPHERIC LINES.

711.1	954.2	964.4	970.5	976.1	988.9	998.1	1008.3	1015.1
918.0	958.8	965.7	972.1	977.4	989.2	999.2	1009.2	1016.4
949.4	959.6	968.7	974.3	977.7	989.6	1000.0	1010.5	1017.7
949.8	961.9	969.0	975.0	982.0	993.1	1001.4	1013.9	1018.2
951.7	963.7	969.6	975.7	982.3	993.4	1005.8		

EXPLANATION OF ÅNGSTRÖM AND THALÉN'S TABLES.[1]

In the following Tables we have registered the principal of the iron lines and of the *new* coincidences which we—*over and above* those observed by Kirchhoff—have discovered between A and G. Their *position* is given in the first column according to the measuring scale adopted in Kirchhoff's and Hofmann's tables, Plates III. and IV. The second column gives the relative *strength* of the lines, according to the gradation employed in the same tables, where 6 indicates the strongest and 1 the weakest lines. The third column contains the *name* of the metal, and the fourth the lines *already* identified by Kirchhoff with corresponding sun lines. The letter *K* indicates that Kirchhoff has observed the coincidence for the same metal as we, but the other signs, as for example *K. Sr*, that he found the coincidence belong to a strontium line, whereas we have found it belong moreover to the metal stated in the third column.

PLATE III. STRIP 1.

468·1	2	Ca		509·1	6	Fe		635·5	2	Fe	
484·4	4	Fe		513·6	2	Ca		661·1	2	Fe	
490·9	5	Ca		528·4	1	Ca		683·6	2	Fe	
503·8	6	Fe		554·3	2	Fe					

PLATE III. STRIP 2.

G 694·1	6	H		802·9	2	Fe		937·5	4	Mn	
719·6	3	Fe		804·5	4	Fe	K	953·3	4	Fe	
721·1	3	Fe	K	820·9	4	Fe		955·1	4	Mn	
744·3	4	Fe		826·0	4	Fe		956·7	4	Mn	
748·1	4	Fe		831·9	4	Fe	K	960·1	3	Fe	
757·3	4	Fe		849·7	3	Fe	K	963·4	3	Fe	
773·8	3	Fe	K. Sr	894·4	1	Na		972·9	3	Fe	
784·9	3	Fe	K	897·6	1	Na		974·3	3	Fe	
789·3	3	Fe		907·8	4	Fe	K	981·8	3	Fe	
793·1	4	Fe		908·1	2	Fe		984·8	3	Fe	
797·8	3	Fe		912·1	3	Fe	K	986·6	3	Fe	
798·9	3	Fe		914·3	2	Fe		991·9	3	Fe	K
801·4	3	Fe		931·2	4	Fe	K				

[1] Abhandl. d. K. Akad. d. Wiss. zu Berlin 1861 u. 1862.

Plate III. Strip 3.

1011.9	3	Fe		1151.1	4	Na		1250.9	4	Fe	K
1025.5	3	Fe		1152.5	2	Fe		1251.6	6	Fe	K
1027.7	3	Fe		1155.8	3	Na		1261.8	4	Fe	K
1040.0	2	Fe		1160.1	3	Fe		1218.4	3	Fe	
1075.5	3	Fe		1159.6	4	Fe		1251.4	3	Fe	
1085.5	2	Fe		1171.2	3	Fe		1257.5	3	Fe	
1086.1	3	Fe	K	1176.6	3	Fe		1261.9	2	Fe	
1102.8	3	Fe		1190.1	2	Fe		1265.0	3	Fe	
1103.3	2	Mn		1191.1	2	Fe		1274.2	3	Fe	K. Ba
1118.0	2	Fe		1201.6	4	Fe	K	1279.2	2	Fe	
1125.3	2	Fe		1202.5	1	Fe	K	1279.9	4	Mn	
1130.9	2	Fe		1217.8	3	Fe	K	1282.6	2	Fe	
1133.9	3	Fe		1221.6	2	Fe	K Ca	1289.9	5	Fe	
1137.1	6	Fe		1224.7	5	Fe	K Ca	1301.5	5	Fe	
1147.8	3	Fe		1220.6	4	Fe		1306.7	5	Fe	
1141.3	8	Fe		1231.3	5	Fe	K	1315.0	4	Fe	

Plate III. Strip 4.

1315.6	4	Fe		1414.0	2	Fe		1530.2	4	Fe	
1320.0	4	Fe	K. Sr	1415.8	2	Mn		1531.8	3	Fe	
1324.4	4	Fe	K. Ni	1421.5	4	Fe	K	1541.2	2	Fe	
1327.7	4	Fe		1421.0	5	Fe	K	1541.5	2	Fe	
1340.0	4	Fe		1425.6	5	Fe	K	1545.2	3	Fe	
1347.6	1	Fe	K	1428.2	5	Fe	K	1550.6	2	Fe	
1349.5	6	Fe	K	1435.3	4	Fe	K. Co	1553.6	2	Fe	
1351.1	5	Fe	K	1441.4	2	Fe		1552.8	3	Fe	
1357.7	5	Fe	K	1450.6	3	Fe. Mn	K. Fe	1560.6	5	Fe	K
1362.9	5	Fe	K	1451.8	3	Fe	K	1563.5	5	Fe	
1367.6	6	Fe	K	1456.6	4	Fe		1572.2	3	Fe	K
1372.6	5	Fe	K	1462.8	5	Fe	K	1580.1	3	Fe	
1377.4	1	Mn		1472.8	3	Fe	K	1589.7	3	Fe	
1380.5	6	Fe	K	1461.8	3	Fe	K	1592.2	3	Fe	
1384.7	4	Fe	K	1462.5	3	Fe	K	1601.7	3	Fe	
1388.1	6	Fe	K	1483.0	4	Fe		1600.4	3	Fe	K. Cr
1390.0	5	Fe	K	1487.2	3	Fe	K	1609.2	5	Fe	
1394.2	4	Mn		1498.3	3	Fe		1612.9	3	Fe	
1397.5	5	Fe	K	1500.8	3	Fe	K	1616.0	4	Fe	
1400.2	3	Mn		1506.6	3	Fe	K	1622.3	5	Fe	K
1401.6	4	Fe	K	1510.0	3	Fe		1623.4	3	Fe	K
1403.1	3	Fe		1523.7	5	Fe	K	1627.2	5	Fe	K. Ca
1405.2	3	Fe		E 1523.3	3	Fe	K				
1410.5	4	Fe	K	1527.7	3	Fe	K				

Plate IV. Strip 1.

1622.3	3	Fe	K	1737.1	5	Fe		1857.9	2	Ca	
1625.6	6	Fe	K	1741.4	5	Fe		1860.6	2	Ca	
1627.2	3	Fe	K. Ca	1752.8	4	Fe		1861.3	3	Fe	
1633.5	4	Fe		1762.0	3	Fe		1864.2	3	Ca	
b 1635.7	1	Fe		1772.5	3	Fe		1867.1	3	Fe	K
1650.3	6	Fe	K	1777.0	3	Fe		1868.4	3	Fe	K. Ni
1653.7	6	Fe	K	1778.3	3	Fe		1872.4	6	Fe	
1663.6	6	Fe	K	1782.7	3	Fe		1873.4	6	Ca	
1665.0	5	Fe	K	1791.6	4	Fe		1882.5	6	Fe	
1671.5	3	Na		1796.7	2	Fe		1884.8	3	Fe Ca	
1674.7	3	Na	K. Ca	1798.8	4	Fe		1890.8	3	Fe	
1676.5	4	Fe		1798.9	3	Fe		1892.8	5	Fe	
1677.9	4	Fe		1799.6	4	Fe		1894.8	3	Fe	
1681.0	4	Fe		1804.8	3	Ca		1896.2	4	Ca	
1690.0	3	Fe	K. Ni	1818.7	5	Fe		1904.5	4	Fe	
1692.0	5	Fe		1821.4	5	Fe		1906.5	5	Fe Ca	
1695.6	6	Fe	K	1830.1	5	Fe		1911.0	3	Fe	
1698.5	3	Fe		1832.4	5	Fe		1917.5	4	Fe	
1701.8	5	Fe	K	1834.3	5	Fe		1919.8	4	Fe	
1704.9	3	Fe		1835.9	3	Fe. Ca		1921.1	4	Fe Na	
1710.7	5	Fe		1837.5	3	Ca		1922.4	4	Fe	
1713.4	5	Fe		1841.0	4	Fe		1923.5	4	Ca	
1715.0	4	Fe		1841.6	4	Ca		1926.0	4	Na	
1717.9	4	Fe		1850.7	3	Fe		1930.2	3	Fe	
1733.0	5	Fe		1851.3	4	Fe		1932.5	2	Fe	

Plate IV. Strip 2.

[illegible]	[illegible]	Fe		[illegible]	[illegible]	Fe		[illegible]	[illegible]	Fe	
[illegible]	[illegible]	Fe		[illegible]	[illegible]	Fe		[illegible]	[illegible]	Fe	
[illegible]	[illegible]	Fe		[illegible]	[illegible]	Fe		[illegible]	[illegible]	Fe	
[illegible]	[illegible]	Fe		[illegible]	[illegible]	Fe	K	[illegible]	[illegible]	Mn	
[illegible]	[illegible]	Fe	K	[illegible]	[illegible]	Fe	K	[illegible]	[illegible]	Fe	
[illegible]	[illegible]	Fe		[illegible]	[illegible]	Fe		[illegible]	[illegible]	Fe	
[illegible]	[illegible]	Fe		[illegible]	[illegible]	Fe		[illegible]	[illegible]	Mn	
[illegible]	[illegible]	Fe		[illegible]	[illegible]	Fe		[illegible]	[illegible]	Mn	
[illegible]	[illegible]	Fe		[illegible]	[illegible]	Fe	K	[illegible]	[illegible]	Mn	
[illegible]	[illegible]	Fe		[illegible]	[illegible]	Fe	K	[illegible]	[illegible]	Fe	
[illegible]	[illegible]	Fe		[illegible]	[illegible]	Fe	K	[illegible]	[illegible]	Fe	
[illegible]	[illegible]	Fe		[illegible]	[illegible]	Fe		[illegible]	[illegible]	Mn	
[illegible]	[illegible]	Fe		F [illegible]	[illegible]	H		[illegible]	[illegible]	Fe	
[illegible]	[illegible]	Fe	K	[illegible]	[illegible]	Fe	K	[illegible]	[illegible]	Fe	
[illegible]	[illegible]	Fe	K	[illegible]	[illegible]	Fe		[illegible]	[illegible]	Fe	
[illegible]	[illegible]	Fe	K	[illegible]	[illegible]	Mn		[illegible]	[illegible]	Fe	
[illegible]	[illegible]	Fe		[illegible]	[illegible]	Fe		[illegible]	[illegible]	Fe	
[illegible]	[illegible]	Fe		[illegible]	[illegible]	Fe		[illegible]	[illegible]	Mn	

Plate IV. Strip 3.

[illegible]	[illegible]	Fe		[illegible]	[illegible]	Fe		[illegible]	[illegible]	Ca	
[illegible]	[illegible]	Mn		[illegible]	[illegible]	Fe		[illegible]	[illegible]	Ca	K. Co
[illegible]	[illegible]	Fe		[illegible]	[illegible]	Fe		[illegible]	[illegible]	Ca	
[illegible]	[illegible]	Fe		[illegible]	[illegible]	Fe		[illegible]	[illegible]	Fe	
[illegible]	[illegible]	Mn		[illegible]	[illegible]	Fe		[illegible]	[illegible]	Fe	
[illegible]	[illegible]	Fe		[illegible]	[illegible]	Fe		[illegible]	[illegible]	Fe	
[illegible]	[illegible]	Mn		[illegible]	[illegible]	Fe		[illegible]	[illegible]	Ca	
[illegible]	[illegible]	Fe		[illegible]	[illegible]	Fe		[illegible]	[illegible]	Ca	
[illegible]	[illegible]	Mn	K. Zn	[illegible]	[illegible]	Fe		[illegible]	[illegible]	Fe	K. Co
[illegible]	[illegible]	Fe		[illegible]	[illegible]	Fe		[illegible]	[illegible]	Fe	
[illegible]	[illegible]	Fe		[illegible]	[illegible]	Fe, Ca		[illegible]	[illegible]	Fe	
[illegible]	[illegible]	Fe		[illegible]	[illegible]	Fe		[illegible]	[illegible]	Fe	
[illegible]	[illegible]	Fe, Ca		[illegible]	[illegible]	Fe					
[illegible]	[illegible]	Fe		[illegible]	[illegible]	Fe					

Plate IV. Strip 4.

[illegible]	[illegible]	Fe		[illegible]	[illegible]	Fe		[illegible]	[illegible]	Fe	
[illegible]	[illegible]	Fe		[illegible]	[illegible]	Fe		[illegible]	[illegible]	Fe	
[illegible]	[illegible]	Fe		[illegible]	[illegible]	Fe, Ca		[illegible]	[illegible]	Fe	
[illegible]	[illegible]	Fe		[illegible]	[illegible]	Fe		[illegible]	[illegible]	Fe	
[illegible]	[illegible]	Fe		[illegible]	[illegible]	Fe, Mn	K. Fe	[illegible]	[illegible]	H	
[illegible]	[illegible]	Fe		[illegible]	[illegible]	Fe		[illegible]	[illegible]	Fe	
[illegible]	[illegible]	Fe		[illegible]	[illegible]	Fe		[illegible]	[illegible]	Fe	K
[illegible]	[illegible]	Fe		[illegible]	[illegible]	Fe	K	[illegible]	[illegible]	Fe	
[illegible]	[illegible]	Ca		[illegible]	[illegible]	Fe		[illegible]	[illegible]	Fe	
[illegible]	[illegible]	Fe	K. Ca	[illegible]	[illegible]	Fe		G [illegible]	[illegible]	Fe	K
[illegible]	[illegible]	Fe		[illegible]	[illegible]	Fe		[illegible]	[illegible]	Fe	
[illegible]	[illegible]	Fe		[illegible]	[illegible]	Fe		[illegible]	[illegible]	Fe	K. Ca
[illegible]	[illegible]	Fe		[illegible]	[illegible]	Fe					
[illegible]	[illegible]	Ca		[illegible]	[illegible]	Fe	K				

DESCRIPTION OF BROWNING'S NEW AUTOMATIC SPECTROSCOPE.

This instrument is furnished with a battery of six equilateral prisms of dense flint glass. All the prisms are linked together like a chain by their respective corners; the bases being in this manner linked together. This chain of prisms is then bent round, so as to form a circle with the apices outwards. The centre of the base of each prism is attached to a radial rod. All these rods pass through a common centre. The prism nearest the collimator, that is, the first prism of the train, is a fixture. The movement of the other prisms is then in the proportion of 1, 2, 3, 4, 5, the last, or 6th prism, moving five times the amount of the 2nd. All these motions are communicated by the simple revolution of the micrometer screw, which is used for measuring the position of the lines in the spectrum, and the amount of motion of each and of the telescope is so arranged, that the prisms are automatically adjusted to the minimum angle of deviation for the ray under examination. It is easy to test the efficiency of the instrument in this respect. On taking the lenses out from the eyepiece of the telescope, the whole field of view is found to be filled with light of the colour of that portion of the spectrum which the observer wishes to examine; while in a spectroscope of the usual construction, at the extreme ends of the spectrum, just where the light is most required, only a lens-shaped line of light would be found in the field of view. As a consequence of this peculiarity, the violet and deep-red ends of the spectrum are greatly elongated, or, rather, much more of them can be seen than in an ordinary spectroscope, and the H lines, which are generally seen only with difficulty, come out in a marked manner.

LIST OF THE PRINCIPAL

MEMOIRS, ETC. ON SPECTRUM ANALYSIS.

I.

LECTURES OR MEMOIRS RELATING TO THE SUBJECT OF SPECTRUM ANALYSIS GENERALLY.

BREWSTER, SIR D.:

Data towards a History of Spectrum Analysis. Compt. Rend. lxii. 17.

DELAUNAY:

Notice sur la Constitution de l'Univers. Première Partie: Analyse Spectrale. Annuaire (1869) publié par le Bureau des Longitudes. Paris: Gauthier-Villars.

A most masterly and complete essay on the subject.

DIBBITS, H. C.:

De Spectraal-Analyse. Academisch Proefschrift. Rotterdam: Tamelmeyer, 1863.

A complete treatise on Spectrum Analysis, giving an historical sketch of the discoveries, with chromoliths of the Carbon and other Spectra.

GRANDEAU, L.:

Instruction pratique sur l'Analyse Spectrale. Paris: Mallet-Bachelier. 1863. I. Description des Appareils.—II. Leur Application aux Recherches chimiques.—III. Leur Application aux Observations physiques.—IV. La Projection des Spectres. Avec 2 planches sur cuivre et 1 planche chomolithographiée.

HERSCHEL, ALEX. S.:

On the Methods and recent Progress of Spectrum Analysis. Chem. News, xix. 157.

HUGGINS, WILLIAM:

Lecture on the Physical and Chemical Constitution of the Fixed Stars and Nebulæ. Royal Institution of Great Britain, May 19, 1865. Chemical News xi. 270.

HUGGINS, WILLIAM:

On some further Results of Spectrum Analysis as applied to the Heavenly Bodies. Printed *in extenso* in Report of British Association, 1868, p. 152.

On the Results of Spectrum Analysis as applied to the Heavenly Bodies. A Lecture delivered before the British Association at the Nottingham Meeting, August 24, 1866. Published, with photographs of the Stellar Spectra, by William Ladd, Beak Street, London. Chemical News, xiv. 173, 190, 200, 215.

On some Recent Spectroscopic Researches. Quarterly Journal of Science, No. xxii. April 1869.

JAMIN:

Lectures on Spectrum Analysis. Journ. Pharm. Third Series, xlii. 9, 1862.

KIRCHHOFF, G.:

On the Solar Spectrum and the Spectra of the Chemical Elements. Parts I. and II. Macmillan, 1861–62.

These Memoirs are translations of the original communications to the Academy of Sciences of Berlin. They contain Kirchhoff's theory of the chemical and physical constitution of the Sun, and are accompanied by four plates of the fixed dark lines in the Solar Spectrum from A to G, and the bright lines of the Metals, showing their coincidences. Reduced copies of these plates are given facing Lecture V., and copies of the Tables at the end of this Volume.

LOCKYER, J. N.:

On Recent Discoveries in Solar Physics made by means of the Spectroscope. Royal Institution Proceedings, May 28, 1869. Phil. Mag. [4] xxxviii. 142.

Giving an abstract of Lockyer's own researches on Solar Physics.

On Spectrum Analysis. Lectures delivered before the Society of Arts. Journal of the Society of Arts, 1870.

MILLER, W. A.:

Lectures on Spectrum Analysis (1862). Pharmaceutical Journal, Second Series, iii. 399. Chemical News, v. 201–214.

A Course of Four Lectures on Spectrum Analysis, with its Applications to Astronomy. Delivered at the Royal Institution of Great Britain. May—June 1867. Chemical News, xv. 259, 276; xvi. 8, 20, 47, 71.

Exeter Lecture, 1869. Popular Science Review, Oct. 1869.

MOUSSON, A.:

Résumé de nos Connaissances actuelles sur le Spectre. Archives des Sciences Naturelles de Genève, tome x. mars 1861.

ROSCOE, H. E.:

Lectures on Spectrum Analysis. Delivered at the Royal Institution of Great Britain (1861). Chemical News, iv. 118.

ROSCOE, H. E.:

Lectures on Spectrum Analysis. Ditto (1862). Chemical News, v. 218, 261, 287.

SCHELLEN, A.:

Die Spectral Analyse in ihre Anwendung auf die Stoffe der Erde und die Natur der Himmelskorper. Braunschweig: Westermann, 1870.

A valuable and luminous account of the recent discoveries in Celestial Chemistry and Physics, fully and accurately illustrated by engravings and chromoliths.

SECCHI:

Résumé of the Results of Spectrum Analysis applied to Astronomy. N. Arch. Ph. Nat. xxiii. 145.

STEWART, BALFOUR:

On the Sun as a Variable Star. Lecture at Royal Institution, April 12, 1867.

THALÉN, R.:

Spektralanalys exposé och Historik, med en Spektralkarta. Upsala, 1866.

TYNDALL, J.:

On the Basis of Solar Chemistry. June 7, 1861. Phil. Mag. Fourth Series, xxii. 147.

II.

MEMOIRS RELATING TO THE APPLICATION OF SPECTRUM ANALYSIS TO TERRESTRIAL CHEMISTRY.

ALLEN, O. D.:

Observations on Cæsium and Rubidium. Silliman's Journal, November 1862. Phil. Mag. xxv. 189.

The new alkalies shown to be contained in lepidolite from Hebron.

ALLEN, O. D., & JOHNSON, S. W.:

On the Equivalent and Spectrum of Cæsium, showed that the Equivalent of Cæsium is 133. Silliman's Journal, January 1863. Phil. Mag. xxv. 196.

ÅNGSTRÖM, A. J.:

Optical Researches. Pogg. Ann. xciv. 141. Phil. Mag. Fourth Series, ix. 327.

In this he shows that a twofold spectrum is always seen when we examine the Electric Spark; one set of lines being due to the ignition of the particles of air or gas through which the sparks pass, whilst the second set is caused by the incandescence of the metallic particles themselves.

ATTFIELD:

On the Carbon Spectrum. Phil. Trans. 1862, p. 221.

He obtained results similar to those of Swan, but noticed a larger number of lines, and attributes the lines to the glowing vapour of Carbon.

BABINET:

Sur la Parogénie. Cosmos, xxv. 303 et seq.

BECQUEREL:

On the Spectra of Phosphorescent Bodies. La Lumière, vol. i. p. 207. (See Lecture V.)

BRASSACK:

On the Electric Spectra of the Metals. Zeitschr. f. d. ges. Naturw. ix. 185.

BREWSTER, SIR D.:

On the Action of various Coloured Bodies on the Spectrum. Phil. Mag. Fourth Series, xxiv. 441.

On the Monochromatic Lamp. Edin. Royal Soc. Trans. 1822.

On Paragenic Spectra. Phil. Mag. January 1860.

BUNSEN:

Discoveries of the New Alkaline Metals. Berlin Acad. Ber., 10th May, 1860, p. 221. Chemical News, iii. 132.

On Cæsium. Phil. Mag. xxvi. 241.

Confirms the atomic weight of Cæsium to be 133.

On the Presence of Lithium in Meteorites. Phil. Mag. Fourth Series, xxiii. 474.

On the Preparation of the Rubidium Compounds. Phil. Mag. Fourth Series, xxiv. 46.

On the Inversion of the Bands in the Didymium Absorption Spectra. Phil. Mag. Fourth Series, xxviii. 246; xxxii. 177. (See Lecture IV. Appendix F.)

BUNSEN & BAHR:

On the Erbium Spectrum. Ann. Ch. Pharm. cxxxvii. 1.

CHAUTARD:

Spectra of Rarefied Gases. Phil. Mag. Nov. 1864.

CHRISTOFLE & BEILSTEIN:

On the Phosphorus Spectrum. With a Chromolith of the Spectrum. Ann. Chem. Phys. Fourth Series, iii. 280.

COOKE, J. P.:

On the Construction of Spectroscopes. Am. Journ. Sc. and Arts, vol. xl. Nov. 1865.

CROOKES, W.:

On a Means of increasing the Intensity of Metallic Spectra. Chemical News, v. 234 (1862).

Thallium, Discovery of. Chemical News, iii. 193.

On Thallium and its Compounds. Chem. Soc. Journ. xvii. 112.

DANIEL:

On the Spectra of the Induction Spark. Compt. Rend. lvii. 98.

DEBRAY, M. H.:

Sur la Projection des Raies brillantes des Flames colorées par les Métaux. Ann. Chim. Phys. Troisième Série, lxv. 331.

DELAFONTAINE:

Note on the Absorption Spectra of Erbium, Didymium, and Terbium. Ann. Ch. Pharm. cxxxv. 194. Chemical News, xi. 253.

DIACON, M. E.:

Recherches sur l'Influence des Éléments électronégatifs sur le Spectre des Métaux. Ann. Chim. Phys. Quatrième Série, vi. 1.

With Drawings of the Spectra of Copper Chloride and Bromide, &c.

DIBBITS:

Pogg. Ann. 1864, cxxii. 497.

Observed continuous spectra by combustion of hydrogen in oxygen and chlorine.

FEUSSNER:

On the Absorption of Light at different Temperatures. Phil. Mag. Fourth Series, xxix. 471.

FIZEAU:

On the Spectrum of Burning Sodium. Compt. Rend. liv. 493.

A figure of the phenomena here observed is given in Fig. 62.

FOUCAULT:

Institut, 1849, p. 45.

Observed the dark double line D in the Spectrum from the Electric Arc.

FRANKLAND:

On the Combustion of Hydrogen and Carbonic Oxide in Oxygen under great pressure. Proc. Roy. Soc. xvi. p. 419. (See Lecture II. Appendix D.)

FRASER, W.:

On the Spectrum of Osmium. Chemical News, viii. 34.

GAMGEE, ARTHUR:

On the Action of Nitrites on the Blood. Phil. Trans. 1868, p. 589.

The author shows that the Colour as well as the Absorption Spectrum of Blood undergoes change when acted on by nitrites. The two sharply-defined absorption bands of the oxidised colouring matter become very faint, and an additional though faint band appears in the red. If the blood thus altered be acted upon by ammonia, the colour changes from chocolate-brown to blood-red again, and the absorption band in the red disappears, and the two bands between D and E again become visible.

GASSIOT:

Description of a Large Spectroscope. Proc. Roy. Soc. 1863, xii. 536.

Spectroscope with Eleven Prisms. Phil. Mag. Fourth Series, xxviii. 69.

GIBBS, WOLCOTT:

Description of Large Spectroscope. Silliman's Journal, Second Series, xxxv. 110.

GLADSTONE, J. H.:

On an Optical Test for Didymium. Chem. Soc. Journ. 1858, x. 219.

In this paper the existence of the Didymium Absorption Lines was first pointed out.

On the Use of the Prism in Qualitative Analysis. Chem. Soc. Journ. 1858, x. 79.

In this paper the Absorption Spectra of many coloured metallic salts are given.

On the Violet Flame of many Chlorides. Phil. Mag. Fourth Series, xxiv. 417.

GRANDEAU, L.:

Recherches sur la Présence de Rubidium et Cæsium dans les Eaux naturelles, les Minéraux et les Végétaux. Ann. de Chim. et de Phys. Troisième Série, lxvii.

HEINRICHS:

On the Distribution of Lines in Spectra. Silliman's Journal, July 1864.

HERAPATH, W. BIRD:

On the Use of the Micro-Spectroscope in the Discovery of Bloodstains. Chem. News, xvii. 113, 123.

HOPPE, F.:

On the Absorption Lines in the Blood Spectrum. Schmidt's Jahrbuch d. ges. Med. cxiv. 3 1862.

HUGGINS, WILLIAM:

On the Spectra of some of the Chemical Elements. Phil. Trans. 1864, p. 139, and Poggendorff's Ann.

Giving exact measurement of the lines of twenty-four metals, with maps. For reduced copy of these maps see Plates I. and II. end of Lecture III.

On the Prismatic Examination of Microscopic Objects. Transactions of the Royal Microscopic Society. Quarterly Journal of Microscopical Science, July 1865.

Description of a Hand Spectrum-Telescope. Proc. Roy. Soc. xvi 241.

KETTELER:

Wave-length Measurement of Metallic Lines. Monatsbericht d. K. Pr. Akad. d. Wiss. zu Berlin, 1864, p. 632.

KIRCHHOFF & BUNSEN:

Chemical Analysis by Spectrum Observations. First Memoir. Pogg. Ann. cx. 161. Phil. Mag. Fourth Series, xx. 89. See Lecture II. Appendix A.)

This Memoir contains the exposition of the method of experiment, and a description of the Spectra of the metals of the alkalies and the alkaline earths.

Description of the Properties of the new Metals Cæsium and Rubidium. Second Memoir. Pogg. Ann. cxiii. 337. Phil. Mag. Fourth Series, xxii. 329, 498. (See Lecture III. Appendix A.)

KUNDT:

Spectrum of Phosphorescent Light, with Dark Lines. Phil. Mag. Dec. 1867.

LAMY, A.:

De l'Existence d'un nouveau Métal, le Thallium. Ann. de Chim. et de Phys. Troisième Série, lxvii. 385.

LIELEGG, A.:

On the Spectrum of the Bessemer Flame. Phil. Mag. Fourth Series, xxxiv. 302.

Contributions to our Knowledge of the Spectra of the Flames of Gases containing Carbon. Phil. Mag. Fourth Series, xxxvii. 208.

MASCART:

On the Wave-length of the Lines of certain Metals, directly determined by a Diffraction Grating. Annales Scientifiques de l'Ecole Normale Supérieure, tome iv. Paris, 1866.

MASSON:

On the Nature of the Electric Spark. Giving Drawings of the Electric Spectra of several Metals. Ann. de Chim. et de Phys. Troisième Série, xxxi. 295.

MELDE, F.:

On the Absorption of Light by Coloured Liquids. Pogg. Ann. cxxiv. 91; cxxvi. 264.

MELVILL, THOMAS:

On the Examination of Coloured Flames by the Prism. Edinburgh Phys. and Lit. Essays, ii. 12 (1752).

To the writer of this paper the remarkable phenomena exhibited by coloured lights when examined by the prism were well known.

MILLER, WILLIAM ALLEN:

Experiments and Observations on some cases of Lines in the Prismatic Spectrum produced by the Passage of Light through coloured Vapours and Gases, and from certain coloured Flames. Read June 21, 1845, at British Association. Printed in Phil. Mag. Third Series, xxvii. 81.

On the Photographic Transparency of various Bodies, and on the Photographic Effects on Metallic and other Spectra obtained by means of the Electric Spark. Phil. Trans. 1862, p. 861.

Note on the Spectrum of Thallium. Proc. Roy. Soc. xii. 407.

Heated in the Electric Spark, Thallium exhibits five lines in addition to the green line Tl α.

MILLER, W. HALLOWS:

On the Absorption Bands of Nitrous Acid Gas, &c. Phil. Mag. Third Series, ii. 381.

MITSCHERLICH, ALEXANDER:

On the Spectra of Compounds and Simple Substances. With Two Plates. Phil. Mag. xxviii. 169.

He concludes that compounds whose vapours can be heated up to incandescence without undergoing decomposition yield spectra different to those of their elementary constituents. He adds some singular speculations concerning a supposed relation between the atomic weights of the haloid compounds of Barium and the distance, on an arbitrary scale, read off between the chief lines of their spectra.

MORREN, M. A.:

De la Flamme de quelques Gaz Carburés, et en particulier de celle de l'Acetylène et du Cyanogen. Ann. Chem. Phys. (4), iv. 305. With figure of the Carbon Spectrum.

On the Spectrum of the Non-luminous Gas-flame. Chemical News, ix. 135.

MULDER, E.:

On the Spectra of Phosphorus, Sulphur, and Selenium. Journ. Pr. Chem. xci. 111.

MÜLLER, J.:

Determination of the Wave-length of certain Bright Lines in the Spectrum. Phil. Mag. Fourth Series, xxvi. 259.

Wave-lengths of the Metallic Lines. Fortsch. d. Physik. Jahrg. 1863, p. 191; 1865, p. 229.

On the Wave-length of the Blue Indium Line. Phil. Mag. (4), xxx. 76.

PASTEUR:

On the Spectrum of the Phosphorescent Light emitted by certain Animals. Compt. Rend. lix. 509.

He found that the light from a Mexican *Pyrophorus* gave a continuous spectrum.

PICKERING:

Comparative Efficiency of different Forms of Spectroscopes. Silliman's Journal, May 1868.

PISANI:

On Pollux, a Silicate of Cæsium. Compt. Rend. lviii. 714.

PLÜCKER:

On the Measurements of the Wave-lengths of the Metallic Lines. Quoted in Wiedemann, Lehre von Galvanismus, ii. 875, Taf. i.

On the Nature of the Electric Discharge *in vacuo*. Pogg. Ann. civ. March, August, 1858; cv. May 1859; cvii. 497—638.

PLÜCKER & HITTORFF:

On the Spectra of Ignited Gases and Vapours, with especial regard to the different Spectra of the same Elementary Gaseous Substance. Phil. Trans. 1865, p. 1.

These important experiments show that by varying the physical conditions certain of the elementary bodies yield two distinct spectra. Plücker explains this by the assumption of several allotropic conditions of the element existing at various temperatures.

ROBINSON, DR.:

On Electric Spectra. Phil. Trans. 1863.

ROOD, O. N.:

On the Didymium Absorption Spectrum. Silliman's Journal, Second Series, xxxiv. 129.

ROSCOE, H. E.:

On the Spectrum produced by the Flame evolved in the Manufacture of Steel by the Bessemer Process. Proc. Lit. Phil. Soc. Manchester, Feb. 24, 1863; Phil. Mag. Fourth Series, xxv. 318.

ROSCOE & CLIFTON:

On the Effect of Increased Temperature upon the Nature of the Light emitted by the Vapour of certain Metals or Metallic Compounds. Chemical News, v. 233 (1862).

RUTHERFURD, L. M.:

On the Construction of the Spectroscope. Am. Journ. Sc. Arts. vol. xxxix. 1860, p. 120.

SEGUIN, J. M.:

On the Spectrum of Fluoride of Silicium. Compt. Rend. liv. 993; Chemical News, vi. 282.

On the Spectra of Phosphorus and Sulphur. Phil. Mag. Fourth Series, xxiii. 416.

SORBY, H. C.:

On the Application of Spectrum Analysis to Microscopical Investigations, and especially to the Detection of Bloodstains. Chemical News, xi. 186, 194, 232, 256.

On a Definite Method of Qualitative Analysis of Animal and Vegetable Colouring Matters by means of the Spectrum-Microscope. Proc. Roy. Soc. xv. 433.

On a New Micro Spectroscope, and on a New Method of printing a Description of the Spectra seen with the Spectrum-Microscope. Chem. News, xv. 220.

On Jargonium, a new Element accompanying Zirconium. Chemical News, xix. 121. Proc. Roy. Soc. xvii. 511.

On some remarkable Spectra of Compounds of Zirconia and the Oxides of Uranium. Proc. Roy. Soc. xviii. 107.

STOKES, G. G.:

On the Long Spectrum of the Electric Light. Phil. Trans. 1862, p. 599.

He shows that rays exist in the ultra-violet portion at a distance from the last visible rays equal to six times the length of the whole visible spectrum. Each metal possesses a peculiar series of these bands, which may be rendered visible by allowing the rays to fall on a fluorescent body.

On the Discrimination of Organic Bodies by their Optical Properties. Roy. Inst. March 4, 1864. Phil. Mag. Fourth Series, xxvii. 388.

On the Reduction and Oxidation of the Colouring Matter of the Blood. Proc. Roy. Soc. xiii. 353.

SWAN:

On the Blue Lines of the Spectrum of the Non-luminous Gas-flame. Ed. Phil. Trans. iii. 370, and xxi. 411.

These lines remained unaffected by alteration in composition of the burning body in hydrogen or oxygen. Swan carefully measured the position of these lines, and he first explained the delicacy of the Sodium reaction (1857).

TALBOT, H. FOX:

Some Experiments on Coloured Flames. Brewster's Journal of Science, v. 1826.

On a Method of obtaining Homogeneous Light of great Intensity. Phil. Mag. Third Series, 1833, iii. 35.

On the Flame of Lithia. Phil. Mag. Third Series, 1834, iv. 11.

On Prismatic Spectra. Phil. Mag. 1836, ix. 3.

THALÉN, R.:

On the Determination of the Wave-lengths of the Lines of the Metals. Nova Acta Reg. Soc. Sc. Upsal, Third Series, vi. Upsala, 1868.

Taking as his starting-point the determination of wave-lengths of the principal Fraunhofer's Lines by Angstrom (Recherches sur le Spectre Solaire, par A. J. Ångström. I.—Spectre Normal du Soleil. Upsal. 1868), the author, by graphical interpolation, obtains, from Kirchhoff's and Ångström's tables, the wave-lengths of the bright metallic lines. A large plate accompanies the Memoir, giving the lines and their wave-lengths of forty-five metals. The following twenty-three were examined in the metallic state:- K, Na, Mg, Al, Fe, Co, Ni, Zn, Cd, Pb, Tl, Bi, Cu, Hg, Ag, Au, Sn, Pt, Pd, Os, Sb, Te, In. The remaining twenty-two were examined as chlorides:—Li, Cs, Rb, Ba, Sr, Ca, Gl, Zr, Er, Y, Th, Mn, Cr, Ce, D, L, U, Ti, Wo, Mo, V, As.

TYNDALL & FRANKLAND:

On the Blue Band of the Lithium Spectrum. Phil. Mag. Fourth Series, xxii. 151. 472.

VAN DER WILLIGEN:

On the Spark of the Induction Coil. Pogg. Ann. cvi. 615.

WALTENHOFEN, A. VON:

Spectra of Electric Spark in rarefied Gases. Dingl. Pol. J. clxxvii. 38.

WATTS, W. M.:

On the Spectrum of the Bessemer Flame. Phil. Mag. Fourth Series, xxxiv. 437.

On the Spectra of Carbon. Phil. Mag. Fourth Series, xxxviii. 249.

WHEATSTONE, C.:

On the Prismatic Decomposition of the Electric, Voltaic, and Electro-magnetic Sparks. Read August 12, 1835. British Assoc. Dublin. Chemical News, iii. 198.

WÜLLNER:

On the Spectra of the Gases under different Pressures. Phil. Mag. Fourth Series, xxxvii. p. 405; xxxix. p. 365.

III.

MEMOIRS RELATING TO THE APPLICATION OF SPECTRUM ANALYSIS TO CELESTIAL CHEMISTRY.

AIRY, G. B., Astronomer Royal:

Measurements of Stellar Lines. Monthly Notices of the Roy. Astron. Soc. xxiii. 190.

Wave-lengths of Lines in Kirchhoff's Maps. Phil. Trans. 1868, p. 29.

ÅNGSTRÖM, A. J.:

Wave-length Measurements. Aus Oefversigt af K. Vetensk. Acad. Förh. No. 2. Pogg. Ann. cxxiii. 489.

On the Fraunhofer Lines visible in the Solar Spectrum, and on the Coincidence of these Lines with the Bright Lines of certain Metals. Phil. Mag. Fourth Series, xxiv. 1.

Optical Researches. Vetensk. Acad. February 1853. Phil. Mag. Fourth Series, ix. 327.

The nature of the Electric Spectrum pointed out.

Observations on certain Lines of the Solar Spectrum. Phil. Mag. Fourth Series, xxiii. 76.

Ångström believes that the groups A and B, as well as a group lying between B and C, are due to absorption in the earth's atmosphere, caused by the presence of some compound gas—perhaps carbonic acid—but not due to aqueous vapour.

Recherches sur le Spectre normal du Soleil, avec Atlas de six planches. Upsal: W. Schultz, 1868.

The Solar Lines from A to H mapped according to their wave-lengths. A most valuable and interesting memoir. See Lecture V. Appendix A.

BREWSTER, SIR D.:

Observations of the Lines of the Solar Spectrum, and on those produced by the Earth's Atmosphere, and by the Action of Nitrous Acid Gas. Phil. Mag. Third Series, viii. 384.

BREWSTER & GLADSTONE:

On the Lines of the Solar Spectrum. Map of the Solar Spectrum, giving the Absorption Lines of the Earth's Atmosphere. Phil. Trans. 1860, p. 149.

BROWNING, J.:

On the Spectra of the Meteors of November 13–14, 1866. Phil. Mag. Fourth Series, xxxiii. 234.

CHACORNAC:

Physical Constitution of the Sun. Compt. Rend. lx. 170.

COOKE, J. P.:

On the Aqueous Lines of the Solar Spectrum. Phil. Mag. Fourth Series, xxxi. 337.

When the air was damp Cooke saw seven lines and a nebulous band, besides the Nickel line, between the D lines. These seven lines disappeared when the air was dry.

DITSCHEINER, L.:

Wave-length of Fraunhofer's Lines measured. Wien. Acad. Ber. 2 Abthl. l. 296.

DONATI:

Intorno alle Strie degli Stellari. Il Nuovo Cimento, xv. 292.

DRAPER, WILLIAM:

On the Variation in Intensity of the Fixed Lines of the Solar Spectrum. Phil. Mag. (4), xxv. 342.

FAYE:

On the Physical Constitution of the Sun. Compt. Rend. lx. 89, 138, 468.

FRANKLAND & LOCKYER:

Preliminary Note of Researches on Gaseous Spectra in relation to the Physical Constitution of the Sun. Proc. Roy. Soc. xvii. 288.

Researches on Gaseous Spectra in relation to the Physical Constitution of the Sun, Stars, and Nebulæ. Proc. Roy. Soc. xvii. 453; xviii. 79.

FRAUNHOFER:

Denkschriften der Münchener Academie. 1814 and 1815. Giving exact measurement of the dark Solar Lines.

For fac-simile of Fraunhofer's map see p. 27.

GIBBS, WOLCOTT:

On the Normal Solar Spectrum. Silliman's Journal, Jan. 1867.

This paper gives the wave-lengths of the principal lines of the Solar Spectrum.

On Wave-lengths. Silliman's Journal, March 1869.

GLADSTONE, J. H.

Notes on the Atmospheric Lines of the Solar Spectrum, and on certain Spectra of Gases. Proc. Roy. Soc. xi. 305.

GRANDEAU, L.:

On the Spectrum of Lightning. Chemical News, ix. 66.

HERSCHEL, ALEX.:

On the Spectra of Meteors. Intellectual Observer, Oct. 1868.

On the Total Eclipse of the Sun of 18th August, 1868. Proc. Roy. Inst. 1868–9.

HERSCHEL, LIEUT. JOHN:

On Spectra of Southern Nebulæ, and Spectrum of Lightning. Proc. Roy. Soc. xvi. 416, 451; xvii. pp. 58—61.

Additional Observations of Southern Nebulæ. Proc. Roy. Soc. xvii. 303.

On the Solar Eclipse of 1868 seen at Jamkandi.

Spectroscopic Observations of the Solar Prominences. Proc. Roy. Soc. xviii. 62.

HUGGINS, WILLIAM:

On the Disappearance of the Spectrum of ε Piscium at its Occultation of January 4, 1865. With Conclusions as to the Non-existence of a Lunar Atmosphere. Monthly Notices, Roy. Astron. Soc. xxv. 60. Chem. News, xi. 175.

On the Spectra of the Nebulæ. Phil. Trans. 1864, p. 437. Phil. Mag. June 1866.

On the Nebula in the Sword-handle of Orion. Proc. Roy. Soc. 1865, p. 39.

Further Observations on the Spectra of some of the Nebulæ, with a Mode of determining the Brightness of the Bodies. Phil. Trans. 1866, pp. 381—397.

On the Spectrum of Comet I. 1866. Proc. Roy. Soc. xv. 5.

On the Spectrum of Comet II. 1867. Monthly Notices, Royal Astron. Soc. xvii. 288.

On the Spectrum of Brorsen's Comet, 1868. Proc. Roy. Soc. xvi. 386.

Further Observations on the Spectra of some of the Stars and Nebulæ, with an Attempt to determine therefrom whether these Bodies are moving towards or from the Earth; also Observations of the Spectra of the Sun and of Comet II. 1868. Phil. Trans. 1868, p. 529.

On the Spectrum of Mars. Monthly Notices, Royal Astron. Soc. xxvii. 178.

Description of a Hand Spectrum-Telescope. Proc. Roy. Soc. xvi. 241.

This instrument is suitable for an observation of Meteors, and was successfully used at the Solar Eclipse of 1868.

Note on a Method of viewing the Solar Prominences without an Eclipse. Proc. Roy. Soc. xvii. 302.

Note on the Heat of the Stars. Proc. Roy. Soc. xvii. 309.

On some further Results of Spectrum Analysis as applied to the Heavenly Bodies. Lecture at Roy. Inst. Reported in Chem. News, xix. 187.

Huggins, William, & Miller, W. A.:

On the Lines of the Spectra of some of the Fixed Stars. Proc. Roy. Soc. xii. 444.

On the Spectra of some of the Fixed Stars. Phil. Trans. 1864, p. 413; Phil. Mag. June 1866.

The first complete and accurate investigation of the Stellar Spectra.

On the Spectrum of the Variable Star α Orionis. Monthly Notices, xxvi. p. 215.

On the Spectrum of a New Star in Corona Borealis. Proc. Roy. Soc. xv. 146, May 17, 1866.

Janssen:

On the Terrestrial Rays of the Solar Spectrum. Phil. Mag. Fourth Series, xxx. 78.

Reply to Ångström's Observations on the Solar Lines. Phil. Mag. Fourth Series, xxiii. 78.

On the Spectrum of Aqueous Vapour. Phil. Mag. Fourth Series, xxxii. 315.

According to Janssen the line A, a great part of B, C, and two lines between C and D, are caused by aqueous vapour. See Fig. 60.

Absorption Spectra of Aqueous Vapour. Compt. Rend. lvi. 538; lx. 213.

On the Terrestrial Atmospheric Absorption Lines. Compt. Rend. liv. 1280.

On the Solar Protuberances. Proc. Roy. Soc. xvii. 276.

Report on the Solar Eclipse of 1868 as seen at Guntoor. Annuaire du Bureau des Longitudes, 1869, p. 584.

Kirchhoff, G.:

On the Relation between the Radiating and Absorbing Powers of different Bodies for Light and Heat. Phil. Mag. (4), xx. 1.

This paper contains a discussion of the Mathematical Theory of the Law of Exchanges. It is followed by a Postscript by the author, on the history of the subject.

Ueber den Zusammenhang zwischen Emission und Absorption von Licht und Wärme. Monats-berichte d. Berliner Acad. 27th Oct. 1859. Phil. Mag. (4), xix. 193.

This contains the statement of the Law of Exchanges, and the first announcement of the discovery of the cause of Fraunhofer's lines.

On the History of Spectrum Analysis. Phil. Mag. (4), xxv. 250.

An interesting historical survey of the early researches on Spectrum Analysis, and remarks on the history of the chemical analysis of the Solar Atmosphere.

LE SUEUR:

On the Nebulæ of Argo and Orion, and on the Spectrum of Jupiter. Proc. Roy. Soc. xviii.

On Spectroscopic Observations with the .215 Great Melbourne Telescope, xviii. 242.

LOCKYER, J. N.:

Spectroscopic Observations of the Sun. No. I. Proc. Roy. Soc. xvii. 91; No. II. xvii. 128–131; No. III. xvii. 350; No. IV. xvii. 415; No. V. xviii. 74. (See also Lecture V. Appendix C.)

Notice of an Observation of the Spectrum of a Solar Prominence. Proc. Roy. Soc. xvii. pp. 91, 104.

Supplementary Note on a Spectrum of a Solar Prominence, xvii. p. 128.

Remarks on the recent Eclipse of the Sun as observed in the United States. Proc. Roy. Soc. xviii. 179.

Reply to some Remarks of Father Secchi on the recent Solar Discoveries. Phil. Mag. Jan. 1870.

Solar Eclipse 1869. American Observations, Report on. Nature, vol. i. p. 14.

Spectroscopic Observations of the Sun, No. II. Phil. Trans. 1869. Part I. p. 425.

MASCART:

On the Rays of the Ultra-violet Solar Spectrum. Compt. Rend. Nov. 1863. Phil. Mag. Fourth Series, xxvii. 159.

MERZ, S.:

On the Dark Lines in the Spectra of Stars. Pogg. Ann. cxvii. 654.

RAYET:

On the Solar Eclipse of 1868. Roy. Astron. Soc. Report, 1868-9, p. 152.

On the Refrangibility of the Brilliant Yellow Ray of the Solar Atmosphere. Chem. News, xix. 158.

ROSCOE, H. E.:

On the Opalescence of the Atmosphere for the Chemically Active Rays. Roy. Inst. June 1, 1866. Chem. News, xiv. 28.

RUTHERFURD, L. M.:

Measurement of Stellar Spectra. Silliman's Journ. xxxv. 71.

SECCHI:

On the Spectrum of the Great Nebula in Orion. Chem. News, xi. 136. Read before the French Academy, March 5, 1865.

On the Spectral Rays of the Planet Saturn. Phil. Mag. Fourth Series, xxx. 73.

Measurement of a few Stellar Lines. Astron. Nachrichten, 3 Marz, 1863.

Spectrum Observations, made at the Roman Observatory. Volumi dell' Academia dei Roma, xl.

SECCHI:

On a Continuous Solar Spectrum. Compt. Rend. Mars 1869.

On Stellar Spectrometry. Chem. News, xviii. 107.

On the Spectrum of the Planet Neptune. Compt. Rend. Nov. 22, 1869.

Spectrum of α Orionis. Monthly Notices, xxvi. 214.

Spettri prismatici delle Stelle fisse atti della Soc. Italian. Roma, 1868.

Catalogo delle Stelle, &c. Parigi: Gauthier Villars, 1867.

Memoria Seconda, 1868.

STEWART, BALFOUR:

Report on the Theory of Exchanges. Brit. Assoc. Reports, 1861.

On the Nature of Light emitted by heated Tourmaline. Phil. Mag. Fourth Series, xxi. 391.

Reply to Kirchhoff on the History of Spectrum Analysis. Phil. Mag. Fourth Series, xxv. 354.

On the Theory of Exchanges. Trans. Roy. Soc. Edin. 1858.

Enunciated this principle completely for the heating rays, previously proposed by Provostaye and Desains. In Feb. 1860, Stewart applied this theory to the luminous rays, subsequently to, but independently of, Kirchhoff.

STOKES, G. G.:

On the Change of Refrangibility of Light. Phil. Trans. 1852, Part II. p. 463. With drawing of the fixed lines in Solar Spectrum in the extreme violet, and in the invisible region beyond.

STONEY:

On the Physical Constitution of the Sun and Stars. Proc. Roy. Soc. xvi. 25; xvii. 1.

On the Bearing of recent Observations upon Solar Physics. Phil. Mag. Fourth Series, xxxvi. 447.

STRUVE, OTTO VON:

Beobachtung eines Nordlichtspectrum (Aurora Borealis). Bull. de l'Acad. Imp. des Sciences de St. Petersbourg, tome xiii. 49, 50.

TENNANT, MAJOR:

Report of the Indian Eclipse, 1868. Royal Astron. Soc. Memoirs, vol. vii. Nature, vol. i. 536.

TYNDALL, J.:

On Cometary Theory. Phil. Mag. Fourth Series, xxxvii. 241.

WEISS, A.:

On the Changes produced in the Position of the Fixed Lines in the Spectrum of Hyponitric Acid by Changes in Density. Phil. Mag. Fourth Series, xxii. 80.

On Fraunhofer's Lines seen in Sunlight at low Altitudes. Phil. Mag. Fourth Series, xxiv. 407.

WOLF & RAYET:

On Three Small Stars with Bright Lines. Compt. Rend. August 1867.

WOLLASTON, W. H.:

A Method of examining Refractive and Dispersive Powers by Prismatic Reflexion. Containing the first discovery of the dark Solar Lines. Phil. Trans. 1802, p. 365.

YOUNG, C. A.:

Spectroscopic Notes on the Sun. Chem. News, xx. 271.

ZÖLLNER, F.:

On a New Spectroscope, with Contributions to the Spectral Analysis of the Stars. Phil. Mag. Fourth Series, xxxviii. 380.

INDEX.

A.

B.

C.

D.

M.

N.

O.

P.

R.

S.

T.

U.

V.

W.

Z.

www.ingramcontent.com/pod-product-compliance
Lightning Source LLC
Chambersburg PA
CBHW051726300426
44115CB00007B/484

* 9 7 8 3 7 4 1 1 7 3 6 2 2 *